D1544357

MOLECULAR ELECTRONIC STRUCTURES
OF TRANSITION METAL COMPLEXES

McGRAW-HILL
INTERNATIONAL
BOOK COMPANY

New York
St. Louis
San Francisco
Auckland
Bogotá
Guatemala
Hamburg
Johannesburg
Lisbon
London
Madrid
Mexico
Montreal
New Delhi
Panama
Paris
San Juan
São Paulo
Singapore
Sydney
Tokyo
Toronto

C. J. BALLHAUSEN

Department of Physical Chemistry
University of Copenhagen

Molecular Electronic Structures of Transition Metal Complexes

This book was set in Times Roman Series 327

British Library Cataloging in Publication Data

Ballhausen, C. J.
Molecular electronic structures of transition metal complexes.
1. Coordination compounds 2. Transition metal compounds 3. Electrons
I. Title
546′.6 QD474

ISBN 0-07-003495-8

MOLECULAR ELECTRONIC STRUCTURES OF TRANSITION METAL COMPLEXES

1 2 3 4 5 WJM 8 1 0 7 9
Printed and bound in Great Britain
by W & J Mackay Limited, Chatham

CONTENTS

PREFACE

The electronic structures of inorganic complexes are to a large extent decided by the fact that these ions and molecules have a transition-metal atom situated at a center of high symmetry. The high degeneracy of the metal d- and f-valence orbitals, so characteristic of central symmetry, is therefore only removed in part. Since, furthermore, the degenerate molecular orbitals of the complex may only be partially filled with electrons, the electronic-term problem bears many similarities to that encountered in the theory of atomic spectra.

The analogies between the theory of atomic structures and the electronic structures of complexes are therefore many. Nevertheless, the all-important fact is that inorganic complexes are not just pure metal ions in disguise. As realized by Van Vleck in 1935 they are true molecular entities, and should be treated as such. The only existing theory which can give a simple consistent picture of the ground and excited molecular electronic states is the molecular orbital theory. Consequently, in this book the theory of electronic structures of complexes is developed solely from this point of view. Furthermore, the aim has been to characterize rather than to calculate.

Most chemists working with the electronic structures of inorganic transition-metal complexes have arrived at the subject via an education in inorganic chemistry. Using a variant of a simplified electronic theory, their aim is usually to extract parameters for a class of complexes and then to arrange these in some series. On the other hand, the solid-state physicists working in this field will often be content to write down an "effective hamiltonian" for one compound and add terms to this until the experimental numbers can be reproduced. Both groups therefore strive to express their experimental findings using some parameterized theory. Unfortunately the temptation is to elaborate an approximate theory and to introduce an increasing number of loosely defined "effects" in order to "explain" the movements of the parameters. For the uninitiated this can easily lead to great confusion, and there is little reason to expect that deeper insight can be gained in this way.

It is well known that a model may correctly describe certain features in one area of a research field but be hopelessly inadequate in others. As an obvious example the crystal field concept springs to mind. In order to avoid the many pitfalls which await a molecular spectroscopist when interpreting experimental results or when building theoretical models, it is therefore imperative to possess a sound knowledge of quantum chemistry.

I have tried to write such a text which twenty years of experience have taught me might be of help to a chemist wishing to work with the electronic structures of inorganic complexes. Rigorous mathematical derivations have not been stressed. Hopefully, enough details have been given so that the essential lines of development are transparent. I have furthermore assumed that the reader commands a working knowledge of group theory. Calculational methods, exemplified by *ab initio* and semi-empirical procedures have not been dealt with. No attempts have been made to cover the literature, either from a theoretical or experimental point of view. I have merely picked out what I considered to be the important ideas and developments, and I have illustrated them by using simple examples.

I wish to thank my colleagues Professor Dr. Jens Peder Dahl and Lektor Dr. Aage E. Hansen for many helpful discussions on the material in this book, and I am greatly indebted to Mrs. E. Ottonello for the preparation of the manuscript.

C.J.B
March 1977

CHAPTER

ONE

THE ELECTRONIC STATES

1-1 THE BORN-OPPENHEIMER APPROXIMATION

The whole of chemistry is a reflection of the behavior of electrons in various potential fields. Experience has shown us that the conduct of electrons cannot be described using the classical laws of motion. Hence chemistry is one huge manifestation of quantum phenomena. Unfortunately, even for a small molecule a reasonably exact quantum-mechanical solution of the molecular Schrödinger equation is a formidable task. In order to gain some insight into electronic behavior, it is therefore necessary to introduce simplifying approximations. The nature of such approximations must of course be adjusted to the phenomena in which we are interested. We must take care not to introduce artificial features into the system, and at the same time not to throw away important traits.

The molecules and ions which we shall be interested in here contain so many nuclei that the rotational structure is washed out. On the other hand, in many interesting cases, the electronic transitions will show some vibrational structure. The wave functions for a molecule are dependent both on the nuclear and electronic coordinates. The first and fundamental approximation we shall introduce is the separation of the electronic and nuclear motions. This is performed using the so-called Born–Oppenheimer approximation.[1,2]

Let us designate the positions and masses of the nuclei μ by $(\mathbf{Q}_\mu, M_\mu)$ and the positions and masses for the electrons by $(\mathbf{r}_i, m)$, where all the vectors $\mathbf{r}$ and $\mathbf{Q}$ are referred to a laboratory-fixed coordinate system. The total non-relativistic hamiltonian for a molecular system of N nuclei and n electrons is

$$\mathscr{H} = -\sum_\mu \frac{\hbar^2}{2M_\mu} \nabla_\mu^2 - \sum_i \frac{\hbar^2}{2m} \nabla_i^2 + V(\mathbf{r}, \mathbf{Q}) \tag{1-1}$$

$V(\mathbf{r}, \mathbf{Q})$ is the potential energy term equal to

$$V(\mathbf{r}, \mathbf{Q}) = -\sum_{\mu,i} \frac{Z_\mu e^2}{|\mathbf{Q}_\mu - \mathbf{r}_i|} + \sum_{\mu<\nu} \frac{Z_\mu Z_\nu e^2}{|\mathbf{Q}_\mu - \mathbf{Q}_\nu|} + \sum_{i<j} \frac{e^2}{|\mathbf{r}_i - \mathbf{r}_j|} \tag{1-2}$$

Let us now suppose that all the nuclear masses are infinite. Clearly, this will quench the nuclear motions, leading to an "electronic" hamiltonian of the form

$$\mathscr{H}_E = -\sum_i \frac{\hbar^2}{2m} \nabla_i^2 + V(\mathbf{r}, \mathbf{Q}) \tag{1-3}$$

The "electronic" Schrödinger equation is then defined as

$$\mathscr{H}_E \Psi_t(\mathbf{r}, \mathbf{Q}) = W_t(\mathbf{Q}) \Psi_t(\mathbf{r}, \mathbf{Q}) \tag{1-4}$$

Both $W_t(\mathbf{Q})$ and $\Psi_t(\mathbf{r}, \mathbf{Q})$ are seen to contain the nuclear positions $\mathbf{Q}$ as parameters, and are indeed continuous functions of the $3N$ nuclear coordinates. In principle, we could solve Eq. (1-4) for all values of the $3N$ nuclear coordinates, and for each nuclear arrangement we would obtain a complete set of electronic wave functions $\Psi_t(\mathbf{r}, \mathbf{Q})$ and eigenvalues $W_t(\mathbf{Q})$.

Introducing the "electronic" hamiltonian into the molecular hamiltonian of Eq. (1-1), this may be rewritten

$$\mathscr{H} = \mathscr{H}_E - \sum_\mu \frac{\hbar^2}{2M_\mu} \nabla_\mu^2 \tag{1-5}$$

We now use the variational principle to solve the molecular Schrödinger equation, and as our variational wave function we take the finite expansion

$$\Psi = \sum_{t=1}^{p} \chi_t(\mathbf{Q}) \Psi_t(\mathbf{r}, \mathbf{Q}) \tag{1-6}$$

The use of a trial wave function of this form is inspired by the fact that had the solutions to the electronic Schrödinger equation been independent of the nuclear coordinates, a single product function would be the exact solution to the hamiltonian in Eq. (1-5).

Without loss of generality we can take $\Psi_t(\mathbf{r}, \mathbf{Q})$ to be real and normalized

$$\int |\Psi_t(\mathbf{r}, \mathbf{Q})|^2 \, d\mathbf{r} = 1 \tag{1-7}$$

Notice in particular that $\Psi_t(\mathbf{r}, \mathbf{Q})$ is assumed normalized for all values of $\mathbf{Q}$.

Using Eq. (1-5) together with Eq. (1-6) now leads to

$$-\sum_\mu \frac{\hbar^2}{2M_\mu} \left\{ \sum_t \Psi_t(\mathbf{r}, \mathbf{Q}) \nabla_\mu^2 \chi_t(\mathbf{Q}) + 2 \sum_t \nabla_\mu \Psi_t(\mathbf{r}, \mathbf{Q}) \nabla_\mu \chi_t(\mathbf{Q}) + \sum_t \chi_t(\mathbf{Q}) \nabla_\mu^2 \Psi_t(\mathbf{r}, \mathbf{Q}) \right\}$$
$$+ \mathscr{H}_E \sum_t \chi_t(\mathbf{Q}) \Psi_t(\mathbf{r}, \mathbf{Q}) = W \sum_t \chi_t(\mathbf{Q}) \Psi_t(\mathbf{r}, \mathbf{Q}) \tag{1-8}$$

Multiplying Eq. (1-8) by $\Psi_u(\mathbf{r}, \mathbf{Q})$, and integrating over the electronic coordinates making use of the orthonormal properties of the set $\Psi_t(\mathbf{r}, \mathbf{Q})$, leads to

$$-\sum_{\mu}\frac{\hbar^2}{2M_\mu}\nabla_\mu^2\chi_u(\mathbf{Q}) + W_u(\mathbf{Q})\chi_u(\mathbf{Q}) - W\chi_u(\mathbf{Q})$$

$$-\sum_{\mu}\frac{\hbar^2}{2M_\mu}\sum_t 2\langle\Psi_u(\mathbf{r},\mathbf{Q})|\nabla_\mu|\Psi_t(\mathbf{r},\mathbf{Q})\rangle\nabla_\mu\chi_t(\mathbf{Q})$$

$$-\sum_{\mu}\frac{\hbar^2}{2M_\mu}\sum_t\langle\Psi_u(\mathbf{r},\mathbf{Q})|\nabla_\mu^2|\Psi_t(\mathbf{r},\mathbf{Q})\rangle\chi_t(\mathbf{Q}) = 0 \qquad (1\text{-}9)$$

If we define an operator $\hat{C}_{u,t}$ equal to

$$\hat{C}_{u,t} = \sum_{\mu}\frac{\hbar^2}{2M_\mu}2\langle\Psi_u(\mathbf{r},\mathbf{Q})|\nabla_\mu|\Psi_t(\mathbf{r},\mathbf{Q})\rangle\nabla_\mu + \sum_{\mu}\frac{\hbar^2}{2M_\mu}\langle\Psi_u(\mathbf{r},\mathbf{Q})|\nabla_\mu^2|\Psi_t(\mathbf{r},\mathbf{Q})\rangle \qquad (1\text{-}10)$$

we can write Eq. (1-9) as

$$\left\{-\sum_{\mu}\frac{\hbar^2}{2M_\mu}\nabla_\mu^2 + W_\mu(\mathbf{Q}) - W - \hat{C}_{u,u}\right\}\chi_u(\mathbf{Q}) - \sum_{u\neq t}\hat{C}_{u,t}\chi_t(\mathbf{Q}) = 0 \qquad (1\text{-}11)$$

The set (1-11) of coupled differential equations can be used to determine the expansion coefficients $\chi_t(\mathbf{Q})$. In principle we can get the total wave function for the molecule by solving the electronic Schrödinger Eq. (1-4) for all values of $\mathbf{Q}$, thereby getting all of the electronic wave functions, $\Psi(\mathbf{r},\mathbf{Q})$. These should then be multiplied by the nuclear wave functions, $\chi(\mathbf{Q})$, to obtain Eq. (1-6). The Eqs. (1-11) can then be used to determine the complete wave functions to any degree of accuracy.

This method to derive the total wave function for a molecular system is called the Born–Oppenheimer separation; it is seen to lead to two different sets of equations, one governing the electronic motions, Eq. (1-4), and one governing the nuclear motions, Eq. (1-11), that is the molecular vibrations and rotations.

For orders of magnitude of the electronic, vibrational, and rotational energies we have

$$w_{\text{el}} = \frac{\hbar^2}{ma_0^2}$$

and

$$w_{\text{rot}} = \frac{\hbar^2}{Ma_0^2}$$

where a_0 is a characteristic length of the molecule. With $w_{\text{vib}} = h\nu = \hbar\sqrt{k/M}$ we can also write $w_{\text{el}} = \hbar\sqrt{k/m}$ since at equilibrium the same forces act on the electrons and nuclei. Hence

$$w_{\text{vib}} = \sqrt{\frac{m}{M}}\,w_{\text{el}} \quad \text{and} \quad w_{\text{rot}} = \frac{m}{M}w_{\text{el}}$$

The original Born–Oppenheimer separation used an expansion[3] in the parameter $\eta = (m/M)^{1/4}$. For a harmonic oscillator in its ground state we have for the average value of the displacement, $\langle\xi^2\rangle$, of the system from its equilibrium position

$\langle \xi^2 \rangle = \hbar/2\pi\nu M$, where ν is the classical vibrational frequency. Defining the dimensionless quantity

$$\eta = \frac{\sqrt{\langle \xi^2 \rangle}}{a_0} \equiv \left(\frac{m}{M}\right)^{1/4} \tag{1-12}$$

we can see the physical background for the expansion parameter.

The equations show that for molecules of a certain size the rotational energies are very small indeed. We shall therefore, in what follows, completely neglect the rotational energies and only consider the electronic and vibrational motions.

1-2 THE ADIABATIC APPROXIMATION

The operators $\hat{C}_{u,t}$ contain integrals which depend on the first and second derivatives of the electronic wave functions with respect to the nuclear coordinates. For the diagonal terms $\hat{C}_{u,u}$ we have, recalling the normalization condition of Eq. (1-7)

$$\nabla_\mu \int |\Psi_u(\mathbf{r}, \mathbf{Q})|^2 \, d\mathbf{r} = 0$$

or

$$\langle \Psi_u(\mathbf{r}, \mathbf{Q}) | \nabla_\mu | \Psi_u(\mathbf{r}, \mathbf{Q}) \rangle = 0 \tag{1-13}$$

To estimate $\hbar^2/2M_\mu \langle \Psi_u(\mathbf{r}, \mathbf{Q}) | \nabla_\mu^2 | \Psi_u(\mathbf{r}, \mathbf{Q}) \rangle$ we observe that nuclei and electrons experience roughly the same Coulombic forces, since the interaction between them is determined by the value of $|\mathbf{Q}_\mu - \mathbf{r}_i|$. Hence $\nabla_\mu^2 \Psi_u(\mathbf{r}, \mathbf{Q}) \approx \nabla_i^2 \Psi_u(\mathbf{r}, \mathbf{Q})$. We have then

$$\frac{\hbar^2}{2M_\mu} \langle \Psi_u(\mathbf{r}, \mathbf{Q}) | \nabla_i^2 | \Psi_u(\mathbf{r}, \mathbf{Q}) \rangle \approx \frac{\hbar^2}{2M_\mu a_0^2} \approx W_{\text{rot}}$$

where a_0 is the Bohr radius and W_{rot} is the energy of a rotational quantum of the molecule.

The electronic integrals contained in the off-diagonal operators $\hat{C}_{u,t}$ can be evaluated by differentiating Eq. (1-4) with respect to $\mathbf{Q}_\mu$. This yields

$$\Psi_t(\mathbf{r}, \mathbf{Q}) \nabla_\mu \mathscr{H}_E + \mathscr{H}_E \nabla_\mu \Psi_t(\mathbf{r}, \mathbf{Q}) = \Psi_t(\mathbf{r}, \mathbf{Q}) \nabla_\mu W_t(\mathbf{Q}) + W_t(\mathbf{Q}) \nabla_\mu \Psi_t(\mathbf{r}, \mathbf{Q}) \tag{1-14}$$

Multiplying from the left by $\Psi_u(\mathbf{r}, \mathbf{Q})$ and integrating over the electronic coordinates gives

$$W_t(\mathbf{Q}) \langle \Psi_u(\mathbf{r}, \mathbf{Q}) | \nabla_\mu | \Psi_t(\mathbf{r}, \mathbf{Q}) \rangle = W_u(\mathbf{Q}) \langle \Psi_u(\mathbf{r}, \mathbf{Q}) | \nabla_\mu | \Psi_t(\mathbf{r}, \mathbf{Q}) \rangle + \langle \Psi_u(\mathbf{r}, \mathbf{Q}) | \nabla_\mu V(\mathbf{r}, \mathbf{Q}) | \Psi_t(\mathbf{r}, \mathbf{Q}) \rangle$$

or for the case where Ψ_t and Ψ_u are nondegenerate

$$\langle \Psi_u(\mathbf{r}, \mathbf{Q}) | \nabla_\mu | \Psi_t(\mathbf{r}, \mathbf{Q}) \rangle = \frac{\langle \Psi_u(\mathbf{r}, \mathbf{Q}) | \nabla_\mu V(\mathbf{r}, \mathbf{Q}) | \Psi_t(\mathbf{r}, \mathbf{Q}) \rangle}{W_t(\mathbf{Q}) - W_u(\mathbf{Q})} \tag{1-15}$$

Differentiating Eq. (1-13) we obtain

$$\langle \Psi_u(\mathbf{r}, \mathbf{Q}) | \nabla_\mu^2 | \Psi_u(\mathbf{r}, \mathbf{Q}) \rangle + \langle \nabla_\mu \Psi_u(\mathbf{r}, \mathbf{Q}) | \nabla_\mu | \Psi_u(\mathbf{r}, \mathbf{Q}) \rangle = 0 \tag{1-16}$$

We have found that the first term of this equation is of the order of magnitude a_0^{-2}. We observe that the second term in Eq. (1-16) is just $\langle \nabla_\mu \Psi_u(\mathbf{r}, \mathbf{Q}) | \nabla_\mu \Psi_u(\mathbf{r}, \mathbf{Q}) \rangle$. The norm of $\nabla_\mu \Psi_u(\mathbf{r}, \mathbf{Q})$ is consequently a_0^{-1}. Therefore using the Schwarz inequality

$$|\langle \phi | \psi \rangle| < \sqrt{\langle \phi | \phi \rangle \langle \psi | \psi \rangle}$$

we get

$$\frac{\hbar^2}{2M_\mu} \langle \Psi_u(\mathbf{r}, \mathbf{Q}) | \nabla_\mu^2 | \Psi_t(\mathbf{r}, \mathbf{Q}) \rangle < \frac{\hbar^2}{2M_\mu a_0^2} \approx W_{\text{rot}}$$

Since we are here only interested in large molecules and ions we may neglect the very small rotational energies. We note that under these conditions the operator $\hat{C}_{u,u}$ disappears and that $\hat{C}_{u,t}$ reduces to the first term in Eq. (1-10). Furthermore, provided that there are no close-lying electronic states we notice from Eq. (1-15) that the first term in Eq. (1-10) is also small. For nondegenerate electronic wave functions it may safely be neglected.

When all the off-diagonal operators $\hat{C}_{u,t}$ in Eq. (1-11) are thrown away, the molecular wave function (1-6) is seen to reduce to a single term

$$\Psi = \chi_t(\mathbf{Q}) \Psi_t(\mathbf{r}, \mathbf{Q}) \tag{1-17}$$

This form of Ψ is called the *adiabatic approximation* to the molecular wave function. In this scheme each "electronic" wave function $\Psi_t(\mathbf{r}, \mathbf{Q})$ is determined by the "electronic" Schrödinger Eq. (1-4) and the associated nuclear wave function is determined by the simplified form of Eq. (1-11):

$$\left\{ -\sum_\mu \frac{\hbar^2}{2M_\mu} \nabla_\mu^2 + W_t(\mathbf{Q}) \right\} \chi_{t,v}(\mathbf{Q}) = W_{t,v} \chi_{t,v}(\mathbf{Q}) \tag{1-18}$$

The nuclear functions $\chi_{t,v}(\mathbf{Q})$ are seen in the adiabatic approximation to be given as the solutions to a wave equation in which the "electronic" energy $W_t(\mathbf{Q})$ acts as the potential energy.

It is quite obvious from the form of Eq. (1-15) that had our electronic wave function belonged to a degenerate set, an adiabatic wave function would have been a very bad approximation to a molecular wave function. In this case, the coupling operator $\hat{C}_{u,t}$ assumes great importance, and the total wave function will be of the type shown in Eq. (1-6). This situation and its consequences are dealt with under the heading of the Jahn–Teller effect (section 1-6).

1-3 THE NUCLEAR MOTIONS

In the adiabatic approximation, as leading to Eqs. (1-17) and (1-18), $W_t(\mathbf{Q})$ and $\chi_{t,v}(\mathbf{Q})$ are functions of the $3N$ cartesian coordinates of the N nuclei:

$$\mathbf{Q} = (Q_{1x}, Q_{1y}, Q_{1z}, \ldots, Q_{Nz}) \tag{1-19}$$

The nuclear and electronic positions $(\mathbf{Q}, \mathbf{r})$ are measured in a laboratory-fixed coordinate system. The nuclear kinetic operators depends upon the $3N$ second-order derivatives with respect to these coordinates. In order to treat the vibrations of the nuclei the so-called normal coordinates are introduced.[4] This involves changing from the external coordinates $\mathbf{Q}$ to an internal molecular-coordinate system ξ.

Let us assume that we can find at least one point $\mathbf{Q}^0$ in the $3N$ dimensional coordinate system for which all the first derivatives of $W_t(\mathbf{Q})$, with respect to Q_{3N} are equal to zero. Thus for all μ

$$\left(\frac{\partial W(\mathbf{Q})}{\partial Q_{\mu x}}\right)_{\mathbf{Q}^0} = \left(\frac{\partial W(\mathbf{Q})}{\partial Q_{\mu y}}\right)_{\mathbf{Q}^0} = \left(\frac{\partial W(\mathbf{Q})}{\partial Q_{\mu z}}\right)_{\mathbf{Q}^0} = 0 \tag{1-20}$$

Obviously the function $W_t(\mathbf{Q})$ will have an extremum at the point $\mathbf{Q}^0$. Whether $\mathbf{Q}^0$ represents a minimum or a maximum depends, of course, upon the values of the second derivatives of $W_t(\mathbf{Q})$.

In the case where the conditions of Eq. (1-20) cannot be fulfilled, and at least one of the $3N$ derivatives is different from zero for all values of the nuclear coordinate, this simply means that the potential hypersurface cannot have a minimum with respect to this nuclear coordinate, and that the corresponding molecular state is unstable. However, if for some point $\mathbf{Q}^0$, Eq. (1-20) can indeed be fulfilled, $\mathbf{Q}^0$ will be taken to represent a stable configuration of the molecule in the state $\Psi_t(\mathbf{r}, \mathbf{Q})$.

We now introduce a set of $3N$ mass-weighted displacement coordinates $S_1, \ldots, S_{3N}$ defined as

$$\begin{aligned}
S_1 &= \sqrt{M_1}\,(Q_{1x} - Q_{1x}^0) \\
S_2 &= \sqrt{M_1}\,(Q_{1y} - Q_{1y}^0) \\
S_3 &= \sqrt{M_1}\,(Q_{1z} - Q_{1z}^0) \\
&\;\;\vdots \\
S_{3N} &= \sqrt{M_N}\,(Q_{Nz} - Q_{Nz}^0)
\end{aligned} \tag{1-21}$$

This transformation amounts to a translation of the Q-coordinate system followed by a change in scale on the axes. The first and second derivatives with respect to these displacement coordinates are seen to be

$$\begin{aligned}
\frac{\partial}{\partial S_1} &= \frac{1}{\sqrt{M_1}}\frac{\partial}{\partial Q_{1x}} & \qquad \frac{\partial^2}{\partial S_1^2} &= \frac{1}{M_1}\frac{\partial^2}{\partial Q_{1x}^2} \\
&\;\;\vdots & \text{and} \qquad &\;\;\vdots \\
\frac{\partial}{\partial S_{3N}} &= \frac{1}{\sqrt{M_N}}\frac{\partial}{\partial Q_{Nz}} & \qquad \frac{\partial^2}{\partial S_{3N}^2} &= \frac{1}{M_N}\frac{\partial^2}{\partial Q_{NZ}^2}
\end{aligned}$$

In this coordinate system the point S^0, for which all the first derivatives of the function $W_t(S)$ vanish, obviously lies at the origin $S^0 = (0, 0, \ldots, 0)$, and we can expand $W_t(S)$ in a Taylor series retaining terms up to second order only:

$$W_t(S) = W_t^0 + \frac{1}{2}\sum_{k,l}\left(\frac{\partial^2 W_t(S)}{\partial S_k\,\partial S_l}\right)_0 S_k S_l + \cdots \tag{1-22}$$

where W_t^0 is the electronic energy evaluated at the origin (or equivalently at $\mathbf{Q}^0$) and where all terms containing the first derivatives of $W_t(S)$ are absent by virtue of Eq. (1-20). The second derivatives of $W_t(S)$ are to be evaluated at the origin, as indicated by the subscript 0.

Introducing this transformation of coordinates into the nuclear wave Eq. (1-18) leads to

$$\left[-\frac{\hbar^2}{2}\sum_k \nabla_k^2 + W_t^0 + \frac{1}{2}\sum_{k,l} f_{kl}^t S_k S_l - W_t\right]\chi_t(S) = 0 \tag{1-23}$$

The potential-energy term in Eq. (1-22) is a so-called quadratic form, containing all of the cross terms $S_k S_l$. Introducing a linear combination of all the displacement coordinates of Eqs. (1-21) we can, however, reduce the potential-energy terms to a form in which all of the cross terms have disappeared. Taking

$$\xi_u = \sum_k b_{ku} S_k \tag{1-24}$$

and introducing relative nuclear energies

$$\tilde{W}_t = W_t - W_t^0 \tag{1-25}$$

Eq. (1-23) can be written as

$$\sum_{u=1}^{3N}\left[-\frac{\hbar^2}{2}\frac{\partial^2}{\partial \xi_u^2} + \frac{1}{2}d_u^t \xi_u^2\right]\chi_t(\xi_u) = \sum_{u=1}^{3N}\tilde{W}_t\chi_t(\xi_u) \tag{1-26}$$

The coordinates ξ_u represent all of the possible movements of the nuclei. Three of them will, therefore, describe the translations of the molecule as a whole, and three of them will characterize the rotations of the molecule. The translations and the rotations cannot, however, depend upon the intermolecular distances. The potential energy terms in Eq. (1-26) associated with the translations and the rotations, depending as they do upon the internuclear distances, must therefore be zero. This implies that for nonlinear molecules the coefficients d_1^t to d_6^t are zero. (For linear molecules only d_1^t to d_5^t are zero.) Equation (1-26) is thereby reduced to

$$\left[\sum_{u=1}^{6}\left(-\frac{\hbar^2}{2}\frac{\partial^2}{\partial \xi_u^2}\right) + \sum_{u=7}^{3N}\left(-\frac{\hbar^2}{2}\frac{\partial^2}{\partial \xi_u^2} + \frac{1}{2}d_u^t\xi_u^2\right) - \tilde{W}_t\right]\chi_t(\xi_u) = 0 \tag{1-27}$$

The $3N - 6$ linear combinations ξ_u of the cartesian nuclear displacement coordinates are called the normal coordinates of the molecule. They are of such a nature as to leave the center of mass of the nuclei unaltered, and the principal axes of inertia are likewise left unchanged. It is therefore natural to use, in the description of the molecule, a cartesian coordinate system with origin at the center of mass, and with the coordinate axes directed along the principal axes of inertia for the nuclei in the equilibrium position. Hence, in the so-called *crude adiabatic approximation* we take for the molecular wave function

$$\Psi_{tv}(\mathbf{r}, \xi) = \chi_{tv}(\xi)\psi_t^0(\mathbf{r}) \tag{1-28}$$

where the static electronic wave function $\psi_t^0(\mathbf{r})$ is calculated at the equilibrium position of the nuclei, as indicated by the zero superscript, and v is the vibrational quantum number.

The totality of all the static electronic wave functions span what has been called a Longuet–Higgins space.[5] The completeness of this space implies that we can obtain a dynamic electronic wave function as a superposition of static electronic wave functions, that is

$$\Psi_j(\mathbf{r}, \xi) = \sum_t \psi_t^0(\mathbf{r})c_{tj}(\xi) \tag{1-29}$$

Such an expansion is encountered in the so-called Herzberg–Teller vibronic scheme.

When changing the electronic coordinates from an external to an internal coordinate system we must take care not to introduce translations and rotations into our space-fixed molecule. Calling the linear momentum $\mathbf{P}$ and the angular momentum $\mathbf{L}$ we have in the center-of-mass system with n electrons

$$\mathbf{P}_{\text{nuc}} + \sum_n \mathbf{p}_n = 0 \tag{1-30}$$

$$\mathbf{L}_{\text{nuc}} + \sum_n \mathbf{l}_n = 0 \tag{1-31}$$

The kinetic-energy terms of the hamiltonian are, with $M = \sum_\mu M_\mu$, the total nuclear mass,

$$T = \frac{1}{2M}\mathbf{P}_{\text{nuc}}^2 + \sum_n \frac{1}{2m}\mathbf{p}_n^2 \tag{1-32}$$

Substituting Eq. (1-30) into Eq. (1-32) leads to

$$T = \frac{1}{2M}\left(\sum_n \mathbf{p}_n\right)^2 + \frac{1}{2m}\sum_n \mathbf{p}_n^2 = \frac{1}{2\eta_r}\sum_n \mathbf{p}_n^2 + \frac{1}{M}\sum_{n'>n} \mathbf{p}_n \cdot \mathbf{p}_{n'} \tag{1-33}$$

where $\eta_r = mM/(m + M)$, the reduced mass. The cross terms which appear in Eq. (1-33) constitute the so-called mass-polarization term. Due to the presence of $1/M$ it is very small and may be neglected in the present context. Similar, though more complicated, are the coupling terms which appear in the preservation of the angular momentum. However, these may also be safely ignored.

Using the crude adiabatic approximation the equations which govern the nuclear vibrations are then given by the last part of Eq. (1-27)

$$\sum_{u=1}^{3N-6}\left(-\frac{\hbar^2}{2}\frac{\partial^2}{\partial\xi_{tu}^2} + \frac{1}{2}d_u^t\,\xi_{tu}^2 - w_{vu}^t\right)\chi_{tv}(\xi_{tu}) = 0 \tag{1-34}$$

where v is a vibrational quantum number and t specifies the electronic state. The eigenvalues and eigenfunctions are those of the harmonic oscillator with unit mass and force constant d_u^t:

$$w_{vu}^t = (v + \tfrac{1}{2})\hbar\sqrt{d_u^t} = (v + \tfrac{1}{2})h\nu_u^t \qquad v = 0, 1, 2, \ldots \tag{1-35}$$

$$\chi_{tv}(\xi_{tu}) = C_v \exp\left(-\frac{\xi_{tu}^2}{2}\right) H_v(\xi_{tu}) \tag{1-36}$$

where C_v is a normalization constant and the functions $H_v(x)$ are Hermite polynomials. The first polynomials are

$$H_0(x) = 1$$

$$H_1(x) = 2x$$

$$H_2(x) = 4x^2 - 2$$

$$H_3(x) = 8x^3 - 12x$$

Notice that when v is even, only even powers of x occur in the Hermite polynomials, whereas for odd values of v, the Hermite polynomials are made up solely of odd powers.

1-4 SYMMETRY CLASSIFICATIONS AND COUPLING COEFFICIENTS

In classical mechanics there are constants of motion, such as the energy, the linear momentum **p**, and others. In quantum mechanics any operator represented by $\hat{A}$ which commutes with $\mathscr{H}$ will be a constant of the motion. This we write

$$[\mathscr{H}, \hat{A}] = \mathscr{H}\hat{A} - \hat{A}\mathscr{H} = 0$$

Hence

$$\hat{A}(\mathscr{H}\Psi) = \mathscr{H}(\hat{A}\Psi)$$

$$\|$$

$$\hat{A}(W\Psi) = W(\hat{A}\Psi)$$

Notice that $(\hat{A}\Psi)$ is a solution to the Schrödinger equation whenever Ψ is, and it has the same eigenvalue W.

For any molecule there are a certain number of symmetry operations that commute with $\mathscr{H}$. These include the operations which rotate, reflect, or invert the molecule into itself. Of all of these operations (including that which leaves the molecule as it was, $\hat{E}$) we have that any two of them are equivalent to a third. Such a set of symmetry operations constitutes a group.[6] The eigenfunctions of $\mathscr{H}$ form a basis for the so-called irreducible representations of the group. The molecular states can therefore be characterized by their irreducible representations in the point group of the molecule.

A set of functions spanning an irreducible representation is described by its transformation properties under the symmetry operations of the molecule. Since the traces of the transformation matrices are independent of the choice of coordinate system, it is the various traces, called the *characters*, which are used as indicators. To take an example: in the molecular symmetry O the symmetry operators are $\hat{E}$, $8\hat{C}_3$, $3\hat{C}_2$, $6\hat{C}_4$, and $6\hat{C}_2'$. The set of functions (xz), (yz), and (xy)

have the traces 3, 0, -1, -1, 1 under these symmetry operators. Such a behavior is indicative of an irreducible t_2 representation.[6] In particular, for all molecular systems $\mathscr{H}$ spans a one-dimensional totally symmetric representation in the point group of the molecule.

The symmetry designations represent what are normally called "good quantum numbers." Their importance lies in the fact that a molecular state can be characterized exactly, even if one can only perform an approximate calculation. Only provided two states have exactly the same symmetry characteristics can they interact under the hamiltonian. It is this last feature which enables us to draw the well-known correlation diagrams in molecular-orbital theory.

As long as no spin-coordinates occur in the hamiltonian, the operator $\hat{S}^2$, where S is the total spin-angular momentum, commutes with $\mathscr{H}$. The molecular states can in this case be characterized by their orbital transformation properties and the spin multiplicity $2S+1$. For a molecular hamiltonian also containing a spin-orbit coupling term, $\mathscr{H}$ must be invariant when the same symmetry operations are applied simultaneously in both spin-space and electronic position–space. The molecular states must then be characterized in this augmented space.

For rotating an angle α around the z axis we have with $\psi = \exp(im_s\phi)$, $m_s = \pm\frac{1}{2}$, that the transformation matrix is

$$\begin{pmatrix} \exp(\mathrm{i}\tfrac{1}{2}\alpha) & 0 \\ 0 & \exp(-\mathrm{i}\tfrac{1}{2}\alpha) \end{pmatrix}$$

having a character $\chi(\alpha) = 2\cos\frac{1}{2}\alpha$. For $\alpha = 0$, $\chi(0) = 2$. Rotating the system an angle of 2π we have $\chi(2\pi) = -2$. The functions having half-integral values of angular momentum are therefore seen to be double-valued. However, for $\alpha = \pi$ and 3π the functions are single-valued since $\chi(\pi) = \chi(3\pi) = 0$. To find the double-valued representations, the rotation which brings the system back into itself is taken to be $\alpha = 4\pi$. Furthermore, we introduce a symmetry operator $\hat{R}$, a rotation through an angle 2π about an arbitrary axis, such that $\hat{R}^2 = \hat{E}$. The operator $\hat{R}$ commutes with all other symmetry elements of the group. The number of symmetry operations in the so-called double group[7] is therefore twice as many as in the single group.

The symmetry elements of the octahedral double group, for instance, are the following:

$$\hat{E} \quad \begin{pmatrix} 4\hat{C}_3 \\ 4\hat{C}_3^2\hat{R} \end{pmatrix} \quad \begin{pmatrix} 3\hat{C}_2 \\ 3\hat{C}_2\hat{R} \end{pmatrix} \quad \begin{pmatrix} 3\hat{C}_4 \\ 3\hat{C}_4^3\hat{R} \end{pmatrix} \quad \begin{pmatrix} 6\hat{C}_2' \\ 6\hat{C}_2'\hat{R} \end{pmatrix} \quad \hat{R} \quad \begin{pmatrix} 4\hat{C}_3^2 \\ 4\hat{C}_3\hat{R} \end{pmatrix} \quad \begin{pmatrix} 3\hat{C}_4^3 \\ 3\hat{C}_4\hat{R} \end{pmatrix}$$

The irreducible representations in the octahedral double group are, in addition to the five single-valued representations A_1, A_2, E, T_1, and T_2, the double-valued representations $E_{1/2}$ (twofold degenerate), $E_{5/2}$ (twofold degenerate), and G (fourfold degenerate). The subscripts on $E_{1/2}$ and $E_{5/2}$ indicate functions with respectively $M_s = \pm\frac{1}{2}$ and $M_s = \pm\frac{5}{2}$. The G representation spans the four functions of $S = \frac{3}{2}$.

A point to note is that the double-valued representations are all of even dimension. Hence any system containing an odd number of electrons will have levels which are at least twofold degenerate. This is valid as long as no magnetic field is present and is referred to as Kramers degeneracy.

Consider the case where we have two different sets of functions $\psi_1, \psi_2, \ldots, \psi_j$ and $\phi_1, \phi_2, \ldots, \phi_i$. Each of these two sets spans a certain irreducible representation, Γ, characterized by its transformation properties. By forming the various products $\psi_k \times \phi_l$ we obtain a new set of $(j \times i)$ functions. This set of functions is called the *direct product* of the other two. Very often the new set spans more than one irreducible representation. Tables for the decomposition of direct products into irreducible representations can for instance be found in Wilson, Decius, and Cross.[4] In the O molecular symmetry we have for instance $E \times T_1 = E \times T_2 = T_1 + T_2$.

The two irreducible representations we have multiplied together may in particular be the same.[7] Let the set of functions spanning the irreducible representation Γ be $\psi_1, \psi_2, \ldots, \psi_j$. The direct product set has j^2 functions. It has the character $[\chi(\hat{A})]^2$. The direct product can in this case be decomposed into two sets. The first is the set of symmetric product functions $\psi_i\psi_k + \psi_k\psi_i$, of which there are $\frac{1}{2}j(j+1)$. The remaining $\frac{1}{2}j(j-1)$ functions are the antisymmetric product functions $\psi_i\psi_k - \psi_k\psi_i$, $i \neq k$.

The characters of the antisymmetric functions are called $\{\chi^2\}(\hat{A})$. We have[7]

$$\{\chi^2\}(\hat{A}) = \tfrac{1}{2}([\chi(\hat{A})]^2 - \chi(\hat{A}^2)) \tag{1-37}$$

The characters of the symmetric product $[\chi'^2](\hat{A})$ are given as $[\chi(\hat{A})]^2 - \{\chi^2\}(\hat{A})$ or

$$[\chi^2](\hat{A}) = \tfrac{1}{2}([\chi(\hat{A})]^2 + \chi(\hat{A}^2)) \tag{1-38}$$

Using the characters of the product sets we can then in turn decompose them into a number of basis sets which transform in the same way as irreducible representations. We have for instance in the molecular O symmetry

$$T_1 \times T_1 = T_2 \times T_2 = A_1 + E + \{T_1\} + T_2 \tag{1-39}$$

where the antisymmetric part of the product is indicated by a bracket.

The vibrational states of the molecules can likewise be symmetry-classified. For a nondegenerate normal coordinate ξ, the vibrational wave function for the ground state, $\chi_0(\xi)$, is always totally symmetric. The first excited vibrational wave function $\chi_1(\xi)$, on the other hand, transforms like ξ as can be seen from Eq. (1-36). Generally, the vibrational wave functions for a onefold degenerate normal coordinate are totally symmetric for v even, and have the symmetry of ξ for v odd.

In the case of degenerate normal coordinates, the ground state is still nondegenerate. If one quantum is excited, the corresponding vibrational state will again show the degeneracy and transformational properties of the normal coordinate itself. When more than one quantum is excited, the situation becomes more difficult, and one must use the properties of symmetric direct product representation theory to classify the resulting states. However, if an even number of a given vibration are excited, at least one of the resulting states will be totally symmetric.

We may also want to know the analytical form the decomposition of a direct product takes. For a single product function $|\Gamma_1 a\rangle |\Gamma_2 b\rangle = |\Gamma_1\Gamma_2 ab\rangle$ we can write[9]

$$|\Gamma_1\Gamma_2 ab\rangle = \sum_c \Omega(\Gamma_1\Gamma_2\Gamma c)\,|\Gamma_1\Gamma_2\Gamma c\rangle \tag{1-40}$$

where $|\Gamma_1\Gamma_2\Gamma c\rangle$ is a basis set for the irreducible representation Γ. The coupling coefficients $\Omega(\Gamma_1\Gamma_2\Gamma c)$ will form a unitary matrix and we have

$$\Omega(\Gamma_1\Gamma_2\Gamma c) = \langle \Gamma_1\Gamma_2 ab \mid \Gamma_1\Gamma_2\Gamma c\rangle \tag{1-41}$$

As an example we calculate some of the coupling coefficients in the product $T_1 \times T_1$ for the octahedral group.

With

$$|T_{1z}\rangle = \sqrt{\frac{3}{4\pi}}\, z_1$$

$$|E_{z^2}\rangle = \frac{3}{4\pi}\sqrt{\frac{1}{6}}(2z_1z_2 - x_1x_2 - y_1y_2)$$

$$|A_1\rangle = \frac{3}{4\pi}\sqrt{\frac{1}{3}}(x_1x_2 + y_1y_2 + z_1z_2)$$

we can find for $T_{1z} \times T_{1z} = c_1 E_{z^2} + c_2 A_1$

$$\frac{3}{4\pi} z_1z_2 = c_1 \frac{3}{4\pi}\sqrt{\frac{1}{6}}(2z_1z_2 - x_1x_2 - y_1y_2) + c_2 \frac{3}{4\pi}\sqrt{\frac{1}{3}}(x_1x_2 + y_1y_2 + z_1z_2)$$

or

$$c_1 = \sqrt{\frac{2}{3}}, \qquad c_2 = \sqrt{\frac{1}{3}}$$

Griffith[9] takes $c_1 = -\sqrt{\frac{2}{3}}$, $c_2 = \sqrt{\frac{1}{3}}$, but the phases of E and A can of course be chosen independently. In the same way we can take

$$|T_{1y}\rangle = \frac{3}{4\pi}\sqrt{\frac{1}{2}}(x_1z_2 - z_1x_2)$$

$$|T_{2xz}\rangle = -\frac{3}{4\pi}\sqrt{\frac{1}{2}}(x_1z_2 + z_1x_2)$$

Then for the $T_{1x} \times T_{1z} = c_3 T_{1y} + c_4 T_{2xz}$

$$\frac{3}{4\pi} x_1z_2 = c_3 \frac{3}{4\pi}\sqrt{\frac{1}{2}}(x_1z_2 - z_1x_2) - c_4 \frac{3}{4\pi}\sqrt{\frac{1}{2}}(x_1z_2 + z_1x_2)$$

or

$$c_3 = \sqrt{\frac{1}{2}}, \qquad c_4 = -\sqrt{\frac{1}{2}}$$

in accordance with the phases chosen by Griffith. The coupling coefficients used in this book follow his choice of phases.

The coupling coefficients are very useful in the computation of matrix elements. Consider the matrix elements of some irreducible tensor operator $\hat{V}(\Gamma c)$, $\langle \alpha\Gamma_1 a | \hat{V}(\Gamma c) | \alpha'\Gamma_2 b \rangle$. The matrix elements can be expressed in as many independent constants as the number of times the reducible representation $|\Gamma_1 a\rangle\,|\Gamma_2 b\rangle$ contains the representation Γc. We write

$$\langle \alpha\Gamma_1 a | \hat{V}(\Gamma c) | \alpha'\Gamma_2 b \rangle = (\Gamma_1)^{-1/2} \langle \alpha\Gamma_1 \| \hat{V}(\Gamma) \| \alpha'\Gamma_2 \rangle \langle \Gamma_1\Gamma_2 ab | \Gamma_1\Gamma_2\Gamma c \rangle \quad (1\text{-}42)$$

where $(\Gamma_1)^{1/2}$ is the square root of the dimension of Γ_1, $\langle \alpha\Gamma_1 \| \hat{V}(\Gamma) \| \alpha'\Gamma_2 \rangle$ the so-called reduced matrix element, and $\langle \Gamma_1\Gamma_2 ab | \Gamma_1\Gamma_2\Gamma c \rangle$ the coupling coefficient. The above equation is the mathematical formulation of the Wigner–Eckart theorem. The advantage of defining the reduced matrix elements is that these quantities are independent of a, b, and c. They can therefore be used as very general parameters. The coupling coefficients in octahedral, tetragonal quantization are given in Table 1-1. For real operators one can see that in octahedral symmetry $\langle \alpha\Gamma_1 \| \hat{V}(\Gamma) \| \alpha\Gamma_1 \rangle = 0$ for $\Gamma = A_2$ and T_1. The real coupling coefficients for octahedral symmetry with trigonal base, D_3, have further been given in Table 1-3, and the relations of the irreducible representations of O to the irreducible representations of the D_3 and C_{3v} sub-groups are given in Table 1-2. In Fig. 1-1 we have pictured these octahedral and trigonal coordinate systems.

Table 1-1 Coupling coefficients for the octahedral group, $\langle \Gamma_1\Gamma_2 ab | \Gamma_1\Gamma_2\Gamma_c \rangle$
Phases as in J. S. Griffith: *The Theory of Transition-Metal Ions*

$A_2 \times A_2$		A_1
		a_1
a_2	a_2	1

$A_2 \times E$		E	
		z^2	x^2-y^2
a_2	z^2	0	-1
a_2	x^2-y^2	1	0

$A_2 \times T_1$		T_2		
		yz	xz	xy
a_2	x	1	0	0
a_2	y	0	1	0
a_2	z	0	0	1

$A_2 \times T_2$		T_1		
		x	y	z
a_2	yz	1	·	·
a_2	xz	·	1	·
a_2	xy	·	·	1

E ×	E	A_1	A_2	E	
		a_1	a_2	z^2	x^2-y^2
z^2	z^2	$\frac{1}{\sqrt{2}}$	·	$-\frac{1}{\sqrt{2}}$	·
x^2-y^2	x^2-y^2	$\frac{1}{\sqrt{2}}$	·	$\frac{1}{\sqrt{2}}$	·
z^2	x^2-y^2	·	$\frac{1}{\sqrt{2}}$	·	$\frac{1}{\sqrt{2}}$
x^2-y^2	z^2	·	$-\frac{1}{\sqrt{2}}$	·	$\frac{1}{\sqrt{2}}$

Table 1-1 *Continued*

$E \times T_1$		T_1			T_2		
		x	y	z	yz	xz	xy
z^2	x	$-\frac{1}{2}$	·	·	$-\frac{1}{2}\sqrt{3}$	·	·
z^2	y	·	$-\frac{1}{2}$	·	·	$\frac{1}{2}\sqrt{3}$	·
z^2	z	·	·	1	·	·	·
$x^2 - y^2$	x	$\frac{1}{2}\sqrt{3}$	·	·	$-\frac{1}{2}$	·	·
$x^2 - y^2$	y	·	$-\frac{1}{2}\sqrt{3}$	·	·	$-\frac{1}{2}$	·
$x^2 - y^2$	z	·	·	·	·	·	1

$E \times T_2$		T_1			T_2		
		x	y	z	yz	zx	xy
z^2	yz	$-\frac{1}{2}\sqrt{3}$	·	·	$-\frac{1}{2}$	·	·
z^2	xz	·	$\frac{1}{2}\sqrt{3}$	·	·	$-\frac{1}{2}$	·
z^2	xy	·	·	·	·	·	1
$x^2 - y^2$	yz	$-\frac{1}{2}$	·	·	$\frac{1}{2}\sqrt{3}$	·	·
$x^2 - y^2$	xz	·	$-\frac{1}{2}$	·	·	$-\frac{1}{2}\sqrt{3}$	·
$x^2 - y^2$	xy	·	·	1	·	·	·

$T_1 \times T_1$ or $T_2 \times T_2$				T_1			T_2		
				x	y	z	yz	xz	xy
x	y	yz	xz	·	·	$-\frac{1}{\sqrt{2}}$	·	·	$-\frac{1}{\sqrt{2}}$
x	z	yz	xy	·	$\frac{1}{\sqrt{2}}$	·	·	$-\frac{1}{\sqrt{2}}$	·
y	x	xz	yz	·	·	$\frac{1}{\sqrt{2}}$	·	·	$-\frac{1}{\sqrt{2}}$
y	z	xz	xy	$-\frac{1}{\sqrt{2}}$	·	·	$-\frac{1}{\sqrt{2}}$	·	·
z	x	xy	yz	·	$-\frac{1}{\sqrt{2}}$	·	·	$-\frac{1}{\sqrt{2}}$	·
z	y	xy	xz	$\frac{1}{\sqrt{2}}$	·	·	$-\frac{1}{\sqrt{2}}$	·	·

Table 1-1 *Continued*

$T_1 \times T_1$ or $T_2 \times T_2$				A_1	E	
				a_1	z^2	$x^2 - y^2$
x	x	yz	yz	$\frac{1}{\sqrt{3}}$	$\frac{1}{\sqrt{6}}$	$-\frac{1}{\sqrt{2}}$
y	y	xz	xz	$\frac{1}{\sqrt{3}}$	$\frac{1}{\sqrt{6}}$	$\frac{1}{\sqrt{2}}$
z	z	xy	xy	$\frac{1}{\sqrt{3}}$	$-\frac{2}{\sqrt{6}}$	$\cdot$

$T_1 \times T_2$		A_2	E	
		a_2	z^2	$x^2 - y^2$
x	yz	$\frac{1}{\sqrt{3}}$	$-\frac{1}{\sqrt{2}}$	$-\frac{1}{\sqrt{6}}$
y	xz	$\frac{1}{\sqrt{3}}$	$\frac{1}{\sqrt{2}}$	$-\frac{1}{\sqrt{6}}$
z	xy	$\frac{1}{\sqrt{3}}$	$\cdot$	$\frac{2}{\sqrt{6}}$

$T_1 \times T_2$		T_1			T_2		
		x	y	z	yz	xz	xy
x	xz	$\cdot$	$\cdot$	$-\frac{1}{\sqrt{2}}$	$\cdot$	$\cdot$	$-\frac{1}{\sqrt{2}}$
x	xy	$\cdot$	$-\frac{1}{\sqrt{2}}$	$\cdot$	$\cdot$	$\frac{1}{\sqrt{2}}$	$\cdot$
y	yz	$\cdot$	$\cdot$	$-\frac{1}{\sqrt{2}}$	$\cdot$	$\cdot$	$\frac{1}{\sqrt{2}}$
y	xy	$-\frac{1}{\sqrt{2}}$	$\cdot$	$\cdot$	$-\frac{1}{\sqrt{2}}$	$\cdot$	$\cdot$
z	yz	$\cdot$	$-\frac{1}{\sqrt{2}}$	$\cdot$	$\cdot$	$-\frac{1}{\sqrt{2}}$	$\cdot$
z	xz	$-\frac{1}{\sqrt{2}}$	$\cdot$	$\cdot$	$\frac{1}{\sqrt{2}}$	$\cdot$	$\cdot$

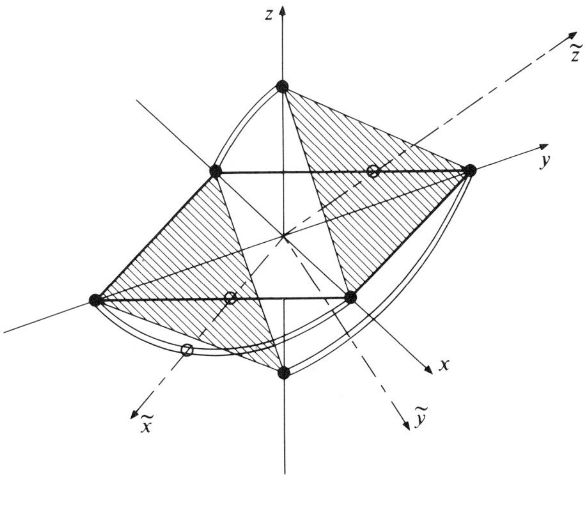

Figure 1-1 The relationship of the C_{3v} *and* D_3 coordinate systems to O_h as represented in Table 1-2.

Table 1-2 The relation of the irreducible representations of O to the irreducible representations of the subgroups D_3 and C_{3v}

O	D_3	C_{3v}
$\frac{1}{\sqrt{3}}(T_2xy + T_2yz + T_2xz)$	T_2a_1	T_2a_1
$\frac{1}{\sqrt{3}}(T_1x + T_1y + T_1z)$	T_1a_2	T_1a_1
(z^2)	Ex	Ex
$(x^2 - y^2)$	Ey	Ey
$\frac{1}{\sqrt{2}}(T_1x - T_1y)$	T_1x	T_1x
$\frac{1}{\sqrt{6}}(T_1x + T_1y - 2T_1z)$	T_1y	T_1y
$-\frac{1}{\sqrt{6}}(T_2yz + T_2xz - 2T_2xy)$	T_2x	T_2x
$\frac{1}{\sqrt{2}}(T_2yz - T_2xz)$	T_2y	T_2y

Table 1-3 Coupling coefficients for octahedral symmetry with trigonal base, D_3, derived from Tables 1-1 and 1-2

$E \times T_1$	T_1x	T_1y	T_1a_2	T_2x	T_2y	T_2a_1
$ex\ \ t_1x$	$-\frac{1}{2}$	·	·	$\frac{1}{2}$	·	$-\frac{1}{\sqrt{2}}$
$ex\ \ t_1y$	·	$\frac{1}{2}$	$-\frac{1}{\sqrt{2}}$	·	$-\frac{1}{2}$	·
$ex\ \ t_1a_2$	·	$-\frac{1}{\sqrt{2}}$	·	·	$-\frac{1}{\sqrt{2}}$	·
$ey\ \ t_1x$	·	$\frac{1}{2}$	$\frac{1}{\sqrt{2}}$	·	$-\frac{1}{2}$	·
$ey\ \ t_1y$	$\frac{1}{2}$	·	·	$-\frac{1}{2}$	·	$-\frac{1}{\sqrt{2}}$
$ey\ \ t_1a_2$	$\frac{1}{\sqrt{2}}$	·	·	$\frac{1}{\sqrt{2}}$	·	·

Table 1-3 *Continued*

$E \times T_2$	T_1x	T_1y	T_1a_2	T_2x	T_2y	T_2a_1
$ex \quad t_2x$	$\frac{1}{2}$	·	·	$\frac{1}{2}$	·	$\frac{1}{\sqrt{2}}$
$ex \quad t_2y$	·	$-\frac{1}{2}$	$-\frac{1}{\sqrt{2}}$	·	$-\frac{1}{2}$	·
$ex \quad t_2a_1$	$-\frac{1}{\sqrt{2}}$	·	·	$\frac{1}{\sqrt{2}}$	·	·
$ey \quad t_2x$	·	$-\frac{1}{2}$	$\frac{1}{\sqrt{2}}$	·	$-\frac{1}{2}$	·
$ey \quad t_2y$	$-\frac{1}{2}$	·	·	$-\frac{1}{2}$	·	$\frac{1}{\sqrt{2}}$
$ey \quad t_2a_1$	·	$-\frac{1}{\sqrt{2}}$	·	·	$\frac{1}{\sqrt{2}}$	·

$T_1 \times T_2$	A_2	Ex	Ey	T_1x	T_1y	T_1a_2	T_2x	T_2y	T_2a_1
$t_1a_2 \quad t_2a_1$	$\frac{1}{\sqrt{3}}$	·	·	·	·	$-\sqrt{\frac{2}{3}}$	·	·	·
$t_1y \quad t_2a_1$	·	·	$-\frac{1}{\sqrt{3}}$	·	$\frac{1}{\sqrt{6}}$	·	·	$-\frac{1}{\sqrt{2}}$	·
$t_1x \quad t_2a_1$	·	$-\frac{1}{\sqrt{3}}$	·	$\frac{1}{\sqrt{6}}$	·	·	$-\frac{1}{\sqrt{2}}$	·	·
$t_1a_2 \quad t_2y$	·	$-\frac{1}{\sqrt{3}}$	·	$\frac{1}{\sqrt{6}}$	·	·	$\frac{1}{\sqrt{2}}$	·	·
$t_1y \quad t_2y$	·	$-\frac{1}{\sqrt{6}}$	·	$-\frac{1}{\sqrt{3}}$	·	·	·	·	$\frac{1}{\sqrt{2}}$
$t_1x \quad t_2y$	$\frac{1}{\sqrt{3}}$	·	$-\frac{1}{\sqrt{6}}$	·	$-\frac{1}{\sqrt{3}}$	$\frac{1}{\sqrt{6}}$	·	·	·
$t_1a_2 \quad t_2x$	·	·	$\frac{1}{\sqrt{3}}$	·	$-\frac{1}{\sqrt{6}}$	·	·	$-\frac{1}{\sqrt{2}}$	·
$t_1y \quad t_2x$	$-\frac{1}{\sqrt{3}}$	·	$-\frac{1}{\sqrt{6}}$	·	$-\frac{1}{\sqrt{3}}$	$-\frac{1}{\sqrt{6}}$	·	·	·
$t_1x \quad t_2x$	·	$\frac{1}{\sqrt{6}}$	·	$\frac{1}{\sqrt{3}}$	·	·	·	·	$\frac{1}{\sqrt{2}}$

Table 1-3 *Continued*

$T_1 \times T_1$		A_1	Ex	Ey	T_1x	T_1y	T_1a_2	T_2x	T_2y	T_2a_1
t_1a_2	t_1a_2	$\frac{1}{\sqrt{3}}$	·	·	·	·	·	·	·	$-\sqrt{\frac{2}{3}}$
t_1y	t_1a_2	·	$\frac{1}{\sqrt{3}}$	·	$-\frac{1}{\sqrt{2}}$	·	·	$-\frac{1}{\sqrt{6}}$	·	·
t_1x	t_1a_2	·	·	$-\frac{1}{\sqrt{3}}$	·	$\frac{1}{\sqrt{2}}$	·	·	$\frac{1}{\sqrt{6}}$	·
t_1a_2	t_1y	·	$\frac{1}{\sqrt{3}}$	·	$\frac{1}{\sqrt{2}}$	·	·	$-\frac{1}{\sqrt{6}}$	·	·
t_1y	t_1y	$\frac{1}{\sqrt{3}}$	$-\frac{1}{\sqrt{6}}$	·	·	·	·	$-\frac{1}{\sqrt{3}}$	·	$\frac{1}{\sqrt{6}}$
t_1x	t_1y	·	·	$-\frac{1}{\sqrt{6}}$	·	·	$-\frac{1}{\sqrt{2}}$	·	$-\frac{1}{\sqrt{3}}$	·
t_1a_2	t_1x	·	·	$-\frac{1}{\sqrt{3}}$	·	$-\frac{1}{\sqrt{2}}$	·	·	$\frac{1}{\sqrt{6}}$	·
t_1y	t_1x	·	·	$-\frac{1}{\sqrt{6}}$	·	·	$\frac{1}{\sqrt{2}}$	·	$-\frac{1}{\sqrt{3}}$	·
t_1x	t_1x	$\frac{1}{\sqrt{3}}$	$\frac{1}{\sqrt{6}}$	·	·	·	·	$\frac{1}{\sqrt{3}}$	·	$\frac{1}{\sqrt{6}}$

$T_2 \times T_2$		A_1	Ex	Ey	T_1x	T_1y	T_1a_2	T_2x	T_2y	T_2a_1
t_2a_1	t_2a_1	$\frac{1}{\sqrt{3}}$	·	·	·	·	·	·	·	$-\sqrt{\frac{2}{3}}$
t_2y	t_2a_1	·	·	$-\frac{1}{\sqrt{3}}$	·	$\frac{1}{\sqrt{2}}$	·	·	$\frac{1}{\sqrt{6}}$	·
t_2x	t_2a_1	·	$-\frac{1}{\sqrt{3}}$	·	$\frac{1}{\sqrt{2}}$	·	·	$\frac{1}{\sqrt{6}}$	·	·
t_2a_1	t_2y	·	·	$-\frac{1}{\sqrt{3}}$	·	$-\frac{1}{\sqrt{2}}$	·	·	$\frac{1}{\sqrt{6}}$	·
t_2y	t_2y	$\frac{1}{\sqrt{3}}$	$\frac{1}{\sqrt{6}}$	·	·	·	·	$\frac{1}{\sqrt{3}}$	·	$\frac{1}{\sqrt{6}}$
t_2x	t_2y	·	·	$\frac{1}{\sqrt{6}}$	·	·	$-\frac{1}{\sqrt{2}}$	·	$\frac{1}{\sqrt{3}}$	·
t_2a_1	t_2x	·	$-\frac{1}{\sqrt{3}}$	·	$-\frac{1}{\sqrt{2}}$	·	·	$\frac{1}{\sqrt{6}}$	·	·
t_2y	t_2x	·	·	$\frac{1}{\sqrt{6}}$	·	·	$\frac{1}{\sqrt{2}}$	·	$\frac{1}{\sqrt{3}}$	·
t_2x	t_2x	$\frac{1}{\sqrt{3}}$	$-\frac{1}{\sqrt{6}}$	·	·	·	·	$-\frac{1}{\sqrt{3}}$	·	$\frac{1}{\sqrt{6}}$

1-5 THE POTENTIAL SURFACES

The transformation, Eq. (1-24), which generates the normal coordinates from the displacement coordinates, is in general not independent of the electronic state of the molecule. In order to investigate this dependence we expand the molecular hamiltonian around the *ground state* equilibrium nuclear geometry point $\mathbf{Q}^0$ with respect to symmetry—adapted displacement coordinates, $\tilde{S}$, rather than the elementary nuclear displacement coordinates of Eq. (1-21). Since these two sets of coordinate systems are related by means of an orthonormal transformation the results are equivalent. We get

$$\mathscr{H} = \mathscr{H}^0 + \sum_{k=1}^{3N} \left(\frac{\partial V}{\partial \tilde{S}_k}\right)_0 \tilde{S}_k + \frac{1}{2}\sum_{i,j}^{3N}\left(\frac{\partial^2 V}{\partial \tilde{S}_i\,\partial \tilde{S}_j}\right)_0 \tilde{S}_i\tilde{S}_j + \cdots \tag{1-43}$$

We are not interested in the translations T and rotations R of the molecule, and shall consequently put $S(T_a)$, $S(T_b)$, $S(T_c)$, $S(R_a)$, $S(R_b)$, and $S(R_c)$ equal to zero. Then

$$\mathscr{H} = \mathscr{H}^0 + \sum_{k=1}^{3N-6} \left(\frac{\partial V}{\partial \tilde{S}_k}\right)_0 \tilde{S}_k + \frac{1}{2}\sum_{i,j}^{3N-6}\left(\frac{\partial^2 V}{\partial \tilde{S}_i\,\partial \tilde{S}_j}\right)_0 \tilde{S}_i\tilde{S}_j + \cdots \tag{1-44}$$

Here $(\partial V/\partial \tilde{S}_k)_0$ and $(\partial^2 V/\partial \tilde{S}_i\,\partial \tilde{S}_j)_0$ depend only upon the electronic coordinates. Since the hamiltonian must transform like a totally symmetric function it follows that $(\partial V/\partial \tilde{S}_k)_0$ transform like $\tilde{S}_k$ and that $(\partial^2 V/\partial \tilde{S}_i\,\partial \tilde{S}_j)_0$ transform like $\tilde{S}_i\tilde{S}_j$ in the point group of the molecule.

We now assume that the crude adiabatic approximation gives an adequate description of the ground and excited electronic states $\psi_t^0(\mathbf{r})$, and we take the states to be nondegenerate. The potential energy for the molecular vibrations are then, according to Eq. (1-18), given by

$$V_t(\tilde{S}) = \langle \psi_t^0(\mathbf{r})\,|\,\mathscr{H}\,|\,\psi_t^0(\mathbf{r})\rangle \tag{1-45}$$

or by insertion of $\mathscr{H}$ from Eq. (1-44)

$$\begin{aligned} V_t(\tilde{S}) = V_t^0 &+ \langle \psi_t^0(\mathbf{r})|\sum_{k=1}^{3N-6}\left(\frac{\partial V}{\partial \tilde{S}_k}\right)_0 \tilde{S}_k\,|\,\psi_t^0(\mathbf{r})\rangle \\ &+ \frac{1}{2}\langle \psi_t^0(\mathbf{r})|\sum_{i,j}^{3N-6}\left(\frac{\partial^2 V}{\partial \tilde{S}_i\,\partial \tilde{S}_j}\right)_0 \tilde{S}_i\tilde{S}_j\,|\,\psi_t^0(\mathbf{r})\rangle + \cdots \end{aligned} \tag{1-46}$$

$\psi_t^0(\mathbf{r}) \times \psi_t^0(\mathbf{r})$ transform like totally symmetric functions. For the integrations over the electronic coordinates to be different from zero $(\partial V/\partial \tilde{S}_k)_0$ and $(\partial^2 V/\partial \tilde{S}_i\,\partial \tilde{S}_j)_0$ must therefore be totally symmetric. It follows from what we said above that $\tilde{S}_k$ behaves like a totally symmetric vibrational coordinate, $\tilde{S}_1$, and that $\tilde{S}_i$ transforms like $\tilde{S}_j$.

For the ground state, $t = 0$, $\langle \psi_0^0(\mathbf{r})\,|\,(\partial V/\partial \tilde{S}_1)_0\,|\,\psi_0^0(\mathbf{r})\rangle$ equals zero since we are at a minimum for the electronic energy. By taking $V_0^0 = 0$ as our zero of energy, Eq. (1-46) is turned into

$$V_0(\tilde{S}) = \frac{1}{2}\sum_{i,j}^{3N-6}\langle \psi_0^0(\mathbf{r})\,|\left(\frac{\partial^2 V}{\partial \tilde{S}_i\,\partial \tilde{S}_j}\right)_0|\,\psi_0^0(\mathbf{r})\rangle\,\tilde{S}_i\tilde{S}_j \tag{1-47}$$

Notice that only the vibrational symmetry coordinates which transform in the same way give cross-terms in Eq. (1-47). A certain linear combination of the $\tilde{S}$ coordinates can then reduce the quadratic form of Eq. (1-47) to an expression in which only the diagonal terms appear, viz.

$$V_0(\xi) = \frac{1}{2} \sum_{i}^{3N-6} a_i^0 \xi_{0i}^2 \tag{1-48}$$

$V_0(\xi)$ is seen to depend upon the normal vibrational coordinates for the ground state.

Consider now the case $t \neq 0$. There is then no reason to expect $\langle \psi_t(\mathbf{r}) | (\partial V/\partial \tilde{S}_1)_0 | \psi_t(\mathbf{r}) \rangle$ to be different from zero. Furthermore, the values of $\langle \psi_t^0(\mathbf{r}) | (\partial^2 V/\partial \tilde{S}_i \, \partial \tilde{S}_j)_0 | \psi_t^0(\mathbf{r}) \rangle$, $t \neq 0$, will certainly differ from those obtained for $t = 0$. The transformation which reduced the quadratic form of $\tilde{S}$ for $t = 0$ will therefore not, in general, lead to a reduction of the quadratic form for a potential surface, $t \neq 0$, of the same molecule. To obtain a quadratic form for $t \neq 0$ a different transformation will have to be used for each state t. The a_i^t coefficients in the expression for $V_t(\xi) = \frac{1}{2} \sum_i a_i^t \xi_{ti}^2$ will therefore be dependent upon which surface we are on. The implication of these observations is that the normal coordinates of an upper adiabatic state in general correspond to a translation and rotation in the space spanned by the ground state normal coordinates. This conclusion was first stated by Duschinky[8] in 1937, and is referred to as the Duschinky effect. Notice that in case the excited state has a different molecular symmetry than the ground state, one should use the highest common point group for the symmetry considerations.

As an example suppose that we have an electronic ground state with two vibrations, v_1^0 and v_2^0, having the same symmetry. Provided the Duschinsky effect is operative, v_1^t and v_2^t will be found also in the electronic excited state t, but having been "mixed" their frequencies will be different from those observed in the ground state. The presence of the Duschinky effect may therefore be detected by comparing the absorption and emission spectra of a molecule, since the "mirror symmetry" of these two properties is destroyed.[10]

In general the electronic energy surface for a molecule in the electronic state t for which the adiabatic approximation is valid will depend parametrically upon the normal coordinates, and we define the potential surface of the molecule as

$$V_t(\xi) = V_t^0 + \frac{1}{2} \sum_{u=1}^{3N-6} a_{tu}(\xi_{tu})^2 + \cdots \tag{1-49}$$

where V_t^0 is the energy at the minimum.

Let $\hat{A}$ be a symmetry operator for a certain molecule. Then $[\hat{A}, \mathscr{H}] = 0$ and with

$$\mathscr{H}_E \psi_t^0(\mathbf{r}) = V_t(\xi) \psi_t^0(\mathbf{r}) \tag{1-50}$$

we easily get by operating on both sides of Eq. (1-50) with $\hat{A}$

$$\mathscr{H}_E(\hat{A}\psi_t^0(\mathbf{r})) = (\hat{A} V_t(\xi))(\hat{A}\psi_t^0(\mathbf{r})) \tag{1-51}$$

It is always possible to construct our wave functions $\psi_t^0(\mathbf{r})$ to be eigenfunctions of the symmetry operator $\hat{A}$. Therefore

$$\hat{A}\psi_t^0(\mathbf{r}) = c\psi_t^0(\mathbf{r}) \tag{1-52}$$

Substituting Eq. (1-52) into Eq. (1-51) gives, after division by the constant c,

$$\mathscr{H}_E\psi_t^0(\mathbf{r}) = (\hat{A}V_t(\xi))\psi_t^0(\mathbf{r}) \tag{1-53}$$

A comparison with Eq. (1-50) shows that

$$V_t(\xi) = \hat{A}V_t(\xi) \equiv V_t(\hat{A}\xi_1, \hat{A}\xi_2, \ldots, \hat{A}\xi_{3N-6}) \tag{1-54}$$

In other words, the potential surface must be invariant under all symmetry operations pertinent to the molecule.

Let us consider as an example an equilateral-triangular arrangement of three nuclei, transforming in the molecular point group D_3 having the symmetry operations $\hat{C}_2$ and $\hat{C}_3$. By standard methods[6] we find that such a molecule possesses three vibrational symmetry coordinates, one of which transforms like a totally symmetric representation α_1, the two others, ε_a and ε_b, spanning a doubly degenerate ε representation.

Expanding the potential surface to third order in the normal coordinates we get

$$V(\xi) = V^0 + \sum_{i,j} c_{ij}\xi_i\xi_j + \sum_{i,j,k} c_{ijk}\xi_i\xi_j\xi_k \tag{1-55}$$

As vibrational coordinates we take $\alpha_1(\xi_1)$ and $\varepsilon(\xi_{2a}, \xi_{2b})$. With

$$\hat{C}_2\begin{pmatrix}\xi_1\\ \xi_{2a}\\ \xi_{2b}\end{pmatrix} = \begin{pmatrix}1 & 0 & 0\\ 0 & 1 & 0\\ 0 & 0 & -1\end{pmatrix}\begin{pmatrix}\xi_1\\ \xi_{2a}\\ \xi_{2b}\end{pmatrix} \tag{1-56}$$

and

$$\hat{C}_3\begin{pmatrix}\xi_1\\ \xi_{2a}\\ \xi_{2b}\end{pmatrix} = \begin{pmatrix}1 & 0 & 0\\ 0 & -\dfrac{1}{2} & -\dfrac{\sqrt{3}}{2}\\ 0 & \dfrac{\sqrt{3}}{2} & -\dfrac{1}{2}\end{pmatrix}\begin{pmatrix}\xi_1\\ \xi_{2a}\\ \xi_{2b}\end{pmatrix} \tag{1-57}$$

the only terms in Eq. (1-55) which are invariant under $\hat{C}_2$ and $\hat{C}_3$ are easily seen to be

$$\begin{aligned}V(\xi) = V^0 &+ \tfrac{1}{2}c_{11}\xi_1^2 + \tfrac{1}{2}c_{22}(\xi_{2a}^2 + \xi_{2b}^2) + \tfrac{1}{6}c_{111}\xi_1^3\\ &+ \tfrac{1}{6}c_{122}\xi_1(\xi_{2a}^2 + \xi_{2b}^2) + \tfrac{1}{6}c_{222}(\xi_{2a}^3 - 3\xi_{2a}\xi_{2b}^2)\end{aligned} \tag{1-58}$$

Denoting the amplitudes of the vibrations ρ we can take $\xi_1 = \rho_1$, $\xi_{2a} = \rho_2 \cos\phi$, and $\xi_{2b} = \rho_2 \sin\phi$. Equation (1-58) is then transformed to

$$V(\xi) = V^0 + \tfrac{1}{2}c_{11}\rho_1^2 + \tfrac{1}{2}c_{22}\rho_2^2 + \tfrac{1}{6}c_{111}\rho_1^3 + \tfrac{1}{6}c_{122}\rho_1\rho_2^2 + \tfrac{1}{2}c_{222}\rho_2^3\cos 3\phi \tag{1-59}$$

Notice that provided c_{222} is small, the parabolic second-order potential surface just exhibits a threefold angular dependence, and the minimum remains at the origin. Should c_{222} turn out to be large, however, the system may not be stable as an equilateral triangle and the molecular equilibrium would correspond to a nonequilateral nuclear conformation.

1-6 THE JAHN–TELLER AND RENNER THEOREMS

Consider now the case where for a certain nuclear arrangement $\mathbf{Q}^0$ we have two or more degenerate solutions to the electronic Schrödinger equation. The degeneracy is therefore symmetry-determined. In order to calculate the potential surfaces we expand $\mathscr{H}$ around $\mathbf{Q}^0$ and use degenerate perturbation theory. As basis functions we take the crude adiabatic n-fold-degenerate electronic functions $\psi_1^0(\mathbf{r}), \psi_2^0(\mathbf{r}), \ldots, \psi_n^0(\mathbf{r})$. To second order in the symmetry-adapted displacement coordinates $\tilde{S}$ the hamiltonian is

$$\mathscr{H} = \mathscr{H}^0 + \sum_i^{3N-6} \left(\frac{\partial V}{\partial \tilde{S}_i}\right)_0 \tilde{S}_i + \frac{1}{2} \sum_{i,j}^{3N-6} \left(\frac{\partial^2 V}{\partial \tilde{S}_i \, \partial \tilde{S}_j}\right)_0 \tilde{S}_i \tilde{S}_j \qquad (1\text{-}60)$$

A displacement of the nuclei which destroys the symmetry-induced degeneracy of the electronic functions will lead to a splitting of the state. To first order in the nuclear-displacement coordinates this splitting is determined by the value of the matrix elements

$$\sum_i \tilde{S}_i \langle \psi_m^0(\mathbf{r}) | \left(\frac{\partial V}{\partial \tilde{S}_i}\right)_0 | \psi_n^0(\mathbf{r}) \rangle \qquad (1\text{-}61)$$

A totally symmetric displacement cannot in the first order destroy the molecular symmetry. Concentrating therefore on the non–totally symmetric displacements we observe that the matrix elements in Eq. (1-61) will be different from zero provided the symmetric product† of the representation which is spanned by $\psi_1^0(\mathbf{r}), \ldots, \psi_n^0(\mathbf{r})$ contains a representation spanned by $\sum_i (\partial V/\partial \tilde{S}_i)_0$. Now $(\partial V/\partial \tilde{S}_i)_0$ must on the other hand transform like the symmetry-adapted displacement coordinate $\tilde{S}_i$ in the point group of the molecule, since otherwise the terms in the hamiltonian $\sum_i (\partial V/\partial \tilde{S}_i)_0 \tilde{S}_i$ cannot transform like a totally symmetric function. In a systematic investigation Jahn and Teller[11] showed that for all molecules, except linear systems, a nuclear arrangement which will lead to electronic degeneracies will also automatically have at least one non–totally symmetric vibrational symmetry coordinate present, such that some of the matrix elements of Eq. (1-61) are different from zero.

The Jahn–Teller theorem therefore tells us that electronic degeneracy in a molecule destroys the symmetry on which it is based. The treatment of this effect is split up into two parts. The elucidation of the potential surfaces is referred to as the *static* Jahn–Teller effect. Solving for the vibrational motions of

† If the electronic wave functions span a double group representation, the antisymmetric product should be used.

the nuclei on the potential surfaces, thereby getting the energy of the system, constitutes the *dynamic* Jahn–Teller effect.[12,13]

It may be that the zero-order electronic states are not degenerate, but are capable of interacting appreciably under the hamiltonian of Eq. (1-60). In such a case we talk about a *pseudo* Jahn–Teller effect.

To clarify the consequences of the Jahn–Teller theorem we shall treat as an example a molecule with the shape of an equilateral triangle. The only electronic degeneracy which can occur here is a twofold one. The symmetric product of the doubly degenerate representation is, in a molecule of D_3 symmetry,

$$[E^2] = E + A_1 \tag{1-62}$$

The vibrational symmetry coordinates of the molecule have previously been found to transform like α_1 and ε. As before, the vibrational coordinates are taken to be $\alpha_1(\xi_1)$ and $\varepsilon(\xi_{2a}, \xi_{2b})$. The two real, degenerate electronic wave functions spanning an E state in D_3 symmetry we call $\psi_A^0(\mathbf{r})$ and $\psi_B^0(\mathbf{r})$. The expanded hamiltonian is to second order given by

$$\mathscr{H} = \mathscr{H}^0 + \sum_i \left(\frac{\partial V}{\partial \xi_i}\right)_0 \xi_i + \frac{1}{2}\sum_{i,j}\left(\frac{\partial^2 V}{\partial \xi_i\, \partial \xi_j}\right)_0 \xi_i \xi_j \tag{1-63}$$

where i and j run over the three indices 1, $2a$, and $2b$.

In order to find the potential surfaces we use degenerate perturbation theory and solve the *secular equation*

$$\begin{vmatrix} H_{AA} - V & H_{AB} \\ H_{BA} & H_{BB} - V \end{vmatrix} = 0 \tag{1-64}$$

where $H_{AA} = \langle\psi_A^0(\mathbf{r})|\mathscr{H}|\psi_A^0(\mathbf{r})\rangle$, $H_{BB} = \langle\psi_B^0(\mathbf{r})|\mathscr{H}|\psi_B^0(\mathbf{r})\rangle$ and $H_{AB} = H_{BA} = \langle\psi_A^0(\mathbf{r})|\mathscr{H}|\psi_B^0(\mathbf{r})\rangle$.

The components of the matrix elements H_{AA}, H_{BB}, and H_{AB} which are invariant under the symmetry operator $\hat{C}_2$ of the molecule can be written down by inspection, using

$$\hat{C}_2\begin{pmatrix}\psi_A^0 \\ \psi_B^0\end{pmatrix} = \begin{pmatrix}1 & 0 \\ 0 & -1\end{pmatrix}\begin{pmatrix}\psi_A^0 \\ \psi_B^0\end{pmatrix} \tag{1-65}$$

This together with Eq. (1-56) leads to

$$H_{AA} = V^0 + a_1\xi_1 + a_2\xi_1^2 + a_3\xi_{2a} + a_4\xi_{2a}^2 + a_5\xi_{2b}^2 + a_6\xi_1\xi_{2a} \tag{1-66}$$

$$H_{BB} = V^0 + a_7\xi_1 + a_8\xi_1^2 + a_9\xi_{2a} + a_{10}\xi_{2a}^2 + a_{11}\xi_{2b}^2 + a_{12}\xi_1\xi_{2a} \tag{1-67}$$

$$H_{AB} = H_{BA} = a_{13}\xi_{2b} + a_{14}\xi_1\xi_{2b} + a_{15}\xi_{2a}\xi_{2b} \tag{1-68}$$

We have here written for instance

$$a_6 = \frac{1}{2}\langle\psi_A^0(\mathbf{r})|\left(\frac{\partial^2 V}{\partial\xi_1\,\partial\xi_{2a}}\right)_0|\psi_A^0(\mathbf{r})\rangle \tag{1-69}$$

and similarly for the other constants $a_1, \ldots, a_{15}$.

Operating now on H_{AA} with the $\hat{C}_3$ symmetry operator of the molecule leads to an elimination of all the constants $a_7, \ldots, a_{15}$. Demanding invariance of the matrix element, we get easily

$$H_{AA} = V^0 + a_1\xi_1 + a_2\xi_1^2 + a_3\xi_{2a} + a_4\xi_{2a}^2 + a_5\xi_{2b}^2 + a_6\xi_1\xi_{2a} \tag{1-70}$$

$$H_{BB} = V^0 + a_1\xi_1 + a_2\xi_1^2 - a_3\xi_{2a} + a_5\xi_{2a}^2 + a_4\xi_{2b}^2 - a_6\xi_1\xi_{2a} \tag{1-71}$$

$$H_{AB} = H_{BA} = -a_3\xi_{2b} - a_6\xi_1\xi_{2b} + (a_4 - a_5)\xi_{2a}\xi_{2b} \tag{1-72}$$

From the form of Eqs. (1-70) and (1-71) we note that it will lead to more symmetrical results if we take

$$\psi_+ = \frac{1}{\sqrt{2}}(\psi_B^0(\mathbf{r}) + i\psi_A^0(\mathbf{r})) \tag{1-73}$$

$$\psi_- = \frac{1}{\sqrt{2}}(\psi_B^0(\mathbf{r}) - i\psi_A^0(\mathbf{r})) \tag{1-74}$$

Then

$$H_{++} = H_{--} = V^0 + a_1\xi_1 + a_2\xi_1^2 + \tfrac{1}{2}(a_4 + a_5)(\xi_{2a}^2 + \xi_{2b}^2) \tag{1-75}$$

$$H_{-+} = (-a_3 - a_6\xi_1)(\xi_{2a} + i\xi_{2b}) - \tfrac{1}{2}(a_4 - a_5)(\xi_{2a} - i\xi_{2b})^2 \tag{1-76}$$

$$H_{+-} = (-a_3 - a_6\xi_1)(\xi_{2a} - i\xi_{2b}) - \tfrac{1}{2}(a_4 - a_5)(\xi_{2a} + i\xi_{2b})^2 \tag{1-77}$$

Substituting Eqs. (1-75, (1-76), and (1-77) into Eq. (1-64) yields for the two potential surfaces

$$\begin{aligned} V = V^0 &+ a_1\xi_1 + a_2\xi_1^2 + \tfrac{1}{2}(a_4 + a_5)(\xi_{2a}^2 + \xi_{2b}^2) \\ &\pm [(\xi_{2a}^2 + \xi_{2b}^2)(a_3 + a_6\xi_1)^2 + \tfrac{1}{4}(a_4 - a_5)^2(\xi_{2a}^2 + \xi_{2b}^2)^2 \\ &+ (a_3 + a_6\xi_1)(a_4 - a_5)\xi_{2a}(\xi_{2a}^2 - 3\xi_{2b}^2)]^{1/2} \end{aligned} \tag{1-78}$$

Making the substitutions:

$(a_4 + a_5) = k_2$, $a_2 = \tfrac{1}{2}k_1$, $(a_4 - a_5) = k_\varepsilon$, $\xi_1 = \rho_1$, $\xi_{2a} = \rho_2 \cos\phi$, and $\xi_{2b} = \rho_2 \sin\phi$, we transform Eq. (1-78) to

$$\begin{aligned} V = V^0 &+ a_1\rho_1 + \tfrac{1}{2}k_1\rho_1^2 + \tfrac{1}{2}k_2\rho_2^2 \\ &\pm \rho_2[(a_3 + a_6\rho_1)^2 + \tfrac{1}{4}k_\varepsilon^2\rho_2^2 + (a_3 + a_6\rho_1)k_\varepsilon\rho_2 \cos 3\phi]^{1/2} \end{aligned} \tag{1-79}$$

Note that the two potential surfaces associated with $\psi_A^0(\mathbf{r})$ and $\psi_B^0(\mathbf{r})$ are split, and that each surface has a threefold angular modulation associated with it. We observe that the threefold angular dependence of the potential surfaces originates from the second-order terms in the expansion of the hamiltonian, whereas for a nondegenerate state such a dependence first turned up in the third order.

Provided $(a_3 + a_6\rho_1)k_\varepsilon$ is positive the lower surface will have its minima at $\cos 3\phi = 1$, or $\phi = 0$, $2\pi/3$, and $4\pi/3$. At these points the quantity in the square parentheses is a perfect square and we get

$$V = V^0 + a_1\rho_1 + \tfrac{1}{2}k_1\rho_1^2 + \tfrac{1}{2}k_2\rho_2^2 - \rho_2[(a_3 + a_6\rho_1) + \tfrac{1}{2}k_\varepsilon\rho_2] \tag{1-80}$$

The minimum value for V is found by differentiating Eq. (1-80) with respect to

ρ_1 and ρ_2. We get

$$\rho_1^{\min} = \frac{a_3 a_6 - a_1(k_2 - k_\varepsilon)}{k_1(k_2 - k_\varepsilon) - a_6^2} \tag{1-81}$$

$$\rho_2^{\min} = \frac{k_1 a_3 - a_1 a_6}{k_1(k_2 - k_\varepsilon) - a_6^2} \tag{1-82}$$

The movement of the nuclei toward the equilibrium configuration is seen to be determined by a coupling between the vibrations α_1 and ε. The distortion $\alpha_1(\rho_1)$ cannot remove the twofold electronic degeneracy, since the threefold rotational symmetry is preserved. Concentrating on the effects of the doubly degenerate vibration ε, we may therefore take the zero-order nuclear configuration $\mathbf{Q}^0$ to be a point where the electronic energy has been minimized with respect to ρ_1. Hence we take $\rho_1^{\min} = 0$, or from Eq. (1-81)

$$a_3 a_6 = a_1(k_2 - k_\varepsilon) \tag{1-83}$$

Substituting Eq. (1-83) into Eq. (1-82) then gives

$$\rho_2^{\min} = \frac{a_3}{k_2 - k_\varepsilon} \approx \frac{a_3}{k_2} \tag{1-84}$$

and

$$V^{\min} = V^0 - \frac{a_3^2}{2(k_2 - k_\varepsilon)} \approx V^0 - \frac{a_3^2}{2k_2} \tag{1-85}$$

The stabilization energy $\Delta V_{J-T} = a_3^2/2k_2$ is referred to as the *Jahn–Teller energy*.

Dissociation will take place provided $k_\varepsilon > k_2$. However, under "normal" conditions $k_\varepsilon \ll k_2$ and we may use the last equalities in Eqs. (1-84) and (1-85). Notice also from Eqs. (1-81) and (1-82) that dissociation will occur provided $a_6 \approx (k_1 k_2)^{1/2}$.

Had we investigated the saddle points of the lower surface where $\cos 3\phi = -1$ we would have found

$$\rho_2^{\text{saddle}} = \frac{a_3}{k_2 + k_\varepsilon} \tag{1-86}$$

$$V^{\text{saddle}} = V^0 - \frac{a_3^2}{2(k_2 + k_\varepsilon)} \tag{1-87}$$

Thus for $k_2 \gg k_\varepsilon$ the lower surface has a trough of approximate depth $a_3^2/2k_2$ at the circle $\rho_2 = a_3/k_2$. The potential at the bottom of the trough has a threefold barrier, with a barrier height of $k_\varepsilon a_3^2/k_2^2$.

Neglecting the ξ_1 coordinate and retaining only the first-order terms in ξ_{2a} and ξ_{2b} in Eqs. (1-76) and (1-77), the secular equation, Eq. (1-64), is simply given by

$$\begin{vmatrix} \frac{1}{2}k_2(\xi_{2a}^2 + \xi_{2b}^2) - V & -a_3(\xi_{2a} - i\xi_{2b}) \\ -a_3(\xi_{2a} + i\xi_{2b}) & \frac{1}{2}k_2(\xi_{2a}^2 + \xi_{2b}^2) - V \end{vmatrix} = 0 \tag{1-88}$$

The transformation $\xi_{2a} = \rho \cos \phi$ and $\xi_{2b} = \rho \sin \phi$ gives $V = \frac{1}{2}k\rho^2 \pm a\rho$. The corresponding wave equations are easily found to be for the lower surface

$$\Psi^{\text{low}} = \frac{1}{\sqrt{2}} \exp(-i\phi/2)\,\psi_+ + \frac{1}{\sqrt{2}} \exp(i\phi/2)\,\psi_- \tag{1-89}$$

and for the higher surface

$$\Psi^{\text{high}} = \frac{1}{\sqrt{2}} \exp(-i\phi/2)\,\psi_+ - \frac{1}{\sqrt{2}} \exp(i\phi/2)\,\psi_- \tag{1-90}$$

The two potential sheets are pictured in Fig. 1-2.

The dynamic Jahn–Teller effect for the D_3 system is now investigated by taking a wave function

$$\Psi = \psi_+\chi_+(\xi) + \psi_-\chi_-(\xi) \tag{1-91}$$

where we can look upon the vibrational functions $\chi_+(\xi)$ and $\chi_-(\xi)$ as variational parameters. For the combined electronic and vibrational (called "vibronic") problem the hamiltonian is given by Eq. (1-1). We consider only the doubly degenerate ε vibration and have to first order in (ξ_{2a}, ξ_{2b})

$$\mathscr{H} = \mathscr{H}^0 + V^{(1)}$$

With p_{2a}, p_{2b} being the momenta canonically conjugate to ξ_{2a}, ξ_{2b} we have

$$\mathscr{H}^0 = \frac{1}{2M_\varepsilon}(p_{2a}^2 + p_{2b}^2) + \tfrac{1}{2}k_2(\xi_{2a}^2 + \xi_{2b}^2) \tag{1-92}$$

$$V^{(1)} = -a_3(\xi_{2a}\sigma_1 + \xi_{2b}\sigma_2) \tag{1-93}$$

where σ_1, σ_2, and σ_3 are the Pauli matrices

$$\sigma_1 = \begin{pmatrix} 0 & 1 \\ 1 & 0 \end{pmatrix} \quad \sigma_2 = \begin{pmatrix} 0 & -i \\ i & 0 \end{pmatrix} \quad \sigma_3 = \begin{pmatrix} 1 & 0 \\ 0 & -1 \end{pmatrix} \tag{1-94}$$

We observe from Eqs. (1-88), (1-89), and (1-90) that keeping only the first-order terms in ξ_{2a} and ξ_{2b}, the problem has cylindrical symmetry. Furthermore, we can

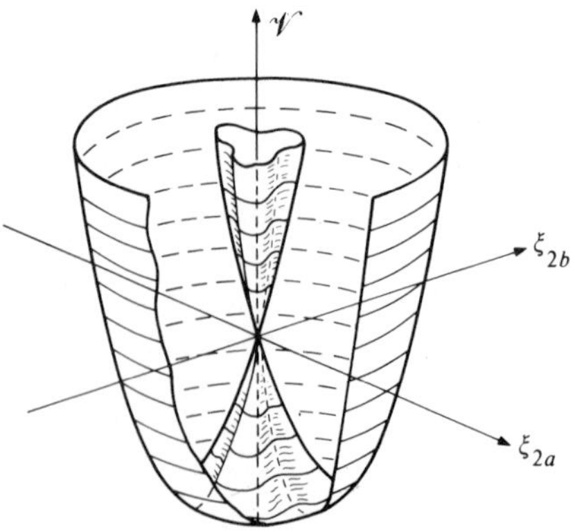

Figure 1-2 Potential sheets for an E level under the influence of a Jahn–Teller effect in a doubly degenerate ε vibration.

associate with ψ_+ and ψ_- an electronic "angular momentum" about the threefold axis of $\pm\hbar$. Without too much trouble it is found that the operator[14]

$$\begin{aligned}\hat{\Lambda} &= 2\hbar^{-1}(\xi_{2a}p_{2b} - \xi_{2b}p_{2a}) + \sigma_3 \\ &= 2\hbar^{-1}\hat{M}_Z + \sigma_3\end{aligned} \tag{1-95}$$

commutes with $\mathscr{H}$ and therefore represents a constant of the motion. This may be looked upon as the total angular momentum, both of the electronic and of the nuclear motions, about the threefold axis. In our definition the eigenvalues of $\hat{\Lambda}$ take only odd integral values, both positive and negative. However, the operator $\hat{O} = \hat{R}\sigma_1$, $\hat{R}$ being a reflection in the ξ_{2a} axis, commutes with $\mathscr{H}$ and anticommutes with $\hat{\Lambda}$. Each energy level is therefore doubly degenerate and characterized by equal and opposite values of Λ.

The vibrational functions $\chi_+(\xi_{2a}, \xi_{2b})$ and $\chi_-(\xi_{2a}, \xi_{2b})$ of Eq. (1-91) are determined by

$$\begin{pmatrix} \mathscr{H}^0 - W & a(\xi_{2a} - i\xi_{2b}) \\ a(\xi_{2a} + i\xi_{2b}) & \mathscr{H}^0 - W \end{pmatrix} \begin{pmatrix} \chi_+ \\ \chi_- \end{pmatrix} = 0 \tag{1-96}$$

with $\mathscr{H}_0$ being given by Eq. (1-92), and where in the off-diagonal elements, given in Eqs. (1-76) and (1-77), we have put $-a_3 = a$. In discussing the solutions to Eq. (1-96) it is convenient to introduce a dimensionless quantity

$$D = \frac{a^2}{2k_2\hbar\omega_2} = \frac{\Delta V_{J-T}}{\hbar\omega} \tag{1-97}$$

D is seen to determine the strength of the interaction between the electronic and vibrational motions.

As our starting point we take $D = 0$. In that case the solutions to Eq. (1-96) are given by

$$\begin{aligned}\mathscr{H}^0\chi_{n,m}(\xi_{2a}, \xi_{2b}) &= (n+1)\hbar\omega\,\chi_{n,m}(\xi_{2a}, \xi_{2b}) \\ m &= -n, -n+2, \ldots, n\end{aligned} \tag{1-98}$$

with $\chi_{n,m}(\xi_{2a}, \xi_{2b})$ being the wave functions for a two-dimensional harmonic oscillator. The degeneracy of each vibronic level is $2(n+1)$, the factor 2 arising from the double electronic degeneracy.

Provided $D \ll 1$ we may find the solutions to Eq. (1-96) using second-order perturbation theory. The matrix elements of interest are,[15] with appropriate phase choices,

$$\langle\chi_{n,m}|\,\xi_{2a} + i\xi_{2b}\,|\chi_{n-1,m-1}\rangle = \sqrt{\frac{\hbar\omega}{2k}}\sqrt{n+m} \tag{1-99}$$

$$\langle\chi_{n,m}|\,\xi_{2a} + i\xi_{2b}\,|\chi_{n+1,m-1}\rangle = \sqrt{\frac{\hbar\omega}{2k}}\sqrt{n-m+2} \tag{1-100}$$

$$\langle\chi_{n,m}|\,\xi_{2a} - i\xi_{2b}\,|\chi_{n+1,m+1}\rangle = \sqrt{\frac{\hbar\omega}{2k}}\sqrt{n+m+2} \tag{1-101}$$

$$\langle \chi_{n,m} | \xi_{2a} - i\xi_{2b} | \chi_{n-1,m+1} \rangle = \sqrt{\frac{\hbar\omega}{2k}} \sqrt{n-m} \tag{1-102}$$

We get easily from Eq. (1-96)

$$W(n, m) = [(n+1) - 2D(m+1)]\,\hbar\omega \tag{1-103}$$

Since for any given value of n, the eigenvalues of $\hat{M}_Z$ are $m\hbar$ ($m = -n, -n+2, \ldots, n$), we observe that Λ takes all odd integral values between $-(2n+1)$ and $+(2n+1)$. We may write

$$|\Lambda| = |2n + 1 - 4\eta| \qquad \eta = 0, 1, 2, \ldots, n \tag{1-104}$$

and

$$W(n, |\Lambda|) = [(n+1) - 2D(n+1-2\eta)]\,\hbar\omega \tag{1-105}$$

The pair of states for which $\Lambda = \pm(2n+1)$ lies lowest, followed at intervals of $4D\hbar\omega$ by those for which $\Lambda = \pm(2n+1-4)$, $\pm(2n+1-8)\cdots\pm(2n+1-4n)$ respectively.

The calculation of the vibrational energies reported above are valid for $D \ll 1$. Let us now consider the solutions to Eq. (1-96) for larger values of D. We use a representation in which $\mathscr{H}^0$ and m are diagonal. For $\Lambda = 2m + 1$ we may expand χ_+ and χ_- on the unperturbed solutions to the two-dimensional harmonic oscillator

$$\chi_+ = c_1\chi_{m,m} + c_3\chi_{m+2,m} + c_5\chi_{m+4,m} + \cdots \tag{1-106}$$

$$\chi_- = c_2\chi_{m+1,m+1} + c_4\chi_{m+3,m+1} + c_6\chi_{m+5,m+1} + \cdots \tag{1-107}$$

Due to the double degeneracy we may restrict ourselves to those solutions for which m is positive.

We now insert the expansions of Eqs. (1-106) and (1-107) in the two coupled Eqs. (1-108) and (1-109)

$$(\mathscr{H}^0 - W)\chi_+ + a(\xi_{2a} - i\xi_{2b})\chi_- = 0 \tag{1-108}$$

$$a(\xi_{2a} + i\xi_{2b})\chi_+ + (\mathscr{H}^0 - W)\chi_- = 0 \tag{1-109}$$

and convert into matrix form by multiplying on the left by each function in the series and integrating. This leads to

$$\begin{pmatrix} m+1-W & \sqrt{D(2m+2)} & 0 & 0 & 0 & \cdot \\ \sqrt{D(2m+2)} & m+2-W & \sqrt{D\times 2} & 0 & 0 & \cdot \\ 0 & \sqrt{D\times 2} & m+3-W & \sqrt{D(2m+4)} & 0 & \cdot \\ 0 & 0 & \sqrt{D(2m+4)} & m+4-W & \sqrt{D\times 4} & \cdot \\ \cdot & \cdot & \cdot & \cdot & \cdot & \cdot \end{pmatrix} \begin{pmatrix} a_1 \\ a_2 \\ a_3 \\ a_4 \\ \cdot \end{pmatrix} = 0 \tag{1-110}$$

No further reduction of the problem seems possible. The eigenvalues and eigenfunctions of Eq. (1-110) can be obtained by numerical methods,[14,16] and we have given in Fig. 1-3 a graphic representation of some of the lowest vibronic levels.

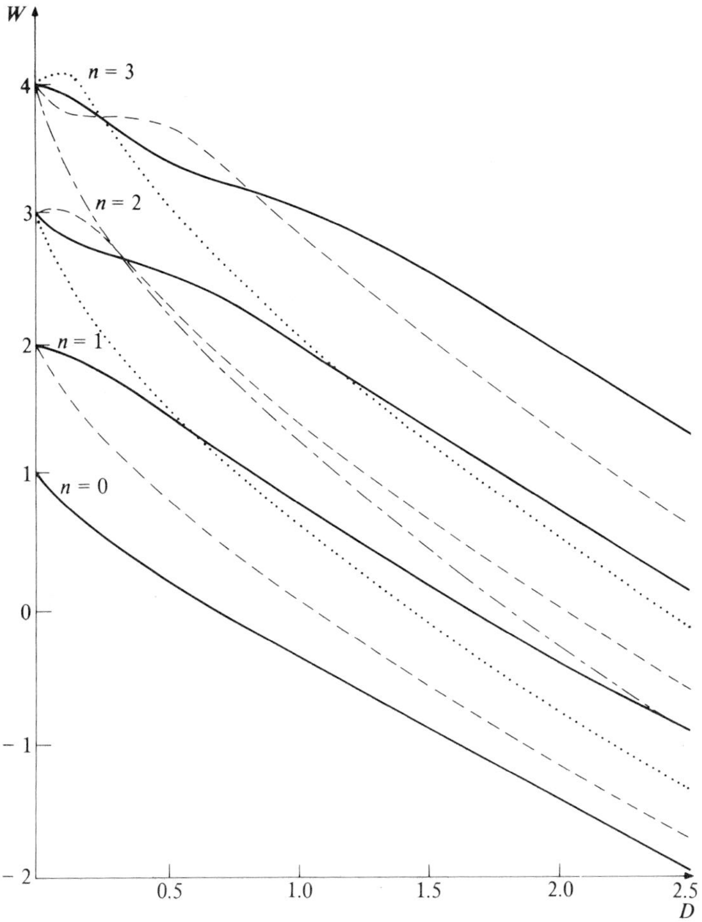

Figure 1-3 The lowest vibronic levels of an E state[14] as a function of the Jahn–Teller coupling parameter D. Energy units of $\hbar\omega$. $|\Lambda| = 1$, solid curve; $|\Lambda| = 3$, dashed curve; $|\Lambda| = 5$, dotted curve; $|\Lambda| = 7$, dash-dotted curve.

We have mentioned that linear molecules form an exception to the Jahn–Teller theorem. This can be proven as follows.[17] In the electronic part of the vibronic wave function a factor $\exp(i\lambda\theta)$ appears where θ is the coordinate conjugate to the axial angular momentum of all the electrons. For an electronic Π state, for instance, $\lambda = \pm 1$.

The vibrational part takes the form

$$\chi_{n,m} = f_{n,m}(\rho)\exp(im\phi) \tag{1-111}$$

The vibronic wave functions can now be written

$$|\lambda, n, m\rangle = \exp(i\lambda\theta)\exp(im\phi)f_{n,m}(\rho) \tag{1-112}$$

where it is understood that $|\lambda, n, m\rangle$ are eigenfunctions of both the electronic and the vibrational angular momentum about the molecular axis.

The terms $\mathscr{H}^{(1)}$ in the hamiltonian which represents the angular coupling between the electronic states and the nuclear framework can only depend upon the relative angle $\alpha = \theta - \phi$ and the amplitude ρ. Also $\mathscr{H}^{(1)}$ must be a symmetrical function of α. Expanding $\mathscr{H}^{(1)}$ we obtain

$$\mathscr{H}^{(1)} = V_0(\rho) + [\exp(\mathrm{i}\alpha) + \exp(-\mathrm{i}\alpha)]\,V_1(\rho) + [\exp(\mathrm{i}2\alpha) + \exp(-\mathrm{i}2\alpha)]\,V_2(\rho) + \cdots \tag{1-113}$$

where $V_k(\rho)$ is of order ρ^k at $\rho = 0$. The matrix elements of $\mathscr{H}^{(1)}$ are given by

$$\langle \exp(\mathrm{i}\lambda\theta)\exp(\mathrm{i}m\phi)|\mathscr{H}^{(1)}|\exp(\mathrm{i}\lambda'\theta)\exp(\mathrm{i}m'\phi)\rangle$$

In this case $\Lambda = \lambda + m$ is a good quantum number, representing the total angular momentum about the molecular axis. With $\Lambda = \lambda + m = \lambda' + m'$ we find easily that in order for the off-diagonal matrix elements to be different from zero we must have

$$|\lambda' - \lambda| = k \tag{1-114}$$

For a Π state $|\lambda' - \lambda| = 2$, and of course for a Δ state $|\lambda' - \lambda| = 4$, and so forth. Hence the first effective coupling term for a linear molecule has $k = 2$, corresponding to a second-order term in the expansion of Eq. (1-113). This effect was first elucidated by Renner[17] in 1934, three years before the paper by Jahn and Teller. It is consequently called the Renner effect.

For given ρ and ϕ the matrix of $\mathscr{H}^{(1)}$ between the two Π functions $\exp(\mathrm{i}\theta)$ and $\exp(-\mathrm{i}\theta)$ is immediately seen to be

$$\begin{pmatrix} V_0(\rho) & V_2(\rho)\exp(-\mathrm{i}2\phi) \\ V_2(\rho)\exp(\mathrm{i}2\phi) & V_0(\rho) \end{pmatrix}$$

The doubly degenerate potential surface is therefore split into two:

$$V = V_0(\rho) \pm V_2(\rho) \tag{1-115}$$

according to whether the electronic wave function is symmetric or antisymmetric about the plane of distortion. The solution of the dynamic equations on such a pair of potential surfaces has been given by Renner[17] and by Pople and Longuet-Higgins.[18] As expected, a splitting of the vibrational levels occurs, dependent upon the strength of the coupling. However, as seen from Fig. 1-4, the linear position of the nuclei in the molecule may be stable, even if the Renner effect is operative.

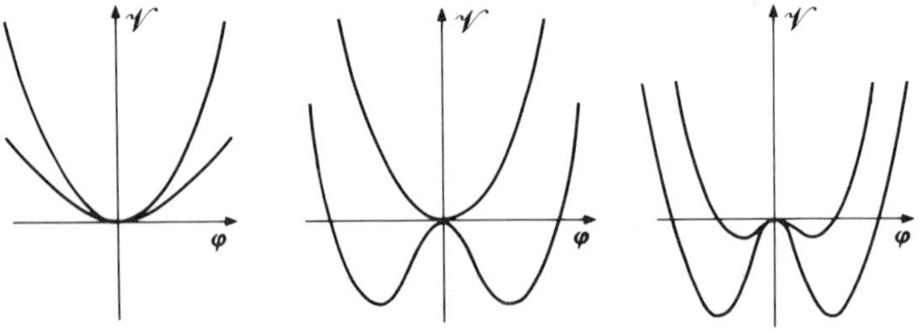

Figure 1-4 Potential surfaces for a linear molecule in a Π state as a function of the angle of bending. Both curves may have positive curvature at the origin, or either or both curves may have negative curvature.

1-7 THE SPIN-ORBIT COUPLING

A more complete treatment of the electronic motions than the one we have given so far introduces a term into the molecular hamiltonian which couples the motion of the electrons to the spin angular momentum. This so-called *spin–orbit coupling* is a relativistic phenomenon;[19] restricting ourselves to consider the interaction between the spin and orbital motions of the same electron, we may write

$$\mathscr{H}^{(1)} = \sum_j \frac{1}{2m^2c^2} (\text{grad } V_j \wedge \mathbf{p}_j) \cdot \mathbf{s}_j \tag{1-116}$$

Here V_j is the potential that electron j experiences, m is the mass of the electron, c the velocity of light, $\mathbf{p}_j$ the linear momentum, and $\mathbf{s}_j$ the spin-momentum operator of electron j.

Let the Bohr radius a_0 be a characteristic length of the molecule. The order of magnitude for the spin–orbit coupling is then seen to be

$$\mathscr{H}^{(1)} \approx \frac{1}{m^2c^2} \frac{V\hbar}{a_0} \frac{\hbar}{a_0} = \left(\frac{e^2}{c\hbar}\right)^2 V$$

Assuming a Coulomb potential $V = -qe^2/r$ and a hydrogenic orbital, the spin–orbit coupling energy is found to be proportional to $\alpha^2 q^2 e^2/a_0$, where $\alpha = e^2/c\hbar$ is the so-called *fine-structure constant*, approximately equal to 1/137. With q small, the spin–orbit coupling energy is therefore small compared with the electronic energy of the system. However, the larger q is, that is the heavier the atoms we have in the molecule, the more important is the spin–orbit coupling term in the molecular hamiltonian.

In order to elucidate the properties of Eq. (1-116) we split the electronic potential V up into two parts;[19] one which is spherically symmetrical, and "the rest" (see Fig. 1-5) as follows:

$$V = V(r) + e^2 \int \frac{\rho(\mathbf{r}')\,d\mathbf{r}'}{|\mathbf{r} - \mathbf{r}'|} \tag{1-117}$$

Here $\rho(\mathbf{r}')$ is the non–spherically symmetric "excess" charge distribution. We may then derive easily for the orbital part of $\mathscr{H}^{(1)}$

$$\text{grad } V \wedge \mathbf{p} = \frac{1}{r}\frac{\partial V}{\partial r} \mathbf{r} \wedge \mathbf{p} - e^2 \int \frac{\rho(\mathbf{r}')\,d\mathbf{r}'}{|\mathbf{r} - \mathbf{r}'|^3} (\mathbf{r} - \mathbf{r}') \wedge \mathbf{p} \tag{1-118}$$

which with $\mathbf{r} \wedge \mathbf{p} = \mathbf{l}_N$, where $\mathbf{l}$ is taken around the center N, becomes

$$\text{grad } V \wedge \mathbf{p} = \frac{1}{r_N}\frac{\partial V_N}{\partial r_N} \mathbf{l}_N - e^2 \int \frac{\rho(x', y', z')\,d\mathbf{r}'}{|\mathbf{r}_N - \mathbf{r}'_N|^3} (\mathbf{r}_N - \mathbf{r}'_N) \wedge \mathbf{p} \tag{1-119}$$

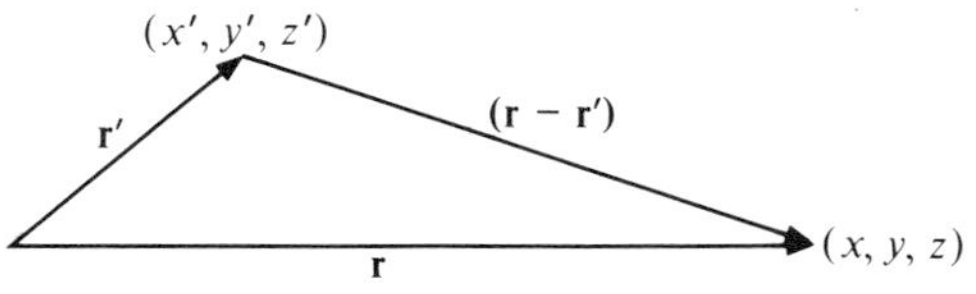

Figure 1-5 Coordinate system for calculating the spin–orbit coupling.

Clearly the orbital part of the spin–orbit coupling operator will transform like the electronic angular momentum **l** in the point group of the molecule. It is then easy to see by symmetry arguments which molecular states will show first-order multiplet separations. A term whose orbital symmetry is Γ, say, will exhibit properties of angular momentum and therefore propensities for spin–orbit coupling only if the direct-product representation $\Gamma \times \Gamma(\mathbf{L}) \times \Gamma$ contains the identity representation.

As already seen in section 1-4 the formal group-theoretical treatment classifies the states simultaneously according to their orbital and spin transformation properties, making use of the "double groups." By virtue of the fact that the spin–orbit coupling operator transforms as the totally symmetric representation in the double group of the molecule, it can only couple together states possessing the same double-group state designation. The spin–orbit coupling term therefore assumes importance in molecular spectroscopy due to its ability to split the spin multiplets and to "mix" the spin states which differ by $\Delta S = 1$. This last feature breaks down the validity of the spin quantum number S.

With N nuclei in the molecule we may write V as a sum of N spherically symmetric potentials located at the N centers, plus the nonspherical part. The spin–orbit coupling operator for a system containing n electrons is therefore

$$\mathscr{H}^{(1)} = \sum_{\mu=1}^{N} \sum_{j=1}^{n} \frac{1}{2m^2c^2}\left(\frac{1}{r_{\mu j}}\frac{\partial V_{\mu j}}{\partial r_{\mu j}}\right)\mathbf{l}_{\mu j}\cdot\mathbf{s}_j - \sum_{j=1}^{n}\frac{e^2}{2m^2c^2}\int\frac{\rho(\mathbf{r}')\,d\mathbf{r}'}{|\mathbf{r}_j-\mathbf{r}'|^3}(\mathbf{r}_j-\mathbf{r}')\wedge\mathbf{p}_j\cdot\mathbf{s}_j \tag{1-120}$$

Neglecting the second part of Eq. (1-120) turns the operator into

$$\mathscr{H}^{(1)} = \sum_{\mu}^{N}\sum_{j}^{n}\frac{1}{2m^2c^2}\left(\frac{1}{r_{\mu j}}\frac{\partial V_{\mu j}}{\partial r_{\mu j}}\right)\mathbf{l}_{\mu j}\cdot\mathbf{s}_j \tag{1-121}$$

Collecting the radial dependence into functions $\zeta_\mu(r)$ characteristic of each atom in the molecule, we get finally

$$\mathscr{H}^{(1)} = \sum_{\mu}^{N}\sum_{j}^{n}\zeta_\mu(r_{\mu j})\,\mathbf{l}_{\mu j}\cdot\mathbf{s}_j \tag{1-122}$$

It must however be remembered that the approximation leading from Eq. (1-120) to Eq. (1-122) is quite drastic since it completely neglects the deviations from spherical symmetry of the electronic clouds of the molecule. The basis of the approximation is that grad V changes the most, close to the nuclei. However, with all overlap terms thrown away one can only expect the above treatment to give an order of magnitude for the molecular spin–orbit coupling. Minor deviations from expected "atomic" values of experimentally determined coupling constants are therefore not to be used for the building of elaborate theories.

For the electronic states we can write in analogy with Eq. (1-121)

$$\mathscr{H}^{(1)} = \lambda(r)\mathbf{L}\cdot\mathbf{S} \tag{1-123}$$

or alternatively

$$\mathscr{H}^{(1)} = \lambda(r)(\tfrac{1}{2}\hat{L}_+\hat{S}_- + \tfrac{1}{2}\hat{L}_-\hat{S}_+ + \hat{L}_z\hat{S}_z) \tag{1-124}$$

where $\hat{L}_+ = \hat{L}_x + i\hat{L}_y$, $\hat{L}_- = \hat{L}_x - i\hat{L}_y$, $\hat{S}_+ = \hat{S}_x + i\hat{S}_y$, and $\hat{S}_- = \hat{S}_x - i\hat{S}_y$.

Consider an octahedral (O_h) or tetrahedral (T_d) molecule. The orbital angular momentum operator transforms under T_1 in both symmetries. For a general $^{2S+1}\Gamma$ term, consultation of the group multiplication table shows that only if $\Gamma = T_1$ or T_2 will the direct product $\Gamma \times \Gamma$ contain a T_1 representation. A first-order effect of the spin–orbit coupling is therefore only found for $^{2S+1}T_1$ and $^{2S+1}T_2$ states, $S \geq \frac{1}{2}$.

Taking a basis set of p-orbitals, $l = 1$, these can be used to span a t_{1u} representation in O_h. Classifying the three components of t_{1u} after their value of m, $\hat{l}_z t_{1,m} = m t_{1,m}$ we have

$$t_{1,1} = -\frac{1}{\sqrt{2}}|x + \mathrm{i}y\rangle$$

$$t_{1,0} = |z\rangle$$

$$t_{1,-1} = \frac{1}{\sqrt{2}}|x - \mathrm{i}y\rangle$$

The components of $\hat{l}$, which are $\hat{l}_+ = \hat{l}_x + \mathrm{i}\hat{l}_y$, $\hat{l}_- = \hat{l}_x - \mathrm{i}\hat{l}_y$, and $\hat{l}_z$, are then represented by the matrices

$$\hat{l}_+\begin{pmatrix} t_{1,1} \\ t_{1,0} \\ t_{1,-1} \end{pmatrix} = 1\begin{pmatrix} 0 & 0 & 0 \\ \sqrt{2} & 0 & 0 \\ 0 & \sqrt{2} & 0 \end{pmatrix}\begin{pmatrix} t_{1,1} \\ t_{1,0} \\ t_{1,-1} \end{pmatrix} \tag{1-125}$$

$$\hat{l}_-\begin{pmatrix} t_{1,1} \\ t_{1,0} \\ t_{1,-1} \end{pmatrix} = 1\begin{pmatrix} 0 & \sqrt{2} & 0 \\ 0 & 0 & \sqrt{2} \\ 0 & 0 & 0 \end{pmatrix}\begin{pmatrix} t_{1,1} \\ t_{1,0} \\ t_{1,-1} \end{pmatrix} \tag{1-126}$$

$$\hat{l}_z\begin{pmatrix} t_{1,1} \\ t_{1,0} \\ t_{1,-1} \end{pmatrix} = 1\begin{pmatrix} 1 & 0 & 0 \\ 0 & 0 & 0 \\ 0 & 0 & -1 \end{pmatrix}\begin{pmatrix} t_{1,1} \\ t_{1,0} \\ t_{1,-1} \end{pmatrix} \tag{1-127}$$

For the threefold degenerate t_{2g} set made up of d-orbitals, $\omega_1 = -1/\sqrt{2}\,|yz + \mathrm{i}xz\rangle$, $\omega_0 = |xy\rangle$, and $\omega_{-1} = 1/\sqrt{2}\,|yz - \mathrm{i}xz\rangle$ we find that the components of $\hat{l}$ are represented by the matrices

$$\hat{l}_+\begin{pmatrix} \omega_1 \\ \omega_0 \\ \omega_{-1} \end{pmatrix} = -1\begin{pmatrix} 0 & 0 & 0 \\ \sqrt{2} & 0 & 0 \\ 0 & \sqrt{2} & 0 \end{pmatrix}\begin{pmatrix} \omega_1 \\ \omega_0 \\ \omega_{-1} \end{pmatrix} \tag{1-128}$$

$$\hat{l}_-\begin{pmatrix} \omega_1 \\ \omega_0 \\ \omega_0 \end{pmatrix} = -1\begin{pmatrix} 0 & \sqrt{2} & 0 \\ 0 & 0 & \sqrt{2} \\ 0 & 0 & 0 \end{pmatrix}\begin{pmatrix} \omega_1 \\ \omega_0 \\ \omega_{-1} \end{pmatrix} \tag{1-129}$$

$$\hat{l}_z\begin{pmatrix} \omega_1 \\ \omega_0 \\ \omega_{-1} \end{pmatrix} = -1\begin{pmatrix} 1 & 0 & 0 \\ 0 & 0 & 0 \\ 0 & 0 & -1 \end{pmatrix}\begin{pmatrix} \omega_1 \\ \omega_0 \\ \omega_{-1} \end{pmatrix} \tag{1-130}$$

We can therefore introduce a fictitious orbital angular momentum $\mathbf{L}'$ where $(\hat{L}')^2\psi(T_i) = 1 \cdot 2\psi(T_i)$, $i = 1, 2$. The quantity 1 before the matrices in Eqs. (1-125), (1-126), and (1-127) can be considered as an "effective" Landé factor, $\alpha = 1$, for a t_{1u} (p) representation and similarly the factor -1 in front of the matrices in Eqs. (1-128), (1-129), and (1-130) is the "effective" Landé factor $\alpha = -1$ for a t_{2g} (d) representation. The advantages of defining the fictitious angular momentum $\mathbf{L}'$ is that $\hat{L}'_z + \hat{S}_z$ is a constant of motion, and its eigenvalues can be used to classify the various states.[20]

Defining the total angular momentum $\mathbf{J} = \mathbf{L}' + \mathbf{S}$, the hamiltonian for the spin–orbit coupling in a ${}^{2S+1}T_1$ or ${}^{2S+1}T_2$ state can be written in general

$$\mathcal{H}^{(1)} = \lambda\alpha\mathbf{L}' \cdot \mathbf{S} = \tfrac{1}{2}\lambda\alpha(\hat{J}^2 - (\hat{L}')^2 - \hat{S}^2)$$

leading to the energies

$$W_{J,S} = \tfrac{1}{2}\alpha\lambda[J(J+1) - 1 \cdot 2 - S(S+1)] \tag{1-131}$$

Consider a ${}^4T_1$ state. With $S = \frac{3}{2}$, J can be $\frac{5}{2}$, $\frac{3}{2}$, and $\frac{1}{2}$. Hence the ${}^4T_1$ is split into three levels: one sixfold degenerate, $W_{5/2,3/2} = \frac{3}{2}\lambda\alpha$; one fourfold degenerate, $W_{3/2,3/2} = -\lambda\alpha$; and one twofold degenerate, $W_{1/2,3/2} = -\frac{5}{2}\lambda\alpha$. The spin–orbit components of the ${}^4T_i$ should therefore occur with energy separations in the ratio 5:3. In general the states ${}^{2S+1}T_i$, $i = 1, 2$, $S \geq 1$, will split into three sublevels. This may leave more degeneracy in the sublevels than expected from group theory. This degeneracy is removed by interactions with other levels transforming in the same way in the double group.

For calculational purposes it is often convenient to adopt cylindrical coordinates whose origin is at the center of the molecule and where the z axis is the axis of highest symmetry (see Fig. 1-6). In this coordinate system we have

$$\text{grad } V = \frac{\partial V}{\partial \rho}\mathbf{i}_1 + \frac{1}{\rho}\frac{\partial V}{\partial \phi}\mathbf{i}_2 + \frac{\partial V}{\partial z}\mathbf{i}_3 \tag{1-132}$$

where $\mathbf{i}_1$, $\mathbf{i}_2$, and $\mathbf{i}_3$ are the unit vectors in the direction of ρ, ϕ, and z respectively. The spin–orbit perturbation operator of Eq. (1-116) is then transformed to

$$\mathcal{H}^{(1)} = \frac{-i\hbar}{2m^2c^2}\left\{\left[\frac{1}{\rho}\frac{\partial V}{\partial \phi}\frac{\partial}{\partial z} - \frac{1}{\rho}\frac{\partial V}{\partial z}\frac{\partial}{\partial \phi}\right]\hat{s}_1 + \left[\frac{\partial V}{\partial z}\frac{\partial}{\partial \rho} - \frac{\partial V}{\partial \rho}\frac{\partial}{\partial z}\right]\hat{s}_2 + \left[\frac{1}{\rho}\frac{\partial V}{\partial \rho}\frac{\partial}{\partial \phi} - \frac{1}{\rho}\frac{\partial V}{\partial \phi}\frac{\partial}{\partial \rho}\right]\hat{s}_3\right\} \tag{1-133}$$

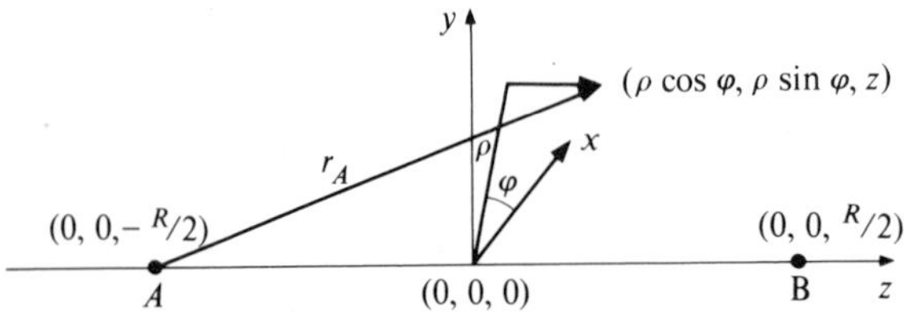

Figure 1-6 Cylindrical coordinates in a linear molecule.

We have further $\hat{s}_3 = \hat{s}_z$ and

$$\hat{s}_1 = \hat{s}_x \cos \phi + \hat{s}_y \sin \phi \tag{1-134}$$

$$\hat{s}_2 = -\hat{s}_x \sin \phi + \hat{s}_y \cos \phi \tag{1-135}$$

Introducing $\hat{s}_+ = \hat{s}_x + i\hat{s}_y$ and $\hat{s}_- = \hat{s}_x - i\hat{s}_y$ we get

$$\mathscr{H}^{(1)} = \frac{-i\hbar}{2m^2 c^2}\left[\tfrac{1}{2}(\hat{A} - i\hat{B}) \exp(-i\phi)\hat{s}_+ + \tfrac{1}{2}(\hat{A} + i\hat{B}) \exp(i\phi)\hat{s}_- + \hat{C}\hat{s}_z\right] \tag{1-136}$$

with

$$\hat{A} = \left[\frac{1}{\rho}\frac{\partial V}{\partial \phi}\frac{\partial}{\partial z} - \frac{1}{\rho}\frac{\partial V}{\partial z}\frac{\partial}{\partial \phi}\right] \tag{1-137}$$

$$\hat{B} = \left[\frac{\partial V}{\partial z}\frac{\partial}{\partial \rho} - \frac{\partial V}{\partial \rho}\frac{\partial}{\partial z}\right] \tag{1-138}$$

$$\hat{C} = \left[\frac{1}{\rho}\frac{\partial V}{\partial \rho}\frac{\partial}{\partial \phi} - \frac{1}{\rho}\frac{\partial V}{\partial \phi}\frac{\partial}{\partial \rho}\right] \tag{1-139}$$

Let us designate our wave functions by $|n\Lambda S M_\Lambda, M_s\rangle$ where Λ is the cylindrical quantum number ($\Lambda = 0, 1, 2, \ldots$). The first two terms in the perturbation will therefore mix wave functions with $\Delta M_\Lambda = 1$. If we do not want to consider off-diagonal terms in the electronic energy we take

$$\mathscr{H}^{(1)} = \sum_j \frac{-i\hbar}{2m^2 c^2}\left[\frac{1}{\rho_j}\frac{\partial V_j}{\partial \rho_j}\frac{\partial}{\partial \phi_j} - \frac{1}{\rho_j}\frac{\partial V_j}{\partial \phi_j}\frac{\partial}{\partial \rho_j}\right]\hat{s}_{zj} \tag{1-140}$$

where the summation is over the j electrons. In true cylindrical symmetry (linear molecules) V_j does not depend upon ϕ_j. Hence Eq. (1-140) reduces to

$$\mathscr{H}^{(1)} = \sum_j \frac{-i\hbar}{2m^2 c^2}\left[\frac{1}{\rho_j}\frac{\partial V_j}{\partial \rho_j}\frac{\partial}{\partial \phi_j}\right]\hat{s}_{zj} \tag{1-141}$$

With $-i\hbar(\partial/\partial\phi_j) = \hat{l}_{zj}$ we get finally

$$\mathscr{H}^{(1)} = \sum_j \frac{1}{2m^2c^2}\frac{1}{\rho_j}\frac{\partial V_j}{\partial \rho_j}\hat{l}_{zj}\hat{s}_{zj} \tag{1-142}$$

The diagonal spin–orbit coupling energy in cylindrical symmetry can evidently be expressed as $D \cdot M_\Lambda \cdot M_s$.

As an example we calculate the fine structure of the ground state $^2\Pi_g$ of the O_2^+ molecule.

Writing down the determinental wave function for $^2\Pi_g(\Lambda = 1, S = \tfrac{1}{2})$ we find easily

$$D(^2\Pi_g) = \frac{\hbar^2}{2m^2c^2}\left\langle \pi_g^x \left| \frac{1}{\rho}\frac{\partial V}{\partial \rho} \right| \pi_g^x \right\rangle \tag{1-143}$$

where in the LCAO approximation

$$\pi_g^x = \frac{1}{\sqrt{2}}(\pi_A^x - \pi_B^x) \tag{1-144}$$

π_A and π_B are the atomic orbitals on the atoms A and B, respectively. Overlap between the atomic orbitals has been neglected since its inclusion is inconsistent with the approximations we shall introduce in the evaluation of D.

Insertion of Eq. (1-144) in (Eq. 1-143) gives

$$D = \frac{\hbar^2}{2m^2c^2}\left[\left\langle \pi_A^x \left| \frac{1}{\rho}\frac{\partial V}{\partial \rho} \right| \pi_A^x \right\rangle - \left\langle \pi_A^x \left| \frac{1}{\rho}\frac{\partial V}{\partial \rho} \right| \pi_B^x \right\rangle\right] \tag{1-145}$$

The second term in Eq. (1-145) is seen to be of the order of magnitude of the π–π overlap times the first term. It is therefore neglected. Further, in the neighborhood of A we take $V \approx V_A(r_A)$. Hence

$$D \approx \frac{\hbar^2}{2m^2c^2}\left\langle \pi_A^x \left| \frac{1}{\rho}\frac{\partial V_A(r_A)}{\partial \rho} \right| \pi_A^x \right\rangle \tag{1-146}$$

With

$$r_A^2 = \rho^2 + (z + R/2)^2 \tag{1-147}$$

(see Fig. 1-6) we get finally

$$D = \frac{\hbar^2}{2m^2c^2}\left\langle 2p_A \left| \frac{1}{r_A}\frac{\partial V_A(r_A)}{\partial r_A} \right| 2p_A \right\rangle \tag{1-148}$$

The expression for D has in this way been reduced to the same form as an atomic spin–orbit coupling constant.[21] The value of D may be estimated as the mean value of ζ in $O(2s)^2(2p)^4$ and ζ in $O^+(2s)^2(2p)^3$. Using atomic tables[22] we get $D \approx 160\ \text{cm}^{-1}$. Experimentally[23] $D \approx 200\ \text{cm}^{-1}$.

REFERENCES

1. M. Born, *Gött. Nachr. math. phys. Kl.* (1951) 1; and M. Born and K. Huang: *Dynamical Theory of Crystal Lattices*, Oxford, 1954, app. VIII.
2. M. Bixon and J. Jortner, *J. Chem. Phys.*, **48**: 715, 1968.
3. M. Born and R. Oppenheimer, *Ann. Phys.*, **84**: 457, 1927.
4. E. B. Wilson, Jr., J. C. Decius, and P. C. Cross, *Molecular Vibrations*, McGraw-Hill, 1955.
5. C. J. Ballhausen and Aa. E. Hansen, *Ann. Rev. Phys. Chem.*, **23**: 15, 1972.
6. F. A. Cotton, *Chemical Applications of Group Theory*, Wiley-Interscience, 1971.
7. L. D. Landau and E. M. Lifshitz, *Quantum Mechanics*, Pergamon Press Ltd, London, 1958.
8. F. Duschinky, *Acta Phys. Chim. URSS*, **7**: 551, 1937.
9. J. S. Griffith, *The Theory of Transition-Metal Ions*, Cambridge University Press, 1961.
10. D. P. Craig and G. J. Small, *J. Chem. Phys.*, **50**: 3827, 1969.
11. H. A. Jahn and E. Teller, *Proc. Roy. Soc.*, **A161**: 220, 1937.
12. H. C. Longuet-Higgins, *Advan. Spectrosc.*, **2**: 429, 1961.
13. R. Englman, *The Jahn–Teller Effect in Molecules and Crystals*, Wiley-Interscience, 1972.
14. W. Moffitt and W. Thorson, *Calcul des Fonctions D'Onde Moleculaire*, Edition du Centre National de la Recherche Scientifique, **82**, Paris, 1958.
15. W. Moffitt and A. D. Liehr, *Phys. Rev.*, **106**: 1195, 1957.
16. H. C. Longuet-Higgins, U. Öpik, M. H. L. Pryce, and R. A. Sack, *Proc. Roy. Soc.*, **A244**: 1, 1958.
17. E. Renner, *Zeits. f. Physik.*, **92**: 172, 1934.
18. J. A. Pople and H. C. Longuet-Higgins, *Mol. Phys.*, **1**: 372, 1958.
19. J. Avery, *Creation and Annihilation Operators*, McGraw-Hill, 1976.

20. A. Abragam and M. H. L. Pryce, *Proc. Roy. Soc.,* **A205**: 135, 1951.
21. E. U. Condon and G. H. Shortley, *The Theory of Atomic Spectra,* Cambridge University Press, 1935.
22. C. E. Moore, *Atomic Energy Levels,* National Bureau of Standards Circular 467, Washington D.C., 1949.
23. M. W. Feast, *Proc. Phys. Soc.,* **A63**: 557, 1950.

CHAPTER

TWO

THE ELECTRONIC ENERGIES

2-1 STATE ENERGIES

The electronic energies of the molecular states are evaluated using the electronic Schrödinger Eq. (1-4) at the molecular configuration $\mathbf{Q}^0$. For a wave function with N electrons $\Psi(\mathbf{r}_1^s, \mathbf{r}_2^s \ldots \mathbf{r}_N^s)$, spin $s = \alpha$ or β, the energy is given by

$$W = \frac{\displaystyle\int \Psi^* \mathscr{H}_E \Psi \, d\tau}{\displaystyle\int \Psi^* \Psi \, d\tau} \tag{2-1}$$

where the integrations are over both electronic space and spin. The antisymmetric wave function Ψ is approximated in the following by a single Slater determinant.

First let us evaluate[1] the normalizing integral $\int |\Psi|^2 \, d\tau$. Each electron in Ψ occupies a molecular spin-orbital ψ_i, $i = 1$ to n. The spin-orbitals are functions of the space and spin coordinates of one electron only, and are assumed to form an orthonormal set. We get

$$\int \Psi^* \Psi \, d\tau = \int \cdots \int |\psi_1^*(1)\psi_2^*(2) \ldots \psi_n^*(N)| \, |\psi_1(1)\psi_2(2) \ldots \psi_n(N)| \, d\tau_1 \ldots d\tau_N$$

Let us now look at one particular term in the expansion of the first determinant, for instance

$$\int \cdots \int \psi_1^*(1)\psi_2^*(N) \ldots \psi_n^*(2) \, |\psi_1(1)\psi_2(2) \ldots \psi_n(N)| \, d\tau_1 \ldots d\tau_N$$

Since the spin-orbitals form an orthonormal set the only way we can make the integral nonvanishing is by selecting from the expansion of the second determinant a term identical to the one we selected from the first determinant. There are $N!$ terms in the expansion of the first determinant and consequently

$$\int \Psi^* \Psi \, d\tau = N! \tag{2-2}$$

Turning to the numerator in Eq. (2-1), the hamiltonian operator can be expanded as a sum of one-electron terms $\hat{h}(i)$ and two-electron terms $g(i, j) = e^2/r_{ij}$. Here $\hat{h}(i)$ represents the kinetic-energy operator plus the nuclear-attraction terms for the ith electron. Hence

$$\mathscr{H} = \sum_{i=1}^{N} \hat{h}(i) + \sum_{i<j}^{N} g(i, j) \tag{2-3}$$

The integral in the numerator can therefore be subdivided into two parts, one depending on the one-electron terms of Eq. (2-3), the other on the two-electron terms. First we shall evaluate the one-electron terms. Take one of the $\hat{h}(i)$ terms in Eq. (2-3) and consider a single term in the expansion of the first wave function in $\int \Psi^* \hat{h}(i) \Psi \, d\tau$

$$\int \cdots \int \psi_1^*(1)\psi_2^*(N)\ldots\psi_n^*(2)\hat{h}(i)\,|\psi_1(1)\psi_2(2)\ldots\psi_n(N)|\, d\tau_1 \ldots d\tau_N$$

The integrals over all the electrons other than i will give zero unless the $N - 1$ spin-orbitals are exactly paired off. What is left is the single term

$$\int \psi_l^*(i)\hat{h}(i)\psi_l(i)\, d\tau_i = \overline{h_{ll}}$$

However, we could have chosen any of the $\hat{h}(i)$ terms in Eq. (2-3) and would have obtained a similar result. Hence

$$\int \cdots \int \psi_1^*(1)\psi_2^*(N)\ldots\psi_n^*(2) \sum_{i=1}^{N} \hat{h}(i)\,|\psi_1(1)\psi_2(2)\ldots\psi_n(N)|\, d\tau_1 \ldots d\tau_N = \sum_{j=1}^{n} \overline{h_{jj}}$$

and since we would have obtained the same result whichever term had been chosen in the expansion of the first determinant, we finally get

$$\int \Psi^* \sum_{i=1}^{N} \hat{h}_i \Psi \, d\tau = N! \sum_{j=1}^{n} \overline{h_{jj}} \tag{2-4}$$

Next we shall evaluate the repulsion terms between the electrons represented by the two-electron operators $g(i, j)$. As before, we look at the integrals coming from one term in the expansion of the first integral

$$\int \cdots \int \psi_1^*(1)\psi_2^*(N)\ldots\psi_n^*(2) g(i, j)\,|\psi_1(1)\psi_2(2)\ldots\psi_n(N)|\, d\tau_1 \ldots d\tau_N$$

Again we match off all spin-orbitals except the two occupied by electrons i and j.

Performing the integrations over the $N - 2$ electrons, we are left with two integrals, namely

$$\iint \psi_m^*(i)\psi_n^*(j)g(i,j)\psi_m(i)\psi_n(j)\,d\tau_i\,d\tau_j$$

$$-\iint \psi_m^*(i)\psi_n^*(j)g(i,j)\psi_m(j)\psi_n(i)\,d\tau_i\,d\tau_j$$

The minus sign in the second integral follows when we remember that we have interchanged two columns in the second determinant when expanding.

It is seen that we may immediately integrate over the spin coordinates in the first integral, but that the second integral will only give a value different from zero provided the spin-orbital ψ_m has the same spin quantum number as the spin-orbital ψ_n. We now define a *Coulomb integral*

$$J_{m,n} = \iint \psi_m^*(i)\psi_m(i)g(i,j)\psi_n^*(j)\psi_n(j)\,d\tau_i\,d\tau_j \tag{2-5}$$

and an *exchange integral*

$$K_{m,n} = \iint \psi_m^*(i)\psi_n(i)g(i,j)\psi_m(j)\psi_n^*(j)\,d\tau_i\,d\tau_j \tag{2-6}$$

Taking the complete sum of the two-electron operators, we observe that each operator will give either a pair of terms $J_{m,n} - K_{m,n}$ or only one term $J_{m,n}$, depending upon whether the spin quantum numbers in the spin-orbitals ψ_m and ψ_n match or do not match. Hence the two-electron operators give the sum

$$\sum_{m<n} J_{m,n} - \sum_{m<n} K_{m,n}\,\delta(s^m, s^n)$$

Once more we notice that we would have obtained exactly the same result whichever term in the expansion of the first determinant we had started with, and the final result is therefore

$$\int \Psi^* \sum_{i<j} g(i,j)\Psi\,d\tau = N!\left[\sum_{m<n} J_{m,n} - \sum_{m<n} K_{m,n}\,\delta(s^m, s^n)\right] \tag{2-7}$$

Inserting Eqs. (2-2), (2-4), and (2-7) in Eq. (2-1) we have

$$W = \sum_n \bar{h}_{n,n} + \sum_{m<n} J_{m,n} - \sum_{m<n} K_{m,n}\,\delta(s^m, s^n)$$

In the special case where we are dealing with a closed shell, the expression for the total electronic energy may be expressed a little more simply. With each orbital being doubly occupied we have the wave function:

$$\Psi = |\psi_1^\alpha(1)\psi_1^\beta(2)\ldots\psi_n^\alpha(2N-1)\psi_n^\beta(2N)| \tag{2-8}$$

Hence for the total state energy

$$W = 2\sum_n \bar{h}_{nn} + \sum_n J_{n,n} + \sum_{n<m} (4J_{n,m} - 2K_{n,m}) \tag{2-9}$$

or noticing that

$$J_{j,j} = K_{j,j} \tag{2-10}$$

we obtain

$$W = 2\sum_{n} \bar{h}_n + \sum_{n,m} (2J_{n,m} - K_{n,m}) \tag{2-11}$$

It is often convenient to define an orbital energy w_n associated with each spin-orbital. Physically this orbital energy is given as the sum of three terms: (1) the kinetic energy of the spin-orbital, (2) the potential energy of attraction between the spin-orbital and the atomic nuclei, and (3) the potential energy of repulsion between the spin-orbital and all of the other spin-orbitals of the system. For the orbital energies in a closed-shell wave function we can therefore take, according to Eq. (2-11),

$$w_n = \bar{h}_n + \sum_{m} (2J_{n,m} - K_{n,m}) \tag{2-12}$$

The total energy of the system can be written either as

$$W = 2\sum (\bar{h}_n + w_n) \tag{2-13}$$

or

$$W = 2\sum_{n} w_n - \sum_{n,m} (2J_{n,m} - K_{n,m}) \tag{2-14}$$

Our definition of the orbital energies w_n has the immediate consequence that we can relate these quantities to the ionization energies of the system. Removing one electron from the orbital ψ_n means that we will lose an energy

$$-\bar{h}_n - \sum_{m} (2J_{m,n} - K_{m,n}) = -w_n$$

Provided that the remaining orbitals are the same before and after the excitation, the ionization energy is therefore equal to the negative of the orbital energy for the electron in question. Photo-ionization spectroscopy, giving us the so-called vertical ionization potentials, can therefore directly be related to w_n (Koopmans' theorem). That the electrons should not relax after an excitation is not usually a very good approximation, however, and deviations from Koopmans' theorem are to be expected.

2-2 THE HARTREE–FOCK EQUATIONS FOR A CLOSED SHELL

In the preceding section we have tacitly assumed that we knew the molecular orbitals ψ_n. In order to find their forms we shall use the variational method.[2,3] We shall minimize the total energy W of the state with respect to a variation of any orbital. Our only condition will be that the molecular orbitals shall form an orthonormal set.

The variational theorem tells us that for the stationary solutions to the molecular Schrödinger equation,

$$\delta W = \int \delta\Psi^* \mathscr{H} \Psi \, d\tau + \int \Psi^* \mathscr{H} \delta\Psi \, d\tau = 0 \tag{2-15}$$

Suppose that in the wave function $\Psi = |\psi_1^\alpha \psi_1^\beta \ldots \psi_n^\alpha \psi_n^\beta|$ where each orbital is doubly occupied, we knew all of the occupied orbitals $\psi_1, \ldots, \psi_n$ together with the empty orbitals $\psi_{n+1} \ldots, \psi_t, \ldots$. These orbitals form a complete orthonormal set, and they are of such a nature that $\delta W = 0$ in Eq. (2-15) for any set of variations, $\delta\psi_1, \delta\psi_2, \ldots, \delta\psi_n$. A variation in an occupied orbital ψ_j can be expanded in terms of the complete set of functions as

$$\delta\psi_j = \sum_{k=1}^{n} c_{jk}\psi_k + \sum_{t=n+1}^{\infty} c_{jt}\psi_t$$

The corresponding change in the state function Ψ is

$$\sum_{t=n+1}^{\infty} c_{jt}[|\psi_1^\alpha \psi_1^\beta \ldots \psi_j^\alpha \psi_t^\beta \ldots \psi_n^\alpha \psi_n^\beta| + |\psi_1^\alpha \psi_1^\beta \ldots \psi_t^\alpha \psi_j^\beta \ldots \psi_n^\alpha \psi_n^\beta|]$$

since all other terms have two columns alike and vanish upon expansion. For a general set of variations of all the orbitals $\psi_1, \ldots, \psi_n$ we therefore have

$$\delta\Psi = \sum_{j=1}^{n} \sum_{t=n+1}^{\infty} c_{jt}[|\psi_1^\alpha \psi_1^\beta \ldots \psi_j^\alpha \psi_t^\beta \ldots \psi_n^\alpha \psi_n^\beta| + |\psi_1^\alpha \psi_1^\beta \ldots \psi_t^\alpha \psi_j^\beta \ldots \psi_n^\alpha \psi_n^\beta|]$$

The energy change is given by Eq. (2-15) as

$$\delta W = 2 \sum_{j=1}^{n} \sum_{t=n+1}^{\infty} c_{jt} \int |\psi_1^\alpha \psi_1^\beta \ldots \psi_j^\alpha \psi_j^\beta \ldots \psi_n^\alpha \psi_n^\beta|^* \mathscr{H} [|\psi_1^\alpha \psi_1^\beta \ldots \psi_j^\alpha \psi_t^\beta \ldots \psi_n^\alpha \psi_n^\beta| + |\psi_1^\alpha \psi_1^\beta \ldots \psi_t^\alpha \psi_j^\beta \ldots \psi_n^\alpha \psi_j^\beta|] \, d\tau_1 \ldots d\tau_{2N} \tag{2-16}$$

The matrix elements of $\mathscr{H}$ are of the type

$$\int \cdots \int |\psi_1^\alpha \psi_1^\beta \ldots \psi_j^\alpha \psi_j^\beta \ldots \psi_n^\alpha \psi_n^\beta|^* \mathscr{H} |\psi_1^\alpha \psi_1^\beta \ldots \psi_j^\alpha \psi_t^\beta \ldots \psi_n^\alpha \psi_n^\beta| \, d\tau_1 \ldots d\tau_{2N}$$

which differs from what we have looked at before. However, they may be expanded, using similar methods as before.

We now define

$$F_{jt} = \int \psi_j^* \hat{h} \psi_t \, dr + \sum_k \left[2 \iint \psi_k^*(2)\psi_k(2) g(1,2) \psi_j^*(1)\psi_t(1) d\tau_1 \, d\tau_2 - \iint \psi_k^*(2)\psi_t(2) g(1,2) \psi_j^*(1)\psi_k(1) \right] d\tau_1 \, d\tau_2 \tag{2-17}$$

where the summation over k runs over all the occupied orbitals, and introduces the Coulomb operator $\hat{J}$ and exchange operator $\hat{K}$

$$\hat{J}_k \psi_t(1) = \int \psi_k^*(2)\psi_k(2) g(1,2) \psi_t(1) \, d\tau_2 \tag{2-18a}$$

$$\hat{K}_k\psi_t(1) = \int \psi_k^*(2)\psi_k(1)g(1,2)\psi_t(2)\,d\tau_2 \tag{2-18b}$$

Hence F_{jt} may be written

$$F_{jt} = \int \psi_j^*(1)\left[\hat{h}(1) + \sum_{k=1}^{n} (2\hat{J}_k - \hat{K}_k)\right]\psi_t(1)\,d\tau_1 \tag{2-19}$$

The matrix element of $\mathscr{H}$ in Eq. (2-16) is found to be equal to $2F_{jt}$. Equation (2-16) demands (since $c_{jt} \neq 0$) that $F_{jt} = 0$ for $j = 1, \ldots, n$ and $t = n + 1, \ldots$.

The operator

$$\hat{F}(1) = \hat{h}(1) + \sum_{k=1}^{n} (2\hat{J}_k - \hat{K}_k) \tag{2-20}$$

can therefore not have any matrix elements between the two sets $\psi_1, \ldots, \psi_n$ and $\psi_{n+1}, \ldots, \psi_t, \ldots$. This implies that we can write for the orbitals inside the set $\psi_1, \ldots, \psi_n$

$$\hat{F}\psi_i = \sum_{k=1}^{n} w_{ki}\psi_k \tag{2-21}$$

The Eqs. (2-21) are the Hartree–Fock equations for the orbitals. Since $\hat{F}$ is hermitian, w_{ki} are the elements of a hermitian matrix, which may be diagonalized by a unitary transformation. The w_{ki}'s are usually called the lagrangian multipliers.

Let us write the variational wave function Ψ as

$$\Psi = |\psi_1^\alpha\psi_2^\alpha \ldots \psi_n^\alpha\psi_1^\beta\psi_2^\beta \ldots \psi_n^\beta|$$

The operator $\hat{F}$ can only couple together orbitals which possess the same spin. Hence the linear transformation which will diagonalize $\psi_1^\alpha, \ldots, \psi_n^\alpha$ will also diagonalize $\psi_1^\beta, \ldots, \psi_n^\beta$. The same set of orbitals $\psi_i^0 = \sum_{k=1}^{n} U_{ik}\psi_k$ therefore emerges from both the α and β sets, and we retain the feature of having doubly occupied orbitals in our wave function. Performing such a linear transformation, the Hartree–Fock equations take on the canonical form

$$\hat{F}\psi_i^0 = w_i\psi_i^0 \qquad (i = 1, \ldots, n) \tag{2-22}$$

Notice that the intuitively derived expression Eq. (2-12) for the orbital energy can be identified with the quantity w_i

$$w_i = \overline{h}_{ii} + \sum_{k=1}^{n} (2J_{ki} - K_{ki}) \tag{2-23}$$

2-3 THE HARTREE–FOCK EQUATIONS FOR AN OPEN SHELL

The concept of doubly occupied orbitals plays an important role in the classification of molecular states according to the operator $\hat{S}^2$. A wave function, say $\Psi = |\psi_1^\alpha\psi_2^\beta\psi_3^\alpha|$, will only be a ^2X state ($S = \frac{1}{2}$, $M_s = \frac{1}{2}$) provided $\psi_1(\mathbf{r}) = \psi_2(\mathbf{r})$.

This is most easily seen by operating with $\hat{S}_+ = \hat{S}_x + i\hat{S}_y$ on Ψ. Having a non-diagonal multiplier between ψ_1 and ψ_3 can assure the spatial equivalence of the ψ_1 and ψ_2 orbitals. However, we must pay for this convenience because ψ_1^α will then no longer have the same "orbital energy" as ψ_1^β. Indeed, the concept "orbital energy" loses its meaning when dealing with wave functions having open shells.

As an example of an open-shell Hartree–Fock-type calculation[3,4] we shall treat the state $\Psi = |\psi_1^\alpha\psi_1^\beta\psi_2^\alpha|$. As in the closed-shell calculation we shall assume that we know the occupied orbitals ψ_1^α and ψ_2^α together with all the empty orbitals $\psi_3^\alpha, \ldots$. Likewise, we assume a knowledge of the occupied ψ_1^β orbital together with the empty orbitals $\psi_2^\beta, \psi_3^\beta, \ldots$. Again we vary the orbitals:

$$\delta\psi_1^\alpha = c_{11}\psi_1^\alpha + c_{12}\psi_2^\alpha + \sum c_{1i}\psi_i^\alpha \qquad i \geq 3$$

$$\delta\psi_1^\beta = \tilde{c}_{11}\psi_1^\beta + \tilde{c}_{12}\psi_2^\beta + \sum \tilde{c}_{1i}\psi_i^\beta \qquad i \geq 3$$

$$\delta\psi_2^\alpha = c_{21}\psi_1^\alpha + c_{22}\psi_2^\alpha + \sum c_{2i}\psi_i^\alpha \qquad i \geq 3$$

The variations in Ψ are then

$$\delta\Psi = \sum_{i=3}^{\infty} c_{1i}\,|\psi_i^\alpha\psi_1^\beta\psi_2^\alpha| + \tilde{c}_{12}\,|\psi_1^\alpha\psi_2^\beta\psi_2^\alpha| + \sum_{i=3}^{\infty} \tilde{c}_{1i}\,|\psi_1^\alpha\psi_i^\beta\psi_2^\alpha| + \sum_{i=3}^{\infty} c_{2i}\,|\psi_1^\alpha\psi_1^\beta\psi_i^\alpha|$$

The variations in W given by $\delta W = 2\int \delta\Psi\mathscr{H}\Psi\, d\tau$ are, writing

$$\iint \psi_i(1)\psi_j(1)\frac{e^2}{r_{12}}\psi_k(2)\psi_l(2)\,d\tau_1\,d\tau_2 = [\psi_i\psi_j\,|\,\psi_k\psi_l]$$

$$\begin{aligned}\delta W = 2\sum_{i=3}^{\infty} c_{1i}\Big[\bar{h}_{i1} + [\psi_i\psi_1\,|\,\psi_2\psi_2] - [\psi_i\psi_2\,|\,\psi_1\psi_2] + [\psi_i\psi_1\,|\,\psi_1\psi_1]\Big]\\
+ \tilde{c}_{12}\Big[\bar{h}_{12} + [\psi_1\psi_2\,|\,\psi_2\psi_2] + [\psi_1\psi_2\,|\,\psi_1\psi_1]\Big]\\
+ 2\sum^{\infty} \tilde{c}_{1i}\Big[\bar{h}_{i1} + [\psi_i\psi_1\,|\,\psi_2\psi_2] + [\psi_i\psi_1\,|\,\psi_1\psi_1]\Big]\\
+ 2\sum_{i=3}^{\infty} c_{2i}\Big[\bar{h}_{i2} + [\psi_i\psi_2\,|\,\psi_1\psi_1] - [\psi_i\psi_1\,|\,\psi_1\psi_2] + [\psi_i\psi_2\,|\,\psi_1\psi_1]\Big]\end{aligned}$$

For the energy to be a minimum, $\delta W = 0$. Hence introducing the Coulomb and exchange operators defined in Eqs. (2-18a) and (2-18b) we must have

$$\int \psi_i^\alpha(\hat{h} + \hat{J}_1 + \hat{J}_2 - \hat{K}_2)\psi_1^\alpha\, d\tau = 0 \qquad i > 3 \tag{2-24a}$$

$$\int \psi_i^\alpha(\hat{h} + 2\hat{J}_1 - \hat{K}_1)\psi_2^\alpha\, d\tau = 0 \qquad i > 3 \tag{2-24b}$$

$$\bar{h}_{12} + [\psi_1\psi_2\,|\,\psi_2\psi_2] + [\psi_1\psi_2\,|\,\psi_1\psi_2] = 0 \tag{2-24c}$$

$$\int \psi_i^\beta (\hat{h} + \hat{J}_1 + \hat{J}_2) \psi_1^\beta \, d\tau = 0 \qquad i > 3 \tag{2-24d}$$

It follows that we can write

$$(\hat{h} + \hat{J}_1 + \hat{J}_2 - \hat{K}_2)\psi_1^\alpha = w_{11}\psi_1^\alpha + w_{12}\psi_2^\alpha \tag{2-25a}$$

$$(\hat{h} + 2\hat{J}_1 - \hat{K}_1)\psi_2^\alpha = w_{21}\psi_1^\alpha + w_{22}\psi_2^\alpha \tag{2-25b}$$

$$(\hat{h} + \hat{J}_1 + \hat{J}_2)\psi_1^\beta = \tilde{w}_{11}\psi_1^\beta \tag{2-26}$$

Therefore

$$w_{11} = \bar{h}_{11} + J_{11} + J_{12} - K_{12} \tag{2-27}$$

$$w_{12} = w_{21} = -[\psi_1\psi_2 \,|\, \psi_2\psi_2] \tag{2-28}$$

$$w_{22} = \bar{h}_{22} + 2J_{12} - K_{12} \tag{2-29}$$

$$\tilde{w}_{11} = \bar{h}_{11} + J_{11} + J_{22} \tag{2-30}$$

Ionizing the electron in ψ_1^α leads to a wave function $|\psi_1^\beta\psi_2^\alpha|$. This is no eigenstate of $\hat{S}^2$, but it can be written as a combination of spin-singlet and a spin-triplet

$$|\psi_1^\beta\psi_2^\alpha| = \frac{1}{\sqrt{2}}\left[\frac{1}{\sqrt{2}}[|\psi_1^\alpha\psi_2^\beta| + |\psi_1^\beta\psi_2^\alpha|] + \frac{1}{\sqrt{2}}[|\psi_1^\beta\psi_2^\alpha| - |\psi_1^\alpha\psi_2^\beta|]\right]$$

We find easily, calling the ground state energy W_0, that

$$-w_{11} = \frac{1}{2}(W^{\text{triplet}} + W^{\text{singlet}}) - W_0 = \frac{I_1^{\text{triplet}} + I_1^{\text{singlet}}}{2}$$

where I^{triplet} is the ionization potential of the triplet state and I^{singlet} the ionization potential of the singlet state, both having the configuration $(\psi_1)^1(\psi_2)^1$.

Ionizing the electron in (ψ_1^β) also leads to the triplet state $|\psi_1^\alpha\psi_2^\alpha|$. We find

$$-\tilde{w}_{11} = I_1^{\text{triplet}}$$

Finally, ionizing out of ψ_2^α leads to the singlet state $|\psi_1^\alpha\psi_1^\beta|$.
We find

$$-w_{22} = I_2^{\text{singlet}}$$

As expected, we notice that w_{11} is not related to a single ionization potential, whereas the other diagonal multipliers are. The more "complicated" the open shell is, the more complex it becomes, of course, to relate the diagonal multipliers to physical quantities.

2-4 THE LCAO MO APPROXIMATION

The Hartree–Fock molecular orbitals ψ_i of Eq. (2-21) are in practice always approximated by a linear combination of atomic orbitals (LCAO MO). Given a set of atomic orbitals $\chi_1, \ldots, \chi_M$ we approximate ψ_i, $i = 1, 2 \ldots n$

$$\psi_i = \sum_{r=1}^{M} c_{ri}\chi_r \qquad M \geq n \tag{2-31}$$

Using a restricted set of given atomic orbitals to approximate ψ_i, this will introduce an error in the calculation of the Hartree–Fock energy. We shall now ask how to calculate the best LCAO MO's for a closed-shell ground state.[2]

Introducing Eqs. (2-18a), (2-18b), and (2-31) into Eq. (2-22) leads to

$$\left\{\hat{h}(1) + \sum_{j=1}^{n} 2 \sum_{t=1}^{M}{}' \sum_{u=1}^{M} c_{tj}^* c_{uj} \int \chi_t^*(2)\chi_u(2) \frac{e^2}{r_{12}} \, d\tau_2 \right\} \sum_{s=1}^{M} c_{si}\chi_s(1)$$
$$- \sum_{j=1}^{n} \sum_{t=1}^{M} \sum_{s=1}^{M} c_{tj}^* c_{si} \int \chi_t^*(2)\chi_s(2) \frac{e^2}{r_{12}} \, d\tau_2 \sum_{u=1}^{M} c_{uj}\chi_u(1) = w_i \sum_{s=1}^{M} c_{si}\chi_s(1) \tag{2-32}$$

We multiply by $\chi_r^*(1)$ and integrate

$$\sum_{s=1}^{M} \left[\bar{h}_{rs} + \sum_{t,u}^{M} \sum_{j=1}^{n} 2c_{tj}^* c_{uj} \{[rs \mid tu] - \tfrac{1}{2}[ru \mid ts]\} \right] c_{si} = w_i \sum_{s=1}^{M} S_{rs} c_{si} \tag{2-33}$$

where we have used $[rs \mid tu] = \int\int \chi_r^*(1)\chi_s(1)e^2/r_{12}\chi_t^*(2)\chi_u(2) \, d\tau_1 \, d\tau_2$ and $S_{rs} = \int \chi_r^*(1)\chi_s(1) \, d\tau_1$.

We define a "charge and bond order matrix" with elements $P_{t,u}$

$$P_{t,u} = 2 \sum_{j=1}^{n} c_{tj}^* c_{uj} \tag{2-34}$$

Equation (2-33) is then transformed to

$$\sum_{s=1}^{M} \left[\bar{h}_{rs} + \sum_{t,u}^{M} P_{t,u} \{[rs \mid tu] - \tfrac{1}{2}[ru \mid ts]\} - w_i S_{rs} \right] c_{si} = 0 \tag{2-35}$$

A further definition

$$G_{rs} = \sum_{t,u}^{M} P_{t,u} \{[rs \mid tu] - \tfrac{1}{2}[ru \mid ts]\} \tag{2-36}$$

turns Eq. (2-35) into

$$\sum_{s=1}^{M} [\bar{h}_{rs} + G_{rs} - w_i S_{rs}] \, c_{si} = 0 \tag{2-37}$$

This set of homogeneous linear equations will only possess nontrivial solutions provided the determinant

$$|\bar{h}_{rs} + G_{rs} - w_i S_{rs}| = 0 \tag{2-38}$$

This secular equation gives us the orbital energies w_i, and Eq. (2-37) determines the corresponding coefficients in the expanded molecular orbitals $\psi_i = \sum_{s=1}^{M} c_{si}\chi_s$. The equations have to be solved by iteration since G_{rs} depends upon c_{si}. That is, we do not know the potential until we have found the appropriate molecular orbitals. A cyclic calculation must be performed. First we assume a certain set of coefficients c_{si} in the molecular orbitals. Then we calculate the potential due to this set of filled orbitals, and in this way construct G_{rs}. We then solve for the energies of Eq. (2-38) and substitute back in Eq. (2-37), finding a set of c_{si} coefficients. These new coefficients c_{si} are then used to construct a new potential, and the whole process repeated over and over until the initial and final coefficients c_{si} agree to within a certain limit. This is called the *self-consistent field* (SCF) limit.

We remind the reader that Eqs. (2-37) and (2-38) were derived under the assumption of doubly filled molecular orbitals. Solving the $M \times M$ secular Eq. (2-38) will yield M solutions. With the number of atomic orbitals M larger than the number of molecular orbitals n, a number of empty orbitals $n + 1$, $n + 2, \ldots, M$ will appear. These last orbitals are called *virtual* orbitals. The set of Eqs. (2-37) and (2-38) are usually referred to as Roothaan's equations.[2]

The Hartree–Fock–Roothaan method determines the "best" single determinant approximation to the ground state. The difference between this Hartree–Fock energy of the ground state and the "true" nonrelativistic energy is usually referred to as the correlation energy. This W_{corr} can be several electron volts. We can improve the calculation of the energies by taking as our trial wave function a sum of several Slater determinants. This procedure is referred to as configuration interaction.

2-5 EXCITATION ENERGIES

The virtual orbitals $\psi_{n+1}, \ldots, \psi_t, \ldots$ arising from a doubly occupied trial wave function are solutions to the equation

$$\left[\hat{h}(1) + \sum_{k=1}^{n} (2\hat{J}_k - \hat{K}_k)\right]\psi_t = w_t\psi_t \tag{2-39}$$

w_t can be found as

$$w_t = \overline{h}_{tt} + \sum_{k=1}^{n} (2J_{kt} - K_{kt}) \tag{2-40}$$

Since $J_{ii} - K_{ii} = 0$, the self-consistent field felt by an electron in the ith occupied orbital consists of the influence of $2n - 1$ electrons. However, in the expression for the energy of a virtual orbital, Eq. (2-39), there is no such self-adjustment since t is not contained in $1, 2, \ldots, n$. Hence an electron in a virtual orbital feels repulsion terms from $2n$ electrons. As shall be shown now by means of an example, Eq. (2-39) indeed produces virtual orbitals more appropriate for a $(2n + 1)$-electron system than for a $2n$-system.

Consider the system $|\psi_1^\alpha \psi_1^\beta|$. The equation for the "virtual orbital" ψ_2^α is

$$(\hat{h} + 2\hat{J}_1 - \hat{K}_1)\psi_2^\alpha = w_2\psi_2^\alpha \tag{2-41}$$

This is to be compared with Eq. (2-25b) which determines the orbital ψ_2 in the system $|\psi_1^\alpha\psi_1^\beta\psi_1^\alpha|$, viz.

$$(\hat{h} + 2\hat{J}_1 - \hat{K}_1)\psi_2^\alpha = w_{21}\psi_1^\alpha + w_{22}\psi_2^\alpha \tag{2-42}$$

The operator has the same form in the two cases, but the presence of the term $w_{21}\psi_1$ in Eq. (2-42) forbids us to identify the operator for the "virtual orbitals" in the $2n$-system with the Hartree–Fock operator of the $(2n + 1)$-electron system. These considerations explain why the calculated virtual orbitals of a neutral molecule often have positive energies; that is, an electron in a virtual orbital will be unbound.

The Hartree–Fock operator for an electron in the occupied orbital ψ_i is

$$\hat{F}_i = \hat{h}(i) + \sum_{k \neq i}^{n} (2\hat{J}_k - \hat{K}_k) + \hat{J}_i \tag{2-43}$$

This operator describes the self-consistent field for the electron and should therefore also be suitable for the excited states of that electron.

Ionizing an electron out of the orbital ψ_i demands an energy equal to

$$W(I) = \bar{h}_{ii} + \sum_{k=1}^{n} (2J_{ki} - K_{ki}) \tag{2-44}$$

Placing the electron back into the virtual orbital t gains the energy

$$W(A) = \bar{h}_{tt} + \sum_{k \neq i}^{n} (2J_{kt} - K_{kt}) + J_{it} \mp K_{it} \tag{2-45}$$

$$= \bar{h}_{tt} + \sum_{k=1}^{n} (2J_{kt} - K_{kt}) - J_{it} + K_{it} \mp K_{it} \tag{2-46}$$

where the $\mp K_{it}$ takes care of the cases where the excited state is a spin-triplet or a spin-singlet. By the use of Eqs. (2-40) and (2-22) we get for the energy differences between the excited $(i \to t)$ ${}^{1,3}\Psi_{ex}$ state and the ground state ${}^1\Psi_0$:

$${}^1W_{ex} - {}^1W_0 = w_t - w_i - J_{it} + 2K_{it} \tag{2-47}$$

$${}^3W_{ex} - {}^1W_0 = w_t - w_i - J_{it} \tag{2-48}$$

For the reasons outlined, one does not get good numerical results by using the virtual orbitals from a Hartree–Fock–Roothaan calculation to assess the positions of the excited states. Reliable results are best obtained by minimizing the energies of the excited states separately from that of the ground state. This leads to the problems of the open shells. Another method of obtaining improved numerical results is to use extended configuration interaction.[5] Here the wave functions are improved by taking a linear combination of Slater determinants which have different electronic configurations. Naturally, the symmetry of the functions which are to be mixed together must be the same.

The actual calculation of the molecular orbitals in the Hartree–Fock–Roothaan scheme runs along two principal roads. Either we include all the electrons of the molecule in the so-called *ab initio* method. Such calculations are always handled on big electronic computers. The LCAO MO wave functions of Eq. (2-31) are usually expanded using a set of gaussian functions as a basis. Each atomic valence orbital is approximated by at least three basis functions. This procedure allows for radial distortion when an atom is at its place in the molecule. An atom in a molecule is indeed very strongly polarized, and its net charge can vary. The valence orbitals expand or contract due to this effect, and good results can be obtained with only a few polarization functions. There is also an angular dependence of the valence orbitals. This is taken care of by an admixture of spherical harmonics not found in the free atom. The reason for using gaussian orbitals as the underlying basis set of functions is that all of the integrals needed for polyatomic molecules can be evaluated fairly easily. The only parameters which are needed are therefore the atomic coordinates.

In the LCAO MO semiempirical methods, the valence electrons are treated separately from the rest of the electrons. It is supposed that the effect of the nonvalence electrons can be lumped into the hamiltonian for the valence electrons. Such an approximation has been widely used for unsaturated hydrocarbons and for inorganic complexes. We write for the N valence electrons

$$\mathscr{H}_{\text{val}}(1, 2, \ldots, N) = \sum_{j=1}^{N} \hat{h}_{\text{core}}(j) + \sum_{i<j}^{N} e^2/r_{ij} \tag{2-49}$$

and we seek wave functions Ψ which are antisymmetric in the valence electrons. This is done by minimizing

$$W = \int \Psi^* \mathscr{H}_{\text{val}} \Psi \, d\tau \tag{2-50}$$

$\hat{h}_{\text{core}}(j)$ is the one-electron part of the hamiltonian for the electron j, consisting of the kinetic-energy operator and the potential-energy "core" operator. The "core potential" is therefore made up of the attractions of the nuclei and of the repulsions arising from the core electrons. The separation of the molecular orbitals into core orbitals and valence orbitals is only strictly valid provided all valence orbitals are kept orthogonal to all core orbitals.

In many semiempirical theories further assumptions are introduced, aiming at simplifying the evaluation of the many molecular integrals. In the so-called ZDO (zero differential overlap) approximation scheme all atomic overlaps are neglected, all two-electron integrals $[ab \mid cd]$ are put equal to $\delta_{ab}\,\delta_{cd}\,[aa \mid cc]$, and $\bar{h}_{a,b}^{(\text{core})}$ integrals are neglected unless the a and b atomic functions are "bonded." The proper application of the ZDO approximation in the calculations of electronic structures of inorganic complexes has been dealt with by Dahl and Ballhausen.[6]

Similar modified types of approximation schemes are the CNDO (complete neglect of differential overlap), the INDO (intermediate neglect of differential overlap), and the NDDO (neglect of diatomic differential overlap) methods. The list could be continued *ad infinitum*. In these types of calculations one should, of course, guard against the complications of the method rising more steeply than the expected reliability of the result. The problem is really that Eq. (2-49) is far from the correct hamiltonian, and it is important to realize under what conditions the hamiltonian in Eq. (2-49) can be expected to produce reliable numbers for molecular properties. We shall not pursue these lines of research here, however, but refer to the literature dealing with these subjects.[7,8,9]

2-6 THE LIGAND-FIELD AND THE CRYSTAL-FIELD METHODS

An inorganic transition metal complex is made up of a metal center surrounded by molecules or ions, the so-called ligands. The valence orbitals of a transition metal with an unfilled nd shell are taken to be nd, $(n+1)$s, and $(n+1)$p. In the first transition series, $n = 3$ (Sc to Zn). The ligand valence orbitals are usually 2s and 2p atomic orbitals.

The energies of the N valence electrons are found in the semiempirical methods as solutions to

$$\mathscr{H}_{\text{val}}\Psi(1, \ldots, N) = W\Psi(1, \ldots, N) \tag{2-51}$$

with

$$\mathscr{H}_{\text{val}} = \sum_{j=1}^{N} \hat{h}_{\text{core}}(j) + \sum_{i<j}^{N} e^2/r_{ij} \tag{2-52}$$

and

$$\hat{h}_{\text{core}}(1) = -\frac{1}{2}\nabla_1^2 - \sum_{\mu} \frac{Z_\mu}{r_{\mu 1}} + \sum_{\substack{\text{core} \\ \text{orbitals}}} (2\hat{J}_k - \hat{K}_k) \tag{2-53}$$

The three terms in $\hat{h}_{\text{core}}$ are the kinetic energy, the potential energy due to the μ nuclei, and the potential stemming from all the filled inner shells of the metal and ligands.

The self-consistent field operator for closed shells of valence orbitals is

$$\hat{F} = \hat{h}_{\text{core}}(1) + \sum_{j=1}^{N} (2\hat{J}_j - \hat{K}_j) \tag{2-54}$$

where

$$\hat{F}\psi_j = w_j\psi_j \tag{2-55}$$

The LCAO molecular orbitals we want to use for the valence electrons are of the type

$$\psi_\gamma = \alpha\chi_{d\gamma} + \beta \sum_{i=1}^{L} c_{i\gamma}\chi_i \tag{2-56}$$

where χ stands for an atomic orbital. The coefficients $c_{i\gamma}$ in front of the $i = 1$ to L ligand orbitals are usually symmetry-determined[10] and the indices γ refer to the symmetry type of the molecular orbital. The coefficients α and β are to be evaluated by minimizing the total energy of the system, using Eqs. (2-54) and

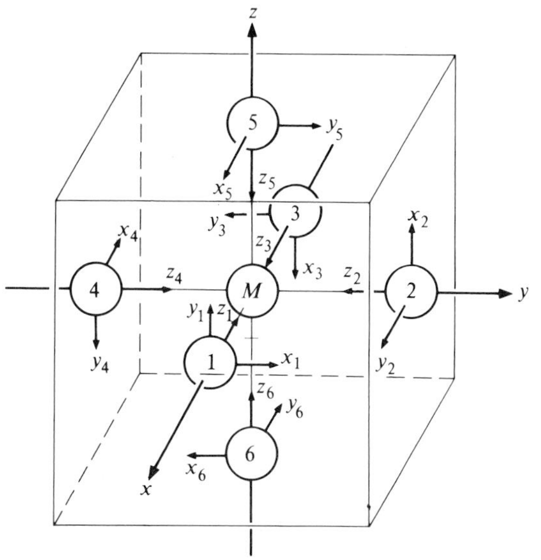

Figure 2-1 Coordinate systems for an octahedral molecule.

(2-55). In Figs. 2-1 and 2-2 we have given the coordinate systems used for octahedral and tetrahedral coordination, and in Tables 2-1 and 2-2 the symmetry-determined metal and ligand orbitals.

To gain some insight into the problem, we shall look at a state $|(\text{core})(\psi_\gamma)^2|$, where

$$\psi_\gamma = \alpha\chi_{M\gamma} + \beta\chi_{L\gamma} \tag{2-57}$$

with $\chi_{L\gamma} = \sum_{i=1}^{L} c_{i\gamma}\chi_i$ being a symmetry-adapted normalized combination of ligand orbitals and $\chi_{M\gamma}$ being a metal orbital.

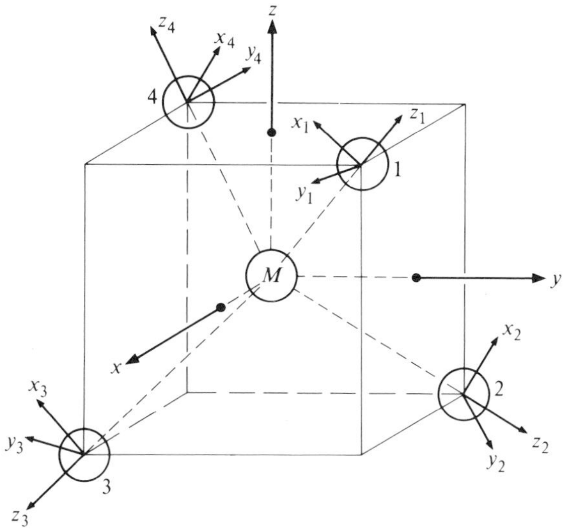

Figure 2-2 Coordinate systems for a tetrahedral molecule.

Table 2-1 Metal and ligand orbitals for the molecular orbitals of an octahedral complex

Representation	Metal orbital	Ligand orbitals σ	Ligand orbitals π
a_{1g}	s	$\frac{1}{\sqrt{6}}(\sigma_1 + \sigma_2 + \sigma_3 + \sigma_4 + \sigma_5 + \sigma_6)$	
e_g	$d_{x^2-y^2}$	$\frac{1}{2}(\sigma_1 - \sigma_2 + \sigma_3 - \sigma_4)$	
	d_{z^2}	$\frac{1}{2\sqrt{3}}(2\sigma_5 + 2\sigma_6 - \sigma_1 - \sigma_2 - \sigma_3 - \sigma_4)$	
t_{1u}	p_x	$\frac{1}{\sqrt{2}}(\sigma_1 - \sigma_3)$	$\frac{1}{2}(p_{y_2} + p_{x_5} - p_{x_4} - p_{y_6})$
	p_y	$\frac{1}{\sqrt{2}}(\sigma_2 - \sigma_4)$	$\frac{1}{2}(p_{x_1} + p_{y_5} - p_{y_3} - p_{x_6})$
	p_z	$\frac{1}{\sqrt{2}}(\sigma_5 - \sigma_6)$	$\frac{1}{2}(p_{y_1} + p_{x_2} - p_{x_3} - p_{y_4})$
t_{2g}	d_{xz}		$\frac{1}{2}(p_{y_1} + p_{x_5} + p_{x_3} + p_{y_6})$
	d_{yz}		$\frac{1}{2}(p_{x_2} + p_{y_5} + p_{y_4} + p_{x_6})$
	d_{xy}		$\frac{1}{2}(p_{x_1} + p_{y_2} + p_{y_3} + p_{x_4})$
t_{1g}			$\frac{1}{2}(p_{y_1} - p_{x_5} + p_{x_3} - p_{y_6})$
			$\frac{1}{2}(p_{x_2} - p_{y_5} + p_{y_4} - p_{x_6})$
			$\frac{1}{2}(p_{x_1} - p_{y_2} + p_{y_3} - p_{x_4})$
t_{2u}			$\pi_\xi = \frac{1}{2}(p_{y_2} - p_{x_5} - p_{x_4} + p_{y_6})$
			$\pi_\eta = \frac{1}{2}(p_{x_1} - p_{y_5} - p_{y_3} + p_{x_6})$
			$\pi_\zeta = \frac{1}{2}(p_{y_1} - p_{x_2} - p_{x_3} + p_{y_4})$

Table 2-2 Basis functions for T_d molecules

Representation	Metal orbitals	Ligand orbitals
a_1	s	$\frac{1}{2}(s_1 + s_2 + s_3 + s_4)$, $\frac{1}{2}(p_{z_1} + p_{z_2} + p_{z_3} + p_{z_4})$
e	d_{z^2}	$\frac{1}{2}(p_{x_1} - p_{x_2} - p_{x_3} + p_{x_4})$
	$d_{x^2-y^2}$	$\frac{1}{2}(p_{y_1} - p_{y_2} - p_{y_3} + p_{y_4})$
t_2	p_x, d_{yz}	$\frac{1}{2}(p_{z_1} - p_{z_2} - p_{z_3} - p_{z_4})$, $\frac{1}{2}(s_1 - s_2 + s_3 - s_4)$, $\frac{1}{4}[p_{x_1} + p_{x_2} - p_{x_3} - p_{x_4} + \sqrt{3}(-p_{y_1} - p_{y_2} + p_{y_3} + p_{y_4})]$
	p_y, d_{xz}	$\frac{1}{2}(p_{z_1} + p_{z_2} - p_{z_3} - p_{z_4})$, $\frac{1}{2}(s_1 + s_2 - s_3 - s_4)$, $\frac{1}{4}[p_{x_1} - p_{x_2} + p_{x_3} - p_{x_4} + \sqrt{3}(p_{y_1} - p_{y_2} + p_{y_3} - p_{y_4})]$
	p_z, d_{xy}	$\frac{1}{2}(p_{z_1} - p_{z_2} - p_{z_3} + p_{z_4})$, $\frac{1}{2}(s_1 - s_2 - s_3 + s_4)$, $-\frac{1}{2}(p_{x_1} + p_{x_2} + p_{x_3} + p_{x_4})$
t_1		$\frac{1}{4}[\sqrt{3}(p_{x_1} + p_{x_2} - p_{x_3} - p_{x_4}) + p_{y_1} + p_{y_2} - p_{y_3} - p_{y_4}]$
		$-\frac{1}{4}[\sqrt{3}(p_{x_1} - p_{x_2} + p_{x_3} - p_{x_4}) - p_{y_1} + p_{y_2} - p_{y_3} + p_{y_4}]$
		$-\frac{1}{2}(p_{y_1} + p_{y_2} + p_{y_3} + p_{y_4})$

Using the Eqs. (2-37) a secular equation is obtained

$$\begin{vmatrix} F_{MM} - w & F_{ML} - wG_\gamma \\ F_{ML} - wG_\gamma & F_{LL} - w \end{vmatrix} = 0 \tag{2-58}$$

where F_{MM}, etc., are the elements of the Hartree–Fock operator, $\hat{F}$, and G_γ is the so-called group-overlap integral defined as

$$G_\gamma = \left\langle \chi_{M\gamma} \middle| \sum_{i=1}^{L} c_{i\gamma}\chi_i \right\rangle \tag{2-59}$$

The "bonding root" is to good approximation given by

$$w_b = F_{LL} - \frac{(F_{ML} - F_{MM}G_\gamma)^2}{F_{MM} - F_{LL}} \tag{2-60}$$

and the "antibonding" root by

$$w_a = F_{MM} + \frac{(F_{ML} - F_{LL}G_\gamma)^2}{F_{MM} - F_{LL}} \tag{2-61}$$

because most inorganic complexes will have

$$F_{MM} - F_{LL} \gg |F_{ML} - F_{LL}G_\gamma| \tag{2-62}$$

Solving for α and β in Eq. (2-57) using Eq. (2-62) leads to the bonding molecular orbital correct to G_γ^2:

$$\psi_b = \chi_{L\gamma} \tag{2-63}$$

and to the antibonding orbital

$$\psi_a = \frac{1}{\sqrt{1 - G_\gamma^2}}(\chi_{M\gamma} - G_\gamma\chi_{L\gamma}) \tag{2-64}$$

Notice that the antibonding orbital is a metal orbital, which is Schmidt orthogonalized on the "pure" bonding ligand function.

Let ψ_a and ψ_b be reasonable approximations to the Hartree–Fock orbitals for the molecular system. Then we have with the effective one-electron molecular hamiltonian $\hat{F}$

$$\hat{F}\psi_a = w_a\psi_a \tag{2-65}$$

and

$$\hat{F}\psi_b = w_b\psi_b \tag{2-66}$$

Substituting Eq. (2-64) into Eq. (2-65) and making use of Eq. (2-66) we get easily

$$\hat{F}\chi_{M\gamma} + (w_{a\gamma} - w_{b\gamma})G_\gamma \sum_{i=1}^{L} c_{i\gamma}\chi_i = w_{a\gamma}\chi_{M\gamma} \tag{2-67}$$

An expansion of $w_{a\gamma}$ in powers of G_γ therefore gives

$$w_{a\gamma} = \langle \chi_{M\gamma} | \hat{F} | \chi_{M\gamma} \rangle + (w_{a\gamma} - w_{b\gamma})G_\gamma^2 \tag{2-68}$$

Notice in particular that there is no linear term in G_γ. With $w_\gamma^0 = w_{a\gamma} - w_{b\gamma} > 0$ we have to second order in G_γ

$$w_{a\gamma} \approx (1 + G_\gamma^2)\langle \chi_{M\gamma} | \hat{F} | \chi_{M\gamma} \rangle - w_{b\gamma} G_\gamma^2 \tag{2-69}$$

Defining a so-called pseudo-potential $\hat{U}$ by

$$\hat{U}\chi_{M\gamma}(1) = w_\gamma^0 \int \chi_{M\gamma}^*(2) \sum_{i=1}^{L} c_{i\gamma}\chi_i(2)\, d\tau_2 \sum_{i=1}^{L} c_{i\gamma}\chi_i(1) \tag{2-70}$$

we can rewrite Eq. (2-67) as[14]

$$(\hat{F} + \hat{U})\chi_{M\gamma} = w_{a\gamma}\chi_{M\gamma}$$

The definition of $\hat{U}$ is seen to run parallel to the definition of an exchange operator. Also $\hat{U}$ plays a role similar to the so-called crystal-field potential V included in $\hat{F}$.

Consider now an octahedral complex of O_h symmetry. The five d-orbitals span representations of t_{2g} and e_g symmetry,[10] and suitable ligand orbital combinations can be made up to go together with $e_g(\sigma)$ and with $t_{2g}(\pi)$. The ligand valence orbitals are filled with electrons. Hence the metal valence electrons will occupy the antibonding molecular $\pi_a(t_{2g})$ and $\sigma_a(e_g)$ orbitals. In the case where we have one antibonding "metal-electron" outside the core, the energy separation Δ between the $\sigma_a(e_g)$ and $\pi_a(t_{2g})$ orbitals will be

$$\Delta = w_a(e_g) - w_a(t_{2g}) \tag{2-71}$$

We get using Eq. (2-69)

$$\Delta \approx (1 + G_e^2)\langle \chi_{Me} | \hat{F} | \chi_{Me} \rangle - w_{be}G_e^2 - (1 + G_{t_2}^2)\langle \chi_{Mt_2} | \hat{F} | \chi_{Mt_2} \rangle + w_{bt_2}G_{t_2}^2 \tag{2-72}$$

Taking $w_{be} \approx w_{bt_2} = w_b$ and expanding $\langle \chi_{M\gamma} | \hat{F} | \chi_{M\gamma} \rangle$ we find, retaining only the leading terms,

$$\Delta \approx (G_e^2 - G_t^2)\, T_{\text{kin}}(d) - (G_e^2 - G_t^2)w_b \tag{2-73}$$

where $T_{\text{kin}}(d)$ is the kinetic energy of a d-electron. The difference $(G_e^2 - G_t^2)$ is likely to be positive since e_g involves σ bonding and t_{2g} the weaker π bonding. With $T_{\text{kin}}(d) \gg w_b$, Δ is found to be positive. Notice that by far the largest contribution to Δ stems from the orthogonalization of the metal and ligand orbitals. It is therefore the Pauli repulsion, exerted on the metal d-electrons by the closed electron shells of the ligands, which governs the value of Δ.

We now make a distinction between crystal-field and ligand-field theory. Crystal-field theory is defined as an electronic theory of inorganic complexes which neglects the overlap of electrons associated with the metal, with electrons associated with the ligands. The corresponding theory, with overlap not neglected, is termed ligand-field theory.

For small-overlap integrals we see from Eq. (2-73) that the crystal-field theory is likely to be a good approximation. With $G_{e_g} = G_{t_{2g}} = 0$ we have from Eq. (2-64) $\psi_a = \chi_{M\gamma}$. For an octahedral compound we can then define the "cubic crystal field parameter" $10Dq$ as

$$10Dq = \langle d_{e_g} | \hat{h}_{\text{core}} | d_{e_g} \rangle - \langle d_{t_{2g}} | \hat{h}_{\text{core}} | d_{t_{2g}} \rangle = \overline{h}(e_g) - \overline{h}(t_{2g}) \qquad (2\text{-}74)$$

We further define our zero of energy by the one-electron energy of a filled $(t_{2g})^6(e_g)^4$ configuration

$$6\overline{h}(t_{2g}) + 4\overline{h}(e_g) = 0 \qquad (2\text{-}75)$$

Solving Eqs. (2-74) and (2-75) we get

$$-4Dq = \overline{h}(t_{2g}) \qquad (2\text{-}76)$$

$$6Dq = \overline{h}(e_g) \qquad (2\text{-}77)$$

The matrix elements of the operator e^2/r_{12} are evaluated in the crystal-field approximation where $\psi_a = nd$ exactly as in atomic theory.[1] They are given for d-orbitals in terms of the atomic Condon–Shortley–Slater integrals F_0, F_2, and F_4 integrals[10] (Table 2-3).

Consider as an example the ground state of an octahedral Co^{3+} complex. The configuration is $(t_{2g})^6$ corresponding to the wave function

$$\Psi(^1A_{1g}) = |d_{xz}^{\alpha} d_{xz}^{\beta} d_{yz}^{\alpha} d_{yz}^{\beta} d_{xy}^{\alpha} d_{xy}^{\beta}| \qquad (2\text{-}78)$$

Exciting an electron will lead to the configuration $(t_{2g})^5(e_g)^1$ producing $^1T_{1g}$, $^1T_{2g}$ states together with the corresponding spin-triplets. Assuming an effective molecular Hartree–Fock operator $\hat{F}$ we take $\hat{F}d_\gamma = w_\gamma d_\gamma$. Utilizing Eq. (2-47) we find, for example, using a descent in symmetry[10] to construct the excited states,

$$W(^1T_{1g}) - W(^1A_{1g}) = w_{x^2-y^2} - w_{xy} - J_{xy,x^2-y^2} + 2K_{xy,x^2-y^2} \qquad (2\text{-}79)$$

$$W(^1T_{2g}) - W(^1A_{1g}) = w_{z^2} - w_{xy} - J_{xy,z^2} + 2K_{xy,z^2} \qquad (2\text{-}80)$$

Table 2-3 Two-electron integrals of the set $t_{2g}(xz, yz, xy)$ and $e_g(x^2 - y^2, z^2)$ according to Ref. 10, and corrected to second order in G_γ

$$J_{xy,xy} = J_{xz,xz} = J_{yz,yz} = (1 - 2G_\pi^2)(F_0 + 4F_2 + 36F_4)$$

$$J_{xz,yz} = J_{xy,yz} = J_{xz,xy} = (1 - 2G_\pi^2)(F_0 - 2F_2 - 4F_4)$$

$$J_{z^2,yz} = J_{z^2,xz} = (1 - G_\sigma^2 - G_\pi^2)(F_0 + 2F_2 - 24F_4)$$

$$J_{z^2,xy} = (1 - G_\sigma^2 - G_\pi^2)(F_0 - 4F_2 + 6F_4)$$

$$J_{x^2-y^2,yz} = J_{x^2-y^2,xz} = (1 - G_\sigma^2 - G_\pi^2)(F_0 - 2F_2 - 4F_4)$$

$$J_{x^2-y^2,xy} = (1 - G_\sigma^2 - G_\pi^2)(F_0 + 4F_2 - 34F_4)$$

$$J_{z^2,z^2} = J_{x^2-y^2,x^2-y^2} = (1 - 2G_\sigma^2)(F_0 + 4F_2 + 36F_4)$$

$$J_{z^2,x^2-y^2} = (1 - 2G_\sigma^2)(F_0 - 4F_2 + 6F_4)$$

$$K_{yz,xz} = K_{xy,yz} = K_{xz,xy} = (1 - 2G_\pi^2)(3F_2 + 20F_4)$$

$$K_{z^2,x^2-y^2} = (1 - 2G_\sigma^2)(4F_2 + 15F_4)$$

$$K_{xy,x^2-y^2} = (1 - G_\sigma^2 - G_\pi^2)35F_4$$

$$K_{z^2,xy} = (1 - G_\sigma^2 - G_\pi^2)(4F_2 + 15F_4)$$

$$K_{z^2,yz} = K_{z^2,xz} = (1 - G_\sigma^2 - G_\pi^2)(F_2 + 30F_4)$$

$$K_{x^2-y^2,yz} = K_{x^2-y^2,xz} = (1 - G_\sigma^2 - G_\pi^2)(3F_2 + 20F_4)$$

$$[(xz)(xy)|(yz)(x^2 - y^2)] = (1 - \tfrac{1}{2}G_\sigma^2 - \tfrac{3}{2}G_\pi^2)(-3)(F_2 - 5F_4)$$

Further

$$w_{x^2-y^2} = \bar{h}(e_g) + 2J_{xz,x^2-y^2} - K_{xz,x^2-y^2} + 2J_{yz,x^2-y^2} - K_{yz,x^2-y^2} + 2J_{xy,x^2-y^2} - K_{xy,x^2-y^2}$$

$$w_{z^2} = \bar{h}(e_g) + 2J_{xz,z^2} - K_{xz,z^2} + 2J_{yz,z^2} - K_{yz,z^2} + 2J_{xy,z^2} - K_{xy,z^2}$$

$$w_{xy} = \bar{h}(t_{2g}) + 2J_{xz,xy} - K_{xz,xy} + 2J_{yz,xy} - K_{yz,xy} + J_{xy,xy}$$

Evaluating the J and K integrals in the crystal-field approximation we find easily, with the help of Table 2-3, putting G_π and G_σ equal to zero,

$$W(^1T_{1g}) - W(^1A_{1g}) = 10Dq - 35F_4 \qquad (2\text{-}81)$$

$$W(^1T_{2g}) - W(^1A_{1g}) = 10Dq + 16F_2 - 115F_4 \qquad (2\text{-}82)$$

Next we shall look at an octahedral Cr^{3+} complex using ligand-field theory. The configuration $|\psi^\alpha_{xz}\psi^\alpha_{yz}\psi^\alpha_{xy}|$ gives rise in O_h symmetry to a $^4A_{2g}$ state, and a component of an excited $^4T_{2g}$ state is given by $|\psi^\alpha_{xz}\psi^\alpha_{yz}\psi^\alpha_{x^2-y^2}|$. We find

$$W(^4T_{2g}) - W(^4A_{2g}) = \Delta + 2J_{xz,x^2-y^2} - 2J_{yz,xz} - 2K_{xz,x^2-y^2} + 2K_{yz,xz} \qquad (2\text{-}83)$$

where Δ is defined in Eq. (2-71). This is as far as the problem can be reduced in octahedral symmetry. Using a symmetry operation which is *not* consistent with an O_h symmetry we can find

$$2J_{xz,x^2-y^2} - 2J_{yz,xz} + 2K_{yz,xz} - 2K_{xz,x^2-y^2} = 0 \qquad (2\text{-}84)$$

This result can of course also be obtained by going to spherical symmetry. Therefore, only by moving outside octahedral symmetry is it possible to identify a measured energy difference with Δ. We do not know how great an error we perpetrate by such an approximation.

Provided Eq. (2-64)

$$\psi_a = \sqrt{\frac{1}{1-G_\gamma^2}}(\chi_{M_\gamma} - G_\gamma\chi_{L_\gamma}) \qquad (2\text{-}85a)$$

where

$$\chi_{L_\gamma} = \sum_{i=1}^{L} c_{i\gamma}\chi_i \qquad (2\text{-}85b)$$

is a good approximation to the molecular orbitals of ligand-field theory, we can expand the two-electron integrals. We have for instance

$$K_{xz,yz} = (1-G_\pi^2)^{-2}\left[\iint d_{xz}(1)d_{yz}(1)\frac{1}{r_{12}}d_{xz}(2)d_{yz}(2)\,d\tau_1\,d\tau_2 - 4G_\pi\iint d_{xz}(1)d_{yz}(1)\frac{1}{r_{12}}d_{xz}(2)\chi_{Lyz}(2)\,d\tau_1\,d\tau_2 + \cdots\right] \qquad (2\text{-}86)$$

Expanding the metal orbital on the ligand-orbital set we get to first order in G_π

$$\chi_{Lyz} = G_\pi d_{yz} + \cdots \qquad (2\text{-}87)$$

Therefore correct to second order in G_π

$$K_{xz,yz} \approx (1 - 2G_\pi^2) K_{xz,yz}^{\text{atomic}} \tag{2-88}$$

A reduction of the molecular K integral over that of the atomic one is therefore expected. In Table 2-3 we have evaluated most of the two-electron integrals, correct to second order in G_γ.

Fifteen different J integrals, ten different K integrals, and nine other two-electron integrals can be met within a $(\psi_{xz}, \psi_{yz}, \psi_{xy}, \psi_{x^2-y^2}, \psi_{z^2})$ set of molecular orbitals. Symmetry of the molecule will, however, impose certain restrictions on the independence of these 34 integrals. In octahedral (O_h) symmetry, for instance, we encounter only 10 independent two-electron integrals.

In the crystal field approximation state energy differences are always independent of the F_0 parameter. This is, however, not true in the ligand field approximation. The orders of magnitude are in the first transition series $F_0 \approx$ 200,000 cm^{-1}, $F_2 \approx 2{,}000$ cm^{-1} and $F_4 \approx 200$ cm^{-1}. The dependence of state energy differences on $(G_\sigma^2 - G_\pi^2)F_0$ is therefore such as to make the other terms insignificant; the conventional definition of $10Dq$ is not operative when we take account of electron delocalization.

The definition of $10Dq$, so simple and appealing in crystal-field theory, is unfortunately rather complex in molecular-orbital theory. In general we can write for an antibonding and bonding orbital

$$\psi_a = N_a\left(\chi_d - \lambda_a \sum_{i=1}^{L} c_i\chi_i\right) \tag{2-89}$$

$$\psi_b = N_b\left(\sum_{i=1}^{L} c_i\chi_i + \lambda_b\chi_d\right) \tag{2-90}$$

Due to orthogonality, with G being the group overlap integral, we must have

$$G - \lambda_a + \lambda_b - \lambda_a\lambda_b G = 0 \tag{2-91}$$

This equation gives a relation between the coefficients in the bonding and antibonding orbitals. These coefficients must be evaluated for a specific electronic configuration in some suitable Hartree–Fock scheme. Therefore, the traditional core orbitals and the antibonding "d-orbitals" will exhibit dependence upon the electronic state under consideration, and each $(t_{2g})^n(e_g)^m$ configuration will possess a different set of $\bar{h}_{aa}^{\text{core}}$ values. As realized by Watson and Freeman,[11] this makes an average over all possible configurations a prerequisite to the extraction of one value of a splitting parameter. This difficulty does not of course exist when all the antibonding orbitals are pure nd atomic orbitals. The quantity $10Dq$ is therefore a parameter specific to crystal-field theory.

One final point. The orbital splittings of conventional crystal-field energy diagrams picture the $\bar{h}_{\text{core}}$ quantities. On the other hand the w_n's of Eq. (2-12) are given by photo-ionization experiments using Koopmans' theorem. $\bar{h}_{\text{core}}$ and w_n are related by means of Eq. (2-12), in which a lot of electronic repulsion terms appear. The crystal-field theory level order cannot therefore *a priori* be assumed to follow the SCF, w_n order.

The most impressive feature of crystal-field theory is the correlation diagrams which picture the excited-state energies as functions of $10Dq$. As an example of how such diagrams are constructed we shall take the simplest case of a many-electron diagram, namely that of the spin-triplet states of two electrons.

Characterizing the triplet states the configurations $(t_{2g})^2$, $(t_{2g})^1(e_g)^1$, and $(e_g)^2$ can give rise to in O_h and D_{4h} we get[10]

	O_h	D_{4h}
$(t_{2g})^2$	${}^3T_{1g}$	${}^3A_{2g} + {}^3E_g$
$(t_{2g})^1(e_g)^1$	${}^3T_{1g}$	${}^3A_{2g} + {}^3E_g$
	${}^3T_{2g}$	${}^3B_{2g} + {}^3E_g$
$(e_g)^2$	${}^3A_{2g}$	${}^3B_{1g}$

Using the classification in D_{4h} it is easy to find the components

$$A_2[{}^3T_{1g}(t_2)^2]\ |(\overset{+}{xz})(\overset{+}{yz})|$$

$$A_2[{}^3T_{1g}(e)^1(t_2)^1]\ |(\overset{+}{xy})(x^2 \overset{+}{-} y^2)|$$

$$B_2[{}^3T_{2g}(e)^1(t_2)^1]\ |(\overset{+}{xy})(\overset{+}{z^2})|$$

and

$$B_1[{}^3A_{2g}(e)^2]\ |(x^2 \overset{+}{-} y^2)(\overset{+}{z^2})|$$

The state energies are given by

$$\begin{array}{cc} {}^3T_{1g}(t_{2g})^2 & {}^3T_{1g}(t_{2g})^1(e_g)^1 \end{array}$$

$$\begin{vmatrix} -8Dq + J_{xz,yz} - K_{xz,yz} - W & 2[(xz)(xy)|(yz)(x^2-y^2)] \\ 2[(xz)(xy)|(yz)(x^2-y^2)] & 2Dq + J_{xy,x^2-y^2} - K_{xy,x^2-y^2} - W \end{vmatrix} = 0 \qquad (2\text{-}92)$$

$$W({}^3T_{2g}) = 2Dq + J_{xy,z^2} - K_{xy,z^2} \qquad (2\text{-}93)$$

$$W({}^3A_{2g}) = 12Dq + J_{x^2-y^2,z^2} - K_{x^2-y^2,z^2} \qquad (2\text{-}94)$$

The ligands are now removed to infinity. Hence the core separation between t_{2g} and e_g vanishes. At the same time G_σ and G_π goes to zero and the two-electron integrals take on the values given in Table 2-3. For $Dq = 0$ we have

$${}^3T_{1g} \qquad \begin{vmatrix} F_0 - 5F_2 - 24F_4 - W & -6F_2 + 30F_4 \\ -6F_2 + 30F_4 & F_0 + 4F_2 - 69F_4 - W \end{vmatrix} = 0 \qquad (2\text{-}95)$$

giving for the A_2 components of the degenerate T_1 state functions

$$W(^3T_{1g}) = F_0 - 8F_2 - 9F_4. \quad \Psi = \sqrt{\frac{4}{5}}|(\overset{+}{xz})(\overset{+}{yz})| + \sqrt{\frac{1}{5}}|(\overset{+}{xy})(x^2 \overset{+}{-} y^2)| \qquad (2\text{-}96)$$

$$W(^3T_{1g}) = F_0 + 7F_2 - 84F_4. \quad \Psi = -\sqrt{\frac{1}{5}}|(\overset{+}{xz})(\overset{+}{yz})| + \sqrt{\frac{4}{5}}|(\overset{+}{xy})(x^2 \overset{+}{-} y^2)| \qquad (2\text{-}97)$$

Further, for the B_2 component of the T_2 state

$$W(^3T_{2g}) = F_0 - 8F_2 - 9F_4. \quad \Psi = |(\overset{+}{xy})(\overset{+}{z^2})| \qquad (2\text{-}98)$$

and

$$W(^3A_{2g}) = F_0 - 8F_2 - 9F_4. \quad \Psi = |(x^2 \overset{+}{-} y^2)(\overset{+}{z^2})| \qquad (2\text{-}99)$$

These wave functions and energies are valid in the limit of the ligands being removed to an infinite distance from the metal center. They are of course nothing but the triplet atomic states of a d^2 atomic configuration. Putting $F_0 - 8F_2 - 9F_4 = 0$ we have a sevenfold degenerate orbital ^3F as ground state, and a threefold degenerate ^3P state placed at $15(F_2 - 5F_4)$ cm^{-1}. This energy separation is known from atomic spectroscopy.[12]

We now assume that the interelectronic repulsion energy is independent of the t_2 and e orbitals' core energy, as measured in units of $10Dq$. Hence we get for the molecular state energies using the above wave function components

$$^3T_{1g} \begin{vmatrix} -6Dq - W & 4Dq \\ 4Dq & 15(F_2 - 5F_4) - W \end{vmatrix} = 0 \qquad (2\text{-}100)$$

$$W(^3T_{2g}) = 2Dq \qquad (2\text{-}101)$$

$$W(^3A_{2g}) = 12Dq \qquad (2\text{-}102)$$

The ground state is therefore $^3T_{1g}$ followed by $^3T_{2g}$. For Dq less than about one-third of $15(F_2 - 5F_4)$, the state $^3A_{2g}$ is topped by the second $^3T_{1g}$ state.

Octahedral complexes having eight electrons in the t_{2g} and e_g shells have state energies which can be related to the formulae valid for two-electron systems. The $(t_{2g})^6(e_g)^2$ configuration produces a $^3A_{2g}$ state with a core energy of $-12Dq$. The configuration $(t_{2g})^5(e_g)^3$ gives rise to $^3T_{1g}$ and $^3T_{2g}$ states, both with a core energy of $-2Dq$. Finally $(t_{2g})^4(e_g)^4$ produces the triplet state $^3T_{1g}$ $(8Dq)$. The electronic repulsions for $Dq = 0$ will again place the two $^3T_{1g}$ states some $15(F_2 - 5F_4)$ apart. We observe that Eqs. (2-100), (2-101), and (2-102) are valid also for a d^8 system provided we take Dq to be negative.

We have pictured in Fig. 2-3 a correlation diagram valid for octahedral complexes having eight electrons in the (t_{2g}) and (e_g) shells. The parameters used correspond to a Ni^{2+} complex. The diagram incorporates the effects of spin-orbit coupling,[13] and we notice the effect of the noncrossing rule for levels possessing the same symmetry. It is important, however, to realize that the pictured functional

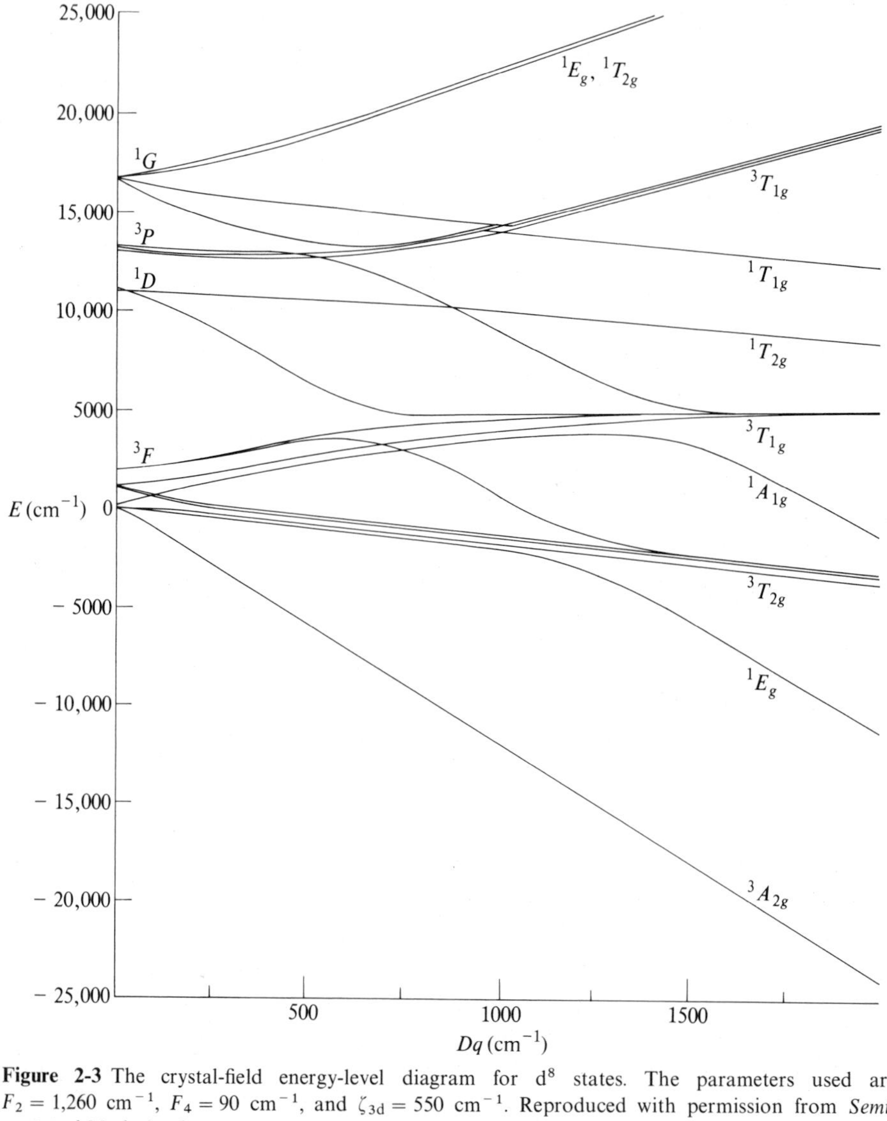

Figure 2-3 The crystal-field energy-level diagram for d^8 states. The parameters used are $F_2 = 1{,}260\ cm^{-1}$, $F_4 = 90\ cm^{-1}$, and $\zeta_{3d} = 550\ cm^{-1}$. Reproduced with permission from *Semi-empirical Methods of Electronic Structure Calculation, Part B*, Plenum Press, 1977.

dependence of the molecular levels on Dq is only qualitative, because in the construction of the diagram the zeroth-order unperturbed states are assumed to rise from a pure $(3d)^8$ configuration. No allowance is therefore made either for atomic configuration interactions or for molecular orbital formation.

For $Dq \to 0$ the molecular states of a transition metal complex correlate with the atomic terms of the free metal ion. In this limit L is a good quantum number, assuming a Russell-Saunders coupling scheme.[1] For small Dq values the molecular ground state will therefore correlate with the ground atomic term. According to

Table 2-4

J	Irreducible representations in O and T_d
0	A_1
$\frac{1}{2}$	$E_{1/2}$
1	T_1
$\frac{3}{2}$	G
2	$E + T_2$
$\frac{5}{2}$	$E_{5/2} + G$
3	$A_2 + T_1 + T_2$
$\frac{7}{2}$	$E_{1/2} + E_{5/2} + G$
4	$A_1 + E + T_1 + T_2$

Hund's rule this term combines the highest spin multiplicity with the highest value of L.

In general the atomic states of the metal ion will form a basis for the irreducible representations which span the molecular states of the complex in question.[10] Table 2-4 accordingly tabulates the various irreducible representations to which each value of the angular momentum J gives rise in O and T_d symmetries. Table 2-5 gives the group multiplication table.

2-7 THE SCF-$X\alpha$-SW METHOD

The self consistent field-$X\alpha$-scattered wave method for investigating the electronic structures of molecules is a technique which arose from developments made by Slater and his group.[15,16] The method differs from the LCAO MO technique in that it does not use atomic orbitals as a starting point. Instead it divides a molecule into three fundamental types of regions: (1) atomic, (2) interatomic, and (3) extra-atomic. Solving the Schrödinger equation separately in these three "boxes", the intersphere wave functions and their first derivatives are required to be continuous with the solutions to the atomic and extra-atomic regions, respectively. In the solutions to the three second-order differential equations the energy occurs as a parameter. It is therefore indirectly determined by the consistency of the linear equations taking care of the behavior of the molecular wave function, and is found by interpolation.

In order to discuss the method it is natural to start out with the so-called density matrices.[17] Let $\Psi(\mathbf{r}_1, \mathbf{r}_2, \ldots, \mathbf{r}_N)$ be a wave function for N electrons. The *first-order density matrix* $\rho_1(\mathbf{r}_1; \mathbf{r}_1')$ is then defined by

$$\rho_1(\mathbf{r}_1; \mathbf{r}_1') = N \int \Psi(\mathbf{r}_1, \mathbf{r}_2, \ldots, \mathbf{r}_N)\Psi^*(\mathbf{r}_1', \mathbf{r}_2, \ldots, \mathbf{r}_N)\, d\tau_2 \ldots d\tau_N \tag{2-103}$$

The *second-order density matrix* $\rho_2(\mathbf{r}_1, \mathbf{r}_2; \mathbf{r}_1', \mathbf{r}_2')$ is defined by

$$\rho_2(\mathbf{r}_1, \mathbf{r}_2; \mathbf{r}_1', \mathbf{r}_2') = N(N-1) \int \Psi(\mathbf{r}_1, \mathbf{r}_2, \ldots, \mathbf{r}_N)\Psi^*(\mathbf{r}_1', \mathbf{r}_2', \ldots, \mathbf{r}_N)\, d\tau_3 \ldots d\tau_N \tag{2-104}$$

Table 2-5 Direct products in O and T_d.
(The antisymmetric products have been put in brackets.)

	A_1	A_2	E	T_1	T_2	$E_{1/2}$	$E_{5/2}$	G
A_1	A_1	A_2	E	T_1	T_2	$E_{1/2}$	$E_{5/2}$	G
A_2		A_1	E	T_2	T_1	$E_{5/2}$	$E_{1/2}$	G
E			$A_1, \{A_2\}, E$	T_1, T_2	T_1, T_2	G	G	$E_{1/2}, E_{5/2}, G$
T_1				$A_1, E, \{T_1\}, T_2$	A_2, E, T_1, T_2	$E_{1/2}, G$	$E_{5/2}, G$	$E_{1/2}, E_{5/2}, 2G$
T_2					$A_1, E, \{T_1\}, T_2$	$E_{5/2}, G$	$E_{1/2}, G$	$E_{1/2}, E_{5/2}, 2G$
$E_{1/2}$						$\{A_1\}, T_1$	A_2, T_2	E, T_1, T_2
$E_{5/2}$							$\{A_1\}, T_1$	E, T_1, T_2
G								$\{A_1\}, A_2, \{E\}, 2T_1, \{T_2\}, T_2$

We now introduce the "natural spin orbitals". Let $\tilde{\psi}_1, \tilde{\psi}_2, \ldots, \tilde{\psi}_i, \ldots$ be a complete, orthonormal set of spin orbitals. We can then expand $\rho_1(\mathbf{r}_1\,;\mathbf{r}'_1)$ on these orbitals

$$\rho_1(\mathbf{r}_1\,;\mathbf{r}'_1) = \sum_{i,j} \gamma_{ij}\tilde{\psi}_i(\mathbf{r}_1)\tilde{\psi}^*_j(\mathbf{r}'_1) \tag{2-105}$$

where the γ_{ij} are constants. Since the density function must be real we may conclude that the matrix $\{\gamma_{ij}\}$ is hermitian. It is therefore possible to change to another orthonormal set $\psi_1, \psi_2, \ldots, \psi_N$, related to the first by a unitary matrix, so that $\{\gamma_{ij}\}$ is diagonal. We have then

$$\rho_1(\mathbf{r}_1\,;\mathbf{r}'_1) = \sum_{i} n_i\psi_i(\mathbf{r}_1)\psi^*_i(\mathbf{r}'_1) \tag{2-106}$$

The functions $\psi_1, \psi_2, \ldots, \psi_i, \ldots$ which make the first-order density matrix diagonal are called the *natural spin-orbitals*. n_i is the occupation number for the spin-orbital ψ_i. Naturally we have $\sum_i n_i = N$. By expanding the wave function Ψ in Slater determinants one may further prove that $0 \leq n_i \leq 1$.

Let the hamiltonian for the N-electron system be

$$\mathscr{H} = \sum_{i=1}^{N} \hat{h}(i) + \frac{1}{2}\sum_{i,j}{}' \frac{1}{r_{ij}} \tag{2-107}$$

where the dash in the last summation means that the term $i = j$ is to be omitted. Using the natural spin-orbitals we have therefore for the electronic energy W

$$W = \sum_{i} n_i\bar{h}_{ii} + \frac{1}{2}\int \frac{1}{r_{12}}\rho_2(\mathbf{r}_1, \mathbf{r}_2)\, d\tau_1\, d\tau_2 \tag{2-108}$$

where

$$\bar{h}_{ii} = \langle\psi_i|\,\hat{h}(i)\,|\psi_i\rangle \tag{2-109}$$

We now express the second-order density matrix as

$$\rho_2(\mathbf{r}_1, \mathbf{r}_2) = \rho(\mathbf{r}_1)\rho(\mathbf{r}_2) + \rho(\mathbf{r}_1)\rho_0(\mathbf{r}_1, \mathbf{r}_2) \tag{2-110}$$

$\rho(\mathbf{r}_1)\rho(\mathbf{r}_2)$ constitutes the "classical" part of $\rho_2(\mathbf{r}_1, \mathbf{r}_2)$. Using Eq. (2-106) we get therefore

$$W = \sum_{i} n_i\bar{h}_{ii} + \frac{1}{2}\sum_{i,j} n_i n_j \cdot J_{ij} + \frac{1}{2}\int d\tau_1\rho(\mathbf{r}_1)\int \frac{1}{r_{12}}\rho_0(\mathbf{r}_1, \mathbf{r}_2)\, d\tau_2 \tag{2-111}$$

with

$$J_{ij} = \langle\psi_i(1)\psi_j(2)|\,\frac{1}{r_{12}}\,|\psi_i(1)\psi_j(2)\rangle \tag{2-112}$$

Defining an exchange potential

$$\hat{U}(\mathbf{r}_1) = \int \frac{1}{r_{12}}\rho_0(\mathbf{r}_1, \mathbf{r}_2)\, d\tau_2$$

we may write Eq. (2-111) as

$$W = \sum_{n_i} n_i\,\bar{h}_{ii} + \frac{1}{2}\sum_{i,j} n_i n_j J_{ij} + \frac{1}{2}\int \rho(\mathbf{r})\hat{U}(\mathbf{r})\, d\tau \tag{2-113}$$

We now assume that $\hat{U}(\mathbf{r})$ can be approximated as the potential at the point $\mathbf{r}$, arising from a spherical distribution of uniform charge density, holding one electron. The distribution is centered at $\mathbf{r}$ with a density equal to $\rho(\mathbf{r})$. With R as the radius of the spherical distribution

$$-\hat{U}(\mathbf{r}) = 4\pi \int_0^R \frac{1}{\tilde{r}} \rho(\mathbf{r})\tilde{r}^2 \, d\tilde{r} = 2\pi\rho(\mathbf{r})R^2 \tag{2-114}$$

Since the sphere of radius R contains one electron we have also

$$\frac{4\pi}{3} R^3 \rho(\mathbf{r}) = 1 \tag{2-115}$$

Eliminating R from Eqs. (2-114) and (2-115) leads to

$$\hat{U}(\mathbf{r}) = -\left[\frac{9\pi}{2}\rho(\mathbf{r})\right]^{1/3} \tag{2-116}$$

Instead of Eq. (2-116) we can use the more flexible exchange potential

$$\hat{U}_{X\alpha}(\mathbf{r}) = -\frac{9}{2}\alpha\left[\frac{3}{4\pi}\rho(\mathbf{r})\right]^{1/3} \tag{2-117}$$

The value $\alpha = 1$ gives the Slater exchange potential.[15,18]

We now want to minimize the expression Eq. (2-113) for W with respect to the occupation numbers n_i. Taking ψ_i and ψ_i^* to be independent we shall vary ψ_i^*, under the restriction

$$\langle \delta\psi_i | \psi_j \rangle = 0 \qquad j = 1, 2, \ldots, N \tag{2-118}$$

Then

$$\delta W = n_i \langle \delta\psi_i | \hat{h} | \psi_i \rangle + \sum_j n_i n_j \langle \delta\psi_i \psi_j | \frac{1}{r_{12}} | \psi_i \psi_j \rangle$$
$$+ \frac{1}{2} n_i \int \delta\psi_i^* \psi_i U(\mathbf{r}) \, d\tau + \frac{1}{2} \int \frac{1}{3} \hat{U}(\mathbf{r}) n_i \psi_i \delta\psi_i^* \, d\tau \tag{2-119}$$

where for the last term we have used Eqs. (2-106) and (2-117). Equation (2-119) can be written

$$\delta W = n_i \langle \delta\psi_i | \hat{h} + \sum_j n_j \hat{J}_j + \frac{1}{2}\left(1 + \frac{1}{3}\right)\hat{U}(\mathbf{r}) | \psi_i \rangle \tag{2-120}$$

where the Coulomb operator $\hat{J}$ has been defined in Eq. (2-18a). The condition $\delta W = 0$ together with Eq. (2-118) leads to

$$\left[\hat{h} + \sum_j n_j \hat{J}_j + \frac{2}{3}\hat{U}(\mathbf{r})\right]\psi_i = \sum_j w_{ji}\psi_i \tag{2-121}$$

For the diagonal lagrangian multipliers we find

$$w_{ii} = \bar{h}_{ii} + \sum_j n_j J_{ij} + \frac{2}{3}\langle \psi_i | \hat{U}(\mathbf{r}) | \psi_i \rangle \tag{2-122}$$

The one-electron energies are also equal to the derivatives of the total energy W with respect to the occupation numbers n_i. This is easily proved by direct differentiation of Eq. (2-113). We get

$$\frac{\partial W}{\partial n_i} = \bar{h}_{ii} + \sum_j n_j J_{ij} + \frac{2}{3}\langle\psi_i|\,\hat{U}(\mathbf{r})\,|\psi_i\rangle \tag{2-123}$$

In solving the set of Eqs. (2-121) for a molecule, we perform the already-mentioned geometrical partitioning of the molecular space. In the atomic region ψ_i is expanded in spherical harmonics times a radial function to be determined by numerical integration. A similar expansion is performed for the extra-atomic region. In the interatomic region the Eqs. (2-121) are solved under the assumption that $\rho(\mathbf{r})$ is a constant and the solutions to Eqs. (2-121) are written as a sum of spherical harmonics times spherical Bessel functions of imaginary arguments (for $w_i < 0$). The wave functions are then joined together, and the energies are found.

Different values for the parameter α may be chosen for different spheres and regions. Also the radii of the spheres may be chosen at will. The method is therefore of a semiempirical nature. It has been stated that the $X\alpha$ method leads to the correct behavior of the energy at large internuclear distances. This would mean that at least some correlation energy must be included in the method. However, the validity of the above assertion is open to some doubt.[19]

The one-electron energies, obtained in the ordinary Hartree–Fock method are, according to Koopmans' theorem, equal to the ionization potentials of the molecule. As already mentioned in this section 2-1, Koopmans' theorem ignores the electronic relaxation which follows ionization, and is only valid provided the other electrons remain "frozen." The relaxation may be quite large for the complexes, so that Koopmans' approximation produces erroneous ionization potentials.

It was shown by Slater[15] that using the SCF-$X\alpha$-SW method, the difference between the total energy of a system and its ionized counterpart to very good approximation is equal to the energy of a spin-orbital containing only half of an electron. Such a spin-orbital, calculated in the SCF-$X\alpha$ approximation, is called a "transition state." The transition states incorporate the effects of orbital relaxation, and we can identify their energies with the ionization potentials.

The excitation energy obtained in this way will, however, be a weighted average of all the transitions to states contained in the direct product $\Gamma(\psi_i) \times \Gamma(\psi_t)$. In particular, having a spin-singlet as ground state the energies of the excited spin-singlets and triplets will be averaged out. By performing a spin-restricted as well as a spin-unrestricted calculation it seems, however, possible to separate the energies of the spin-triplets from the spin-singlets.

REFERENCES

1. E. U. Condon and G. H. Shortley, *The Theory of Atomic Spectra*, Cambridge University Press, 1951.
2. C. C. J. Roothaan, *Rev. Mod. Phys.*, **23**: 69, 1951.

3. J. P. Dahl, *The Independent-Particle Model*, Polyteknisk Forlag, Lyngby (Denmark), 1972.
4. C. C. J. Roothaan, *Rev. Mod. Phys.*, **32**: 179, 1960.
5. R. K. Nesbet, *Proc. Roy. Soc.*, **A230**: 312, 1955.
6. J. P. Dahl and C. J. Ballhausen, *Adv. Quantum Chemistry*, **4**: 170, 1968.
7. J. N. Murrell and A. J. Harget, *Semi-empirical Self-consistent-field Molecular Orbital Theory of Molecules*, Wiley-Interscience, 1972.
8. C. J. Ballhausen and H. B. Gray, *Coordination Chemistry*, Van Nostrand Reinhold, 1971, vol. **1**: 3
9. J. A. Pople and D. L. Beveridge, *Approximate Molecular Orbital Theory*, McGraw-Hill, 1970.
10. C. J. Ballhausen, *Introduction to Ligand Field Theory*, McGraw-Hill, 1962.
11. R. E. Watson and H. J. Freeman, *Phys. Rev.*, **134**: A1526, 1964.
12. C. Moore, *Atomic Energy Levels*, National Bureau of Standards Circular 467, vol. 1, 1949; vol. 2, 1952; and vol. 3, 1958.
13. A. D. Liehr and C. J. Ballhausen, *Ann. of Physics*, **6**: 134, 1959.
14. C. J. Ballhausen and J. P. Dahl, *Theoret. Chim. Acta*, **34**: 169, 1974.
15. J. C. Slater, *Quantum Theory of Molecules and Solids*, McGraw-Hill, vol. IV, 1974.
16. K. H. Johnson, *Adv. Quant. Chem.*, **7**: 143, 1973.
17. R. McWeeny and B. T. Sutcliffe, *Methods of Molecular Quantum Mechanics*, Academic Press, 1969.
18. K. H. Johnson, *Ann. Rev. Phys. Chem.*, **26**: 39, 1975.
19. J. C. Slater, *Intern. J. Quan. Chem. Symp.*, **7**: 533, 1973.

CHAPTER

THREE

THE GROUND STATE. MAGNETIC PROPERTIES

3-1 MOTION OF CHARGED PARTICLES IN AN ELECTROMAGNETIC FIELD

The classical hamilton function for a particle with charge Q moving in an electric or magnetic field is given by[1]

$$\mathscr{H} = \frac{1}{2m}\left[\left(p_x - \frac{Q}{c}A_x\right)^2 + \left(p_y - \frac{Q}{c}A_y\right)^2 + \left(p_z - \frac{Q}{c}A_z\right)^2\right] + Q\phi \qquad (3\text{-}1)$$

p_x, p_y, and p_z are the components of the linear momentum $\mathbf{p}$, $\mathbf{A}$ is a vector potential, and ϕ a scalar potential. $\mathbf{A}$ and ϕ are derived from the electric and magnetic field strengths $\mathbf{E}$ and $\mathbf{H}$ by the relations

$$\mathbf{H} = \nabla \wedge \mathbf{A} \qquad (3\text{-}2)$$

$$\mathbf{E} = -\frac{1}{c}\frac{\partial}{\partial t}\mathbf{A} - \nabla\phi \qquad (3\text{-}3)$$

The construction of the quantum-mechanical hamiltonian is performed by replacing p_x in Eq. (3-1) with $-i\hbar\, \partial/\partial x$, etc. Hence

$$\mathscr{H} = \frac{1}{2m}\left[-\hbar^2\nabla^2 + \frac{i\hbar Q}{c}(\nabla \cdot \mathbf{A} + \mathbf{A} \cdot \nabla) + \frac{Q^2}{c^2}|\mathbf{A}|^2\right] + Q\phi \qquad (3\text{-}4)$$

A closer analysis shows that it is possible to make a different choice with regard to $\mathbf{A}$ and ϕ which leaves $\mathbf{E}$ and $\mathbf{H}$ unchanged. The invariance of $\mathbf{E}$ and $\mathbf{H}$ under such transformations is called *gauge invariance.* In quantum mechanics the Coulomb gauge is particularly important; in this gauge $\nabla \cdot \mathbf{A} = 0$.

With ρ as the charge density of the system, Poisson's equation demands that

$$\nabla^2 \phi = 4\pi\rho \tag{3-5}$$

We now put $\rho = \rho_i + \rho_e(t)$ where ρ_i is the charge density internal to the system which is slowly varying with time, and $\rho_e(t)$ the external charge density. $\rho_e(t)$ may, for example, be associated with a light wave entering the system. Since inside the system $\rho_e(t)$ is equal to zero, Poisson's equation yields

$$\nabla^2 \phi_i = 4\pi\rho_i$$

where ϕ_i is the internal potential. The term $Q\phi$ is therefore equal to the potential energy V of the system. The hamiltonian in Eq. (3-4) is then in the Coulomb gauge given by

$$\mathscr{H} = -\frac{\hbar^2}{2m}\nabla^2 + V + \frac{i\hbar Q}{mc}\mathbf{A}\cdot\nabla + \frac{Q^2}{2mc^2}|\mathbf{A}|^2 \tag{3-6}$$

or

$$\mathscr{H} = -\frac{\hbar^2}{2m}\nabla^2 + V + \mathscr{H}^{(1)} \tag{3-7}$$

For a system containing N charged particles we get therefore

$$\mathscr{H}^{(1)} = \sum_{j=1}^{N}\left(\frac{i\hbar Q_j}{M_j c}\mathbf{A}_j\cdot\nabla_j + \frac{Q_j^2}{2M_j c^2}|\mathbf{A}_j|^2\right) \tag{3-8}$$

Consider now a system of electrons with $Q_j = -|e|$ in a static magnetic field **H**. Writing out the components in Eq. (3-2) we have

$$H_x = \frac{\partial A_z}{\partial y} - \frac{\partial A_y}{\partial z}, \tag{3-9}$$

$$H_y = \frac{\partial A_x}{\partial z} - \frac{\partial A_z}{\partial x} \tag{3-10}$$

$$H_z = \frac{\partial A_y}{\partial x} - \frac{\partial A_x}{\partial y} \tag{3-11}$$

The components of **A** are seen to be given by

$$A_x = \tfrac{1}{2}zH_y - \tfrac{1}{2}yH_z \tag{3-12}$$

$$A_y = \tfrac{1}{2}xH_z - \tfrac{1}{2}zH_x \tag{3-13}$$

$$A_z = \tfrac{1}{2}yH_x - \tfrac{1}{2}xH_y \tag{3-14}$$

or written in vector notation

$$\mathbf{A} = \tfrac{1}{2}\mathbf{H} \wedge \mathbf{r} \tag{3-15}$$

Note that $\nabla \cdot \mathbf{A}$ equals zero; we are in a Coulomb gauge. Substituting Eq. (3-15) in Eq. (3-8) gives

$$\mathscr{H}^{(1)} = \sum_{j=1}^{N} \left(-\frac{i\hbar e}{2mc} \mathbf{H} \wedge \mathbf{r}_j \cdot \nabla_j + \frac{e^2}{8mc^2} |\mathbf{H} \wedge \mathbf{r}_j|^2 \right)$$

Introducing the Bohr magneton $\beta = e\hbar/2mc$, $\mathbf{l} = -i\hbar \mathbf{r} \wedge \nabla$, and using the vector relation $\mathbf{H} \wedge \mathbf{r} \cdot \nabla = \mathbf{H} \cdot \mathbf{r} \wedge \nabla$

$$\mathscr{H}^{(1)} = \hbar^{-1} \beta \mathbf{H} \cdot \sum_{j=1}^{N} (\mathbf{l}_j) + \frac{e^2}{8mc^2} \sum_{j=1}^{N} |\mathbf{H} \wedge \mathbf{r}_j|^2 \qquad (3\text{-}16)$$

Adding to the first term of Eq. (3-16) the spin angular momentum $\mathbf{s}_j$ which enters with the anomalous factor of two and including the spin-orbit coupling term Eq. (1-122), we can finally write for $\mathscr{H}^{(1)}$, $\mathbf{l}$ and $\mathbf{s}$ being measured in units of $\hbar$, with the summation μ over the nuclei,

$$\mathscr{H}^{(1)} = \sum_{j=1}^{N} \beta \mathbf{H} \cdot (\mathbf{l}_j + 2\mathbf{s}_j) + \sum_{i=1}^{\mu} \sum_{j=1}^{N} \zeta_i(r_{ij}) \mathbf{l}_{ij} \cdot \mathbf{s}_j + \frac{e^2}{8mc^2} \sum_{j=1}^{N} |\mathbf{H} \wedge \mathbf{r}_j|^2 \qquad (3\text{-}17)$$

3-2 THE ORBITAL REDUCTION FACTORS

It is only in a centrosymmetric system that $\mathbf{l}$ commutes with $\mathscr{H}$. As we have seen in Eqs. (1-120) to (1-122) the expression for the spin-orbit energies contains the operator $\mathbf{l}_i$ only because the molecular potential has been approximated by a sum of spherical atomic potentials. On the other hand, in the term $\beta \mathbf{H} \cdot \mathbf{l}$ in Eq. (3-17) $\mathbf{l}$ is not associated with any particular center.

We want first to evaluate the matrix elements of $\sum \mathbf{l}_j$ as it occurs in $\beta \mathbf{H} \cdot \sum \mathbf{l}_j$ inside a certain state. As an example we take a $T_{2g}(t_{2g})^1(e_g)^1$ state, spanned by Ψ_{-1}, Ψ_0, and Ψ_1, where

$$\Psi_{-1} = \frac{1}{\sqrt{2}}(\Psi_\xi - i\Psi_\eta),$$

$$\Psi_0 = \Psi_\zeta$$

and

$$\Psi_1 = -\frac{1}{\sqrt{2}}(\Psi_\xi + i\Psi_\eta)$$

with

$$\Psi_\xi = -\frac{1}{2}|(yz)(z^2)| + \frac{\sqrt{3}}{2}|(yz)(x^2 - y^2)|$$

$$\Psi_\eta = -\frac{1}{2}|(xz)(z^2)| - \frac{\sqrt{3}}{2}|(xz)(x^2 - y^2)| \qquad (3\text{-}18)$$

$$\Psi_\zeta = |(xy)(z^2)|$$

We get easily, by putting $\hat{L}_z = \hat{l}_z(1) + \hat{l}_z(2)$ and remembering that e orbitals carry no angular momentum, that for instance

$$\langle\Psi_1|\hat{L}_z|\Psi_1\rangle = \frac{i}{4}\langle xz|\hat{l}_z|yz\rangle - \frac{i}{4}\langle yz|\hat{l}_z|xz\rangle \tag{3-19}$$

Because $\hat{l}_z$ is purely imaginary we find

$$\langle\Psi_1|\hat{L}_z|\Psi_1\rangle = \frac{i}{2}\langle xz|\hat{l}_z|yz\rangle \tag{3-20}$$

Let (xz) and (yz) be pure d-orbitals. Then with the help of Table 3-1, we obtain

$$\langle\Psi_1|\hat{L}_z|\Psi_1\rangle = \tfrac{1}{2}$$

In general we see that operating inside the $T_{2g}(t_{2g})^1(e_g)^1$ set with $\hat{L}_z$, $\hat{L}_+$, and $\hat{L}_-$ the T_{2g} state can be considered to possess an effective Landé factor $\alpha = i/2\langle xz|\hat{l}_z|yz\rangle$. (Compare Eqs. (1-128)–(1-130)).

In the further evaluation of α we take for the two molecular orbitals (xz) and (yz) the linear combinations given in Table 2-1

$$(xz) = N_\pi\left[d_{xz} - \lambda_\pi \tfrac{1}{2}(p_{y_1} + p_{x_5} + p_{x_3} + p_{y_6})\right] \tag{3-21a}$$

$$(yz) = N_\pi\left[d_{yz} - \lambda_\pi \tfrac{1}{2}(p_{x_2} + p_{y_5} + p_{y_4} + p_{x_6})\right] \tag{3-21b}$$

N_π is the normalizing factor and λ_π the mixing coefficient in the molecular orbitals. With the group overlap integral G_π given by

$$G_\pi = \langle d_{xz}|\tfrac{1}{2}(p_{y_1} + p_{x_5} + p_{x_3} + p_{y_6})\rangle \tag{3-22}$$

we have, neglecting ligand–ligand overlap,

$$N_\pi^2(1 - 2\lambda_\pi G_\pi + \lambda_\pi^2) = 1 \tag{3-23}$$

The orbital reduction factor $k_{\pi,\pi}$ is now defined by[2]

$$k_{\pi,\pi}\langle d_{xz}|\hat{l}_z|d_{yz}\rangle = \langle xz|\hat{l}_z|yz\rangle \tag{3-24a}$$

With $\hat{l}_z = -i\hbar(x\,\partial/\partial y - y\,\partial/\partial x)$ being centered on the metal atom and making use of the fact that $\hat{l}_z$ is a purely imaginary operator, we get easily by substituting Eqs. (3-21a) and (3-21b) into Eq. (3-24a)

$$k_{\pi,\pi} = N_\pi^2(1 - 2\lambda_\pi, G_\pi + \tfrac{1}{2}\lambda_\pi^2) \tag{3-24b}$$

Use of Eq. (3-23) turns Eq. (3-24b) into

$$k_{\pi,\pi} = 1 - \tfrac{1}{2}N_\pi^2\lambda_\pi^2 \tag{3-24c}$$

Table 3-1 $\hat{l}$ operating on the d-orbitals

$\hat{l}_x d_{xz} = -i d_{xy}$	$\hat{l}_y d_{xz} = i d_{x^2-y^2} - i\sqrt{3}d_{z^2}$	$\hat{l}_z d_{xz} = i d_{yz}$
$\hat{l}_x d_{yz} = i\sqrt{3}d_{z^2} + i d_{x^2-y^2}$	$\hat{l}_y d_{yz} = i d_{xy}$	$\hat{l}_z d_{yz} = -i d_{xz}$
$\hat{l}_x d_{xy} = i d_{xz}$	$\hat{l}_y d_{xy} - -i d_{yz}$	$\hat{l}_z d_{xy} = -2i d_{x^2-y^2}$
$\hat{l}_x d_{x^2-y^2} = -i d_{yz}$	$\hat{l}_y d_{x^2-y^2} = -i d_{xz}$	$\hat{l}_z d_{x^2-y^2} = 2i d_{xy}$
$\hat{l}_x d_{z^2} = -i\sqrt{3}d_{yz}$	$\hat{l}_y d_{z^2} = i\sqrt{3}d_{xz}$	$\hat{l}_z d_{z^2} = 0.$

With $N_\pi^2\lambda_\pi^2$ giving the fraction f_π of the electron which spends its time on the ligands we get

$$k_{\pi,\pi} = 1 - \tfrac{1}{2}f_\pi \tag{3-24d}$$

$k_{\pi,\pi}$ is thus a measure of the delocalization of the magnetic electron. The effective Landé factor for the $T_{2g}(t_{2g})^1(e_g)^1$ state is seen to be

$$\alpha = \tfrac{1}{2} - \tfrac{1}{4}f_\pi \tag{3-25}$$

As a second example we want to calculate the Landé factor for a T_{1g} state. In a two-electron system the configurations $(t_{2g})^2$ and $(t_{2g})^1(e_g)^1$ both span a T_{1g} representation. We can take

$$\Phi_i = c_1 \,|\, T_{1g}^i(t_{2g})^2\rangle + c_2 \,|\, T_{1g}^i(t_{2g})(e_g)\rangle \tag{3-26}$$

$$i = 1, 0, -1 \text{ and } c_1^2 + c_2^2 = 1.$$

With

$$|\, T_{1g}^1(t_{2g})^2\rangle = -\frac{1}{\sqrt{2}}\left[\,|(xy)(xz)| + \mathrm{i}\,|(yz)(xy)|\,\right] \tag{3-27}$$

and

$$|\, T_{1g}^1(t_{2g})(e_g)\rangle = -\frac{1}{\sqrt{2}}\left[-\frac{\mathrm{i}}{2}|(xz)(x^2-y^2)| + \frac{\mathrm{i}\sqrt{3}}{2}|(xz)(z^2)| - \frac{1}{2}|(yz)(x^2-y^2)| - \frac{\sqrt{3}}{2}|(yz)(z^2)|\right] \tag{3-28}$$

we get for the effective Landé factor α

$$\alpha = \langle\Phi_1|\,\hat{L}_z\,|\Phi_1\rangle = -\mathrm{i}\langle xz|\,\hat{l}_z\,|yz\rangle(c_1^2 - \tfrac{1}{2}c_2^2) - \mathrm{i}\langle x^2-y^2|\,\hat{l}_z\,|xy\rangle c_1c_2 \tag{3-29}$$

Introducing the orbital reduction factor $k_{\sigma,\pi}$ defined by

$$k_{\sigma,\pi}\langle \mathrm{d}_{x^2-y^2}|\,\hat{l}_z\,|\mathrm{d}_{xy}\rangle = \langle x^2-y^2|\,\hat{l}_z\,|xy\rangle \tag{3-30}$$

we can use Table 3-1 and Eq. (3-24a) to transform Eq. (3-29) into

$$\alpha = -(c_1^2 - \tfrac{1}{2}c_2^2)k_{\pi,\pi} - 2c_1c_2k_{\sigma,\pi} \tag{3-31}$$

For $Dq \to 0$ we have for the ground state $c_1 = \sqrt{\tfrac{4}{5}}$ and $c_2 = \sqrt{\tfrac{1}{5}}$ (see Eq. (2-96)). In the limit $k_{\pi,\pi} = k_{\sigma,\pi} = 1$ and we get $\alpha = -\tfrac{3}{2}$. For $Dq \to \infty$, on the other hand, $c_1 = 1$ and $c_2 = 0$. Then $\alpha = -k_{\pi,\pi}$.

The orbital reduction factor $k_{\sigma,\pi}$ is evaluated as follows.[3] We have for the σ orbitals

$$(x^2-y^2) = N_\sigma[\mathrm{d}_{x^2-y^2} - \lambda_\sigma\tfrac{1}{2}(\mathrm{p}_{z_1} - \mathrm{p}_{z_2} + \mathrm{p}_{z_3} - \mathrm{p}_{z_4}) - \lambda_s\tfrac{1}{2}(\mathrm{s}_1 - \mathrm{s}_2 + \mathrm{s}_3 - \mathrm{s}_4)] \tag{3-32}$$

and

$$(xy) = N_\pi\,[\mathrm{d}_{xy} - \lambda_\pi\tfrac{1}{2}(\mathrm{p}_{x_1} + \mathrm{p}_{y_2} + \mathrm{p}_{y_3} + \mathrm{p}_{x_4})] \tag{3-33}$$

For the normalizing constants we get, neglecting ligand–ligand overlap,

$$N_\sigma^2(1 + \lambda_\sigma^2 + \lambda_s^2 - 2\lambda_\sigma G_\sigma - 2\lambda_\pi G_s) = 1 \tag{3-34}$$

$$N_\pi^2(1 + \lambda_\pi^2 - 2\lambda_\pi G_\pi) = 1 \tag{3-35}$$

Denoting by R the distance between the metal atom M and the ligand L, we get by inserting Eqs. (3-34) and (3-35) in Eq. (3-30), after a little algebra,

$$k_{\sigma,\pi} = N_\sigma N_\pi\left(1 - \lambda_\pi G_\pi - \lambda_\sigma G_\sigma - \lambda_s G_s - \frac{1}{2}\lambda_\pi\lambda_\sigma - \frac{1}{2}\lambda_\pi\lambda_s R\left\langle s\left|\frac{\partial p_y}{\partial y}\right.\right\rangle\right) \tag{3-36}$$

Making use of Eqs. (3-34) and (3-35) we can transform Eq. (3-36) to

$$k_{\sigma,\pi} = \frac{1}{2}\left(\frac{N_\pi}{N_\sigma} + \frac{N_\sigma}{N_\pi}\right) - \frac{1}{2}N_\sigma N_\pi\left(\lambda_\sigma^2 + \lambda_s^2 + \lambda_\pi^2 + \lambda_\pi\lambda_\sigma + \lambda_\pi\lambda_s R\left\langle s\left|\frac{\partial p_y}{\partial y}\right.\right\rangle\right) \tag{3-37}$$

To very good approximation we can put the first term in Eq. (3-37) equal to 1 and get

$$k_{\sigma,\pi} = 1 - \frac{1}{2}N_\sigma N_\pi\left(\lambda_\sigma^2 + \lambda_s^2 + \lambda_\pi^2 + \lambda_\pi\lambda_\sigma + \lambda_\pi\lambda_s R\left\langle s\left|\frac{\partial p_y}{\partial y}\right.\right\rangle\right) \tag{3-38}$$

We emphasize that the quantities, $k_{\pi,\pi}$, λ_π, $k_{\sigma,\pi}$, λ_σ, and λ_s all are calculated under the assumption of the same fixed nuclear configuration $\mathbf{Q}^0$ for both ground and excited states.

Let us now calculate the spin-orbit coupling energies of, for instance, a ${}^3T_{2g}(t_{2g})^1(e_g)^1$ state. The $M_J = 2$ state is given by

$$|{}^3T_{2g}, M_L = 1, M_s = 1\rangle = -\frac{1}{\sqrt{2}}(\Psi_\xi + i\Psi_\eta) \tag{3-39}$$

with

$$\Psi_\xi = -\frac{1}{2}|(\overset{+}{yz})(\overset{+}{z^2})| + \frac{\sqrt{3}}{2}|(\overset{+}{yz})(\overset{+}{x^2 - y^2})| \tag{3-40}$$

$$\Psi_\eta = -\frac{1}{2}|(\overset{+}{xz})(\overset{+}{z^2})| - \frac{\sqrt{3}}{2}|(\overset{+}{xz})(\overset{+}{x^2 - y^2})| \tag{3-41}$$

Since all of the electrons carry an α spin we notice from the form of the spin-orbit coupling term that we only have to consider

$$\mathscr{H}^{(1)} = \sum_u\sum_j \zeta_\mu(r_{\mu j})\hat{l}_{\mu jz}\cdot\hat{s}_z \tag{3-42}$$

as active term. Then (compare Eq. (3-20)) $W^{(1)} = \langle{}^3T_{2g}, 1, 1|\mathscr{H}^{(1)}|{}^3T_{2g}, 1, 1\rangle$ reduces to

$$W^{(1)} = \frac{1}{2}\frac{i}{2}\langle xz|\sum_\mu \zeta_\mu(r_\mu)\hat{l}_{\mu z}|yz\rangle \tag{3-43}$$

In the derivation of the form of $\mathscr{H}^{(1)}$ we have taken V to be a sum of spherical potentials where the overlap regions of the molecule have been left out. It would therefore be inconsistent to include the overlap in the normalizing factors of $|xz\rangle$ and $|yz\rangle$. Consequently we take

$$|xz\rangle = \frac{1}{\sqrt{1+\lambda_\pi^2}}(\mathrm{d}_{xz} - \lambda_\pi \chi_L) \tag{3-44}$$

and correspondingly for $|yz\rangle$. Insertion of the molecular orbitals in Eq. (3-43) yields

$$W^{(1)} = \frac{1}{4} \cdot \frac{1}{1+\lambda_\pi^2}\left[\zeta_{3d} + \frac{1}{2}\lambda_\pi^2 \zeta_L\right] \tag{3-45}$$

For $\lambda_\pi = 0$, $W^{(1)} = \frac{1}{4}\zeta_{3d}$. We can therefore utilize an effective molecular spin-orbit coupling constant, $\zeta_{\pi,\pi}$, in the π–π matrix element given by

$$\zeta_{\pi,\pi} = N_\pi^2(\zeta_{3d} + \tfrac{1}{2}\lambda_\pi^2 \zeta_L) \tag{3-46}$$

Similarly can we use an effective spin-orbit coupling constant in a σ–π matrix element equal to

$$\zeta_{\sigma,\pi} = N_\pi N_\sigma(\zeta_{3d} - \tfrac{1}{2}\lambda_\pi \lambda_\sigma \zeta_L) \tag{3-47}$$

Theoretically two spin-orbit constants $\zeta_{\pi,\pi}$ and $\zeta_{\sigma,\pi}$ should therefore be employed. However, in view of the approximations to grad V which have been used, we shall neglect this, and use only one molecular spin-orbit coupling constant ζ_j.

When we deal with an inorganic complex it is natural to take $\mathbf{l}$ to be centered on the metal atom. This choice of center can therefore be compensated for by using in Eq. (3-17) a phenomenological value for the spin-orbit coupling constant and by introducing an orbital reduction factor k in $\beta\mathbf{H}\cdot\mathbf{l}$. We can therefore finally write for $\mathscr{H}^{(1)}$ of Eq. (3-17)

$$\mathscr{H}^{(1)} = \sum_{j=1}^{N} \beta\mathbf{H}\cdot(k\mathbf{l}_j + 2\mathbf{s}_j) + \sum_{j=1}^{N} \zeta_j \mathbf{l}_j \cdot \mathbf{s}_j + \frac{e^2}{8mc^2}\sum_{j=1}^{N} |\mathbf{H} \wedge \mathbf{r}_j|^2 \tag{3-48}$$

3-3 THE SPIN HAMILTONIAN

The idea of a spin hamiltonian is to construct an operator containing a polynomial in the components of the spin vector $\mathbf{S}$, which when operating on a molecular state gives us the corresponding energies.[4] The terms in the molecular hamiltonian which contains $\mathbf{S}$ are given in Eq. (3-48). For a general value of S the spin-orbit coupling is treated as a small perturbation on the zero-order molecular state. Having obtained the wave functions correct to first order in the spin-orbit coupling the term in $\mathbf{S}$ which is proportional to the magnetic field $\mathbf{H}$ is considered.

The zero-order molecular states $|n, M\rangle$ are taken to be eigenfunctions of the molecular hamiltonian $\mathscr{H}_0$, $\hat{S}^2$, and $\hat{S}_z$. Here n specifies the eigenvalues of $\mathscr{H}_0$ and the quantum number S. M is the eigenvalue of $\hat{S}_z$. It is convenient to put $\langle o, M|\mathscr{H}_0|o, M\rangle = 0$.

We now consider a transition metal complex but limit ourselves to the case where we have an orbitally nondegenerate ground state. Furthermore, when the ligands are removed to infinity, the ground state shall move adiabatically into the ground term of the free metal ion. Inside each atomic term the Wigner–Eckart theorem permits the replacement of $\sum_j \zeta_j \mathbf{l}_j \cdot \mathbf{s}_j$ by the operator $\lambda \mathbf{L} \cdot \mathbf{S}$. For ground atomic terms $\lambda = \pm \zeta_j / 2S$ where the plus sign applies when the shell is less than half filled, the minus sign when the shell is more than half filled.[5] Because the commutation rule $[\hat{L}^2, \hat{L}] = 0$ holds true, the atomic states which are eigenfunctions of $\hat{L}^2$ are also eigenfunctions of $\hat{L}$. Hence the only excited molecular states of the same spin multiplicity as $|o, M\rangle$ which have nonzero matrix elements of $\hat{L}$ with the ground state also form parts of the atomic ground term. This permits us to use the form $\lambda \mathbf{L} \cdot \mathbf{S}$ for the spin-orbit coupling term in Eq. (3-48), whose first two terms therefore can be written

$$\mathscr{H}^{(1)} = \beta \mathbf{H} \cdot (k\mathbf{L} + 2\mathbf{S}) + \lambda \mathbf{L} \cdot \mathbf{S} \tag{3-49}$$

The orbital nondegenerate ground state $|o, M\rangle$ cannot carry any orbital momentum because $\hat{L}$ is a purely imaginary operator. Hence $\langle o, M | \hat{L} | o, M' \rangle = 0$. The energies of $\langle o, M|$ are therefore not changed in first order by the spin-orbit interaction. However, the wave function is altered to this order. The first-order modification to $|o, M\rangle$ is given by

$$|\alpha o, M\rangle = |o, M\rangle - \sum_{n,M'} \frac{\langle n, M' | \lambda \mathbf{L} \cdot \mathbf{S} | o, M \rangle}{W_n} |n, M'\rangle \tag{3-50}$$

where we have placed $W_0 = 0$.

Considering one term in the summation over n we have

$$-\sum_{M'} \frac{\langle n, M' | \lambda \mathbf{L} \cdot \mathbf{S} | o, M\rangle}{W_n} |n, M'\rangle$$
$$= -\frac{\lambda}{W_n} \sum_{M'} \langle n, M' | \hat{L}_z \hat{S}_z + \frac{1}{2} \hat{L}_+ \hat{S}_- + \frac{1}{2} \hat{L}_- \hat{S}_+ | o, M \rangle |n, M'\rangle \tag{3-51}$$

We now take $\langle o, M | \hat{L}_a | n, M' \rangle = i\Lambda_a^n \delta_{MM'}$ where Λ_a^n, $a = (x, y, z)$, is a real number. Remembering that $\hat{L}$ is a purely imaginary operator, we get for the right-hand side of Eq. (3-51)

$$\frac{i\lambda}{W_n}\Big[\Lambda_z^n M\,|n, M\rangle + \frac{1}{2}(\Lambda_x^n + i\Lambda_y^n)\sqrt{(S+M)(S-M+1)}\,|n, M-1\rangle$$
$$+ \frac{1}{2}(\Lambda_x^n - i\Lambda_y^n)\sqrt{(S-M)(S+M+1)}\,|n, M+1\rangle\Big]$$

which may be written

$$\frac{i\lambda}{W_n}\left[\Lambda_z^n \hat{S}_z + \frac{1}{2}(\Lambda_x^n + i\Lambda_y^n)\hat{S}_- + \frac{1}{2}(\Lambda_x^n - i\Lambda_y^n)\hat{S}_+\right] |n, M\rangle$$

We have therefore transformed Eq. (3-51) into

$$-\sum_{M'} \frac{\langle n, M' | \lambda \mathbf{L} \cdot \mathbf{S} | o, M \rangle}{W_n} |n, M'\rangle = i\lambda \frac{\mathbf{\Lambda}^n \cdot \mathbf{S}}{W_n} |n, M\rangle$$

The components of the vector $\mathbf{\Lambda}^n$ are pure numbers. The wave function of Eq. (3-50) can therefore be written

$$|\alpha o, M\rangle = |o, M\rangle + i\lambda \sum_n \frac{\mathbf{\Lambda}^n \cdot \mathbf{S}}{W_n} |n, M\rangle \tag{3-52}$$

We now treat the set $|\alpha o, M\rangle$ as a degenerate set of wave functions, using degenerate perturbation theory taking $\mathscr{H}^{(1)} = \beta\mathbf{H} \cdot (k\mathbf{L} + 2\mathbf{S})$. Evaluating the matrix elements of $\mathscr{H}^{(1)}$ we observe that the zero-order contribution is given by the matrix elements $2\beta\mathbf{H} \cdot \langle o, M' | \mathbf{S} | \sigma, M\rangle$. The first-order contributions are seen to come from matrix elements of the form

$$\langle o, M' | k\mathbf{L} + 2\mathbf{S} | n, M\rangle = \langle o, M' | k\mathbf{L} | n, M\rangle$$

since the states $|o\rangle$ and $|n\rangle$ are orthogonal to each other. Evaluating $k\beta \sum_a H_a \hat{L}_a$ to first order inside the set $|\alpha o, M\rangle$ gives the contributions

$$2\sum_n \sum_{a,b} \langle o, M' | k\beta H_a \hat{L}_a \frac{i\lambda}{W_n} \Lambda_b^n \hat{S}_b | n, M\rangle$$

$$= 2i\lambda k\beta \sum_n \frac{1}{W_n} \sum_{a,b} H_a \Lambda_b^n \langle o, M' | \hat{S}_b | o, M\rangle \langle o, M | \hat{L}_a | n, M\rangle$$

$$= -2\lambda k\beta \sum_n \frac{1}{W_n} \sum_{a,b} H_a \Lambda_a^n \Lambda_b^n \langle o, M' | \hat{S}_b | o, M\rangle$$

Introducing the real, symmetric and positive definite tensor

$$\Lambda_{ab} = \sum_n \frac{\langle o | \hat{L}_a | n\rangle \langle n | \hat{L}_b | o\rangle}{W_n} \tag{3-53}$$

the matrix elements of the $|\alpha o, M\rangle$ set which will give us both the zero- and first-order energy contributions are seen to be given by

$$\langle o, M' | 2\beta H_a \hat{S}_b \delta_{ab} - 2\lambda k\beta H_a \Lambda_{ab} \hat{S}_b | o, M\rangle \tag{3-54}$$

The second-order contributions in λ^2 to the ground state are given by matrix elements of the type $\langle \alpha o, M' | \mathscr{H}_0 + \lambda\mathbf{L} \cdot \mathbf{S} | \alpha o, M\rangle$. Using the same method as in the reduction of the first-order term, these are reduced to

$$\langle o, M' | -\lambda^2 \hat{S}_a \Lambda_{ab} \hat{S}_b | o, M\rangle \tag{3-55}$$

If we want to include the energy contributions which are correct to the order of H^2 we have in addition to the term already contained in Eq. (3-48) to include the second-order contribution given by $\mathscr{H}^{(1)} = \beta k\mathbf{H} \cdot \mathbf{L}$ using the $|o, M\rangle$ states. With the first-order wave functions

$$|o, M\rangle + i\beta k \sum_n \frac{\mathbf{H} \cdot \mathbf{\Lambda}^n}{W_n} |n, M\rangle \tag{3-56}$$

the second-order contributions to the energy are given by

$$\langle o, M' | -\beta^2 k^2 H_a \Lambda_{ab} H_b | o, M\rangle \tag{3-57}$$

Collecting the various terms Eqs. (3-54), (3-55), and (3-57) and including the H^2 contribution from Eq. (3-48) we can evidently write, for a spin hamiltonian[4] to use for an orbitally nondegenerate state in an inorganic complex,

$$\mathscr{H} = \beta H_a g_{ab} \hat{S}_b - \lambda^2 \Lambda_{ab} \hat{S}_a \hat{S}_b - \beta^2 k^2 \Lambda_{ab} H_a H_b + \frac{e^2}{8mc^2} \sum_N |\mathbf{H} \wedge \mathbf{r}_N|^2 \quad (3\text{-}58)$$

where the so-called g factor is defined by

$$g_{ab} = 2(\delta_{ab} - \lambda k \Lambda_{ab}) \quad (3\text{-}59)$$

Provided we choose our coordinate axes to be the principal axes of the tensor Λ_{ab}, all the elements Λ_{ab}, $a \neq b$, equal zero. Assuming that we have tetragonal or trigonal symmetry $\Lambda_{xx} = \Lambda_{yy} = \Lambda_\perp$ and $\Lambda_{zz} = \Lambda_\parallel$. In this case we may expand the term $-\lambda^2 \Lambda_{ab} \hat{S}_a \hat{S}_b$ in Eq. (3-58)

$$-\lambda^2 \Lambda_{ab} \hat{S}_a \hat{S}_b = -\lambda^2 [\Lambda_\perp \hat{S}^2 - (\Lambda_\perp - \Lambda_\parallel) \hat{S}_z^2] \quad (3\text{-}60)$$

Defining the quantity D as

$$D = (\Lambda_\perp - \Lambda_\parallel)\lambda^2 \quad (3\text{-}61)$$

we have

$$-\lambda^2 \Lambda_{ab} \hat{S}_a \hat{S}_b = D\hat{S}_z^2 - \lambda^2 \Lambda_\perp S(S+1) \quad (3\text{-}62)$$

or, as this expression is usually written,

$$D[\hat{S}_z^2 - \tfrac{1}{3}S(S+1)] - \tfrac{1}{3}S(S+1)\lambda^2(2\Lambda_\perp + \Lambda_\parallel)$$

Introducing this result into Eq. (3-58) we can write

$$\mathscr{H} = g_\parallel \beta H_z \hat{S}_z + g_\perp \beta (H_x \hat{S}_x + H_y \hat{S}_y) + D[\hat{S}_z^2 - \tfrac{1}{3}S(S+1)] - \tfrac{1}{3}S(S+1)$$
$$\times \lambda^2 (2\Lambda_\perp + \Lambda_\parallel) - \beta^2 k^2 [(H_x^2 + H_y^2)\Lambda_\perp + H_z^2 \Lambda_\parallel] + \frac{e^2}{8mc^2} \sum_N |\mathbf{H} \wedge \mathbf{r}_N|^2 \quad (3\text{-}63)$$

The last three terms in Eq. (3-63) represent a constant shift of all the levels in the lowest spin-multiplet. Retaining only the first three terms in Eq. (3-63) we obtain a spin hamiltonian for an axial molecule

$$\mathscr{H} = g_\parallel \beta H_z \hat{S}_z + g_\perp \beta (H_x \hat{S}_x + H_y \hat{S}_y) + D[\hat{S}_z^2 - \tfrac{1}{3}S(S+1)] \quad (3\text{-}64)$$

If we wish to find the energies of the ground state correct to second order in H we get from Eq. (3-63), disregarding all constant terms and with the magnetic field either parallel or perpendicular to the molecular z axis,

$$\mathscr{H} = D\hat{S}_z^2 + g_\parallel \beta H_z \hat{S}_z - H_z^2 \left[k^2 \beta^2 \Lambda_\parallel - \frac{e^2}{8mc^2} \sum_N (x_N^2 + y_N^2) \right] \quad (3\text{-}65)$$

or

$$\mathscr{H} = D\hat{S}_z^2 + g_\perp \beta H_x \hat{S}_x - H_x^2 \left[k^2 \beta^2 \Lambda_\perp - \frac{e^2}{8mc^2} \sum_N (y_N^2 + z_N^2) \right] \quad (3\text{-}66)$$

Let us consider the g values. The spin hamiltonian for an axial complex can be written

$$\mathscr{H} = DS_z^2 + g_\parallel \beta H_z \hat{S}_z + g_\perp \beta [\tfrac{1}{2}(H_x + iH_y)\hat{S}_- + \tfrac{1}{2}(H_x - iH_y)\hat{S}_+] \quad (3\text{-}67)$$

For a spin state with $S = 1$ we find that the matrix of the spin hamiltonian is

$$
\begin{array}{c} \\ (1,1) \\ (1,0) \\ (1,-1) \end{array}
\begin{array}{c} (1,1) \qquad\qquad (1,0) \qquad\qquad (1,-1) \\
\begin{pmatrix}
D + g_{\parallel}\beta H_z & \frac{\sqrt{2}}{2}\beta g_{\perp}(H_x + iH_y) & 0 \\
\frac{\sqrt{2}}{2}\beta g_{\perp}(H_x - iH_y) & 0 & \frac{\sqrt{2}}{2}\beta g_{\perp}(H_x + iH_y) \\
0 & \frac{\sqrt{2}}{2}\beta g_{\perp}(H_x - iH_y) & D - g_{\parallel}\beta H_z
\end{pmatrix}\end{array}
$$

With **H** being applied in the direction ($H \sin\theta \cos\phi$, $H \sin\theta \sin\phi$, $H \cos\theta$), the solutions to the secular equation derived from the above matrix are

$$-W[(D - W)^2 - g_{\parallel}^2 \beta^2 H^2 \cos^2\theta] - (D - W)\beta^2 g_{\perp}^2 H^2 \sin^2\theta = 0$$

For the two special cases $\theta = 0$ and $\theta = \pi/2$ the solutions are particularly simple. With $\theta = 0$ we get for $H = H_{\parallel}$

$$W = D \pm g_{\parallel}\beta H_{\parallel} \qquad \text{and} \qquad W = 0 \tag{3-68}$$

For $\theta = \pi/2$ we have $H = H_{\perp}$

$$W = \frac{D}{2} \pm \sqrt{\left(\frac{D}{2}\right)^2 + \beta^2 g_{\perp}^2 H_{\perp}^2} \qquad \text{and} \qquad W = D \tag{3-69}$$

In a paramagnetic resonance experiment the sample is placed in a static magnetic field. Plane-polarized radio waves with a fixed wavelength of, say, 3.0 cm are sent through the sample with the magnetic vector perpendicular to the static field. The transitions between the split levels of the spin-multiplet then obey the selection rule $\Delta M = \pm 1$. For a given radio frequency ν two resonance absorption lines will occur for such values of, say, $H_{\parallel}$ that $h\nu = g_{\parallel}\beta H_{\parallel} \pm D$ (compare Fig. 3-1).

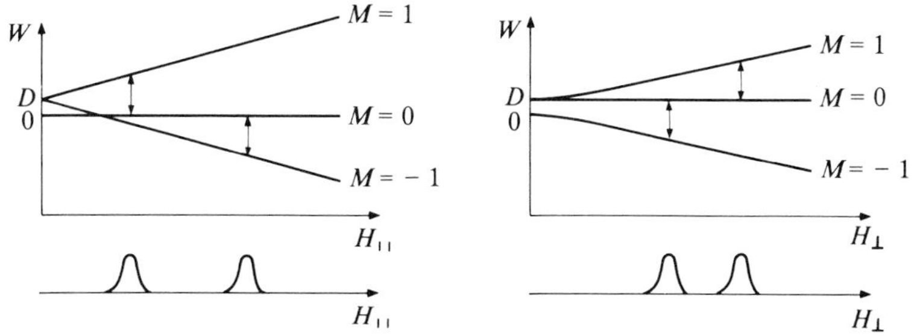

Figure 3-1 Thé energies of a triplet state as a function of the magnetic field. The quantum numbers put on the levels for $H_{\perp}$ are good quantum numbers only when the magnetic field is much greater than the zero-field splitting, i.e., when the magnetic field has overwhelmed the original z axis of quantization.

In 1930 Kramers showed that all electronic systems containing an odd number of electrons must always have at least a twofold degeneracy, provided no magnetic field is present.[6] An orbitally nondegenerate state with $S = \frac{1}{2}$ will therefore behave like a so-called Kramers doublet. With $S \geq 1$ the spin degeneracies may, however, be lifted, apart of course from Kramers degeneracy. The state is said to show zero-field splittings, and we observe from Eq. (3-67) that for axial molecules the splitting parameter is D.

The zero-field splittings due to low symmetry are taken care of by the term $-\lambda^2 \Lambda_{ab} \hat{S}_a \hat{S}_b$ in the spin hamiltonian. With the coordinate axes being the principal axes of the tensor Λ_{ab} we can write in general

$$-\lambda^2 \Lambda_{ab} \hat{S}_a \hat{S}_b = D(\hat{S}_z^2 - \tfrac{1}{3}\hat{S}^2) + E(\hat{S}_x^2 - \hat{S}_y^2) - K\hat{S}^2 \tag{3-70}$$

with

$$D = \lambda^2 \cdot \tfrac{1}{2}(\Lambda_{xx} + \Lambda_{yy} - 2\Lambda_{zz}) \tag{3-71}$$

$$E = \lambda^2 \cdot \tfrac{1}{2}(\Lambda_{yy} - \Lambda_{xx}) \tag{3-72}$$

$$K = \lambda^2 \cdot \tfrac{1}{3}(\Lambda_{xx} + \Lambda_{yy} + \Lambda_{zz}) \tag{3-73}$$

Neglecting all constant terms we have therefore in zero field

$$\mathscr{H} = \tfrac{1}{2}E(\hat{S}_+^2 + \hat{S}_-^2) + D\hat{S}_z^2 \tag{3-74}$$

where the two constants D and E take care of the zero-field splittings.

3-4 MAGNETIC SUSCEPTIBILITIES

Suppose that the energy of a molecular state when the system is subjected to an external magnetic field H can be written as a rapidly convergent series

$$W = W^{(0)} + HW^{(1)} + H^2W^{(2)} + \cdots \tag{3-75}$$

The induced magnetic moment in the field direction H_a is given by

$$\mu_a = -\frac{\partial W}{\partial H_a} = -W^{(1)} - 2H_a W^{(2)} + \cdots \tag{3-76}$$

The induced magnetic moment for one mole of molecules is then given by the statistical average[7]

$$P_a = L_A \frac{\sum \mu_a \exp(-W/k_B T)}{\sum \exp(-W/k_B T)} \tag{3-77}$$

L_A is Avogadro's number, k_B the Boltzmann constant, and T the absolute temperature. The susceptibility χ_a in the field direction is defined by

$$H_a \chi_a = P_a \tag{3-78}$$

We get by insertion of Eq. (3-76) in Eq. (3-77)

$$\chi_a = \frac{L_A}{H_a} \frac{\sum(-W^{(1)} - 2H_a W^{(2)}) \exp(-W/k_B T)}{\sum \exp(-W/k_B T)} \tag{3-79}$$

Expanding the exponential function

$$\exp(-W/k_BT) = \exp(-W^{(0)}/k_BT)\left(1 - \frac{H_a W^{(1)}}{k_BT} + \cdots\right)$$

and assuming the molecule to have no permanent magnetic moment, that is,

$$\sum W^{(1)} = 0$$

we get for the field-independent susceptibility[1]

$$\chi_a = L_A \frac{\sum\left[\dfrac{(W^{(1)})^2}{k_BT} - 2W^{(2)}\right]\exp(-W^{(0)}/k_BT)}{\sum \exp(-W^{(0)}/k_BT)} \tag{3-80}$$

A glance at Eq. (3-69) shows that it is not always practicable to expand W in a power series in H. For $D/2 \langle \beta g H$ we get

$$\sqrt{\left(\frac{D}{2}\right)^2 + \beta^2 g^2 H^2} = \beta g H + \frac{D^2}{8\beta g} H^{-1} + \cdots \tag{3-81}$$

Only by neglecting all terms in H^{-1} is it, however, possible to obtain a field-independent proportionality constant between H_a and P_a, and this demand re-establishes Eq. (3-80).

Let us now calculate the magnetic susceptibility of one mole of molecules having axial symmetry, a $(2S + 1)$ spin manifold and a nondegenerate orbital ground state. All the 6×10^{23} molecules are taken to be aligned with their unique axes parallel to each other. Taking this direction to be the z axis we get immediately from Eq. (3-65) that for H_a parallel to z

$$W = DM^2 + H_{\parallel} g_{\parallel} \beta M - H_{\parallel}^2\left(k^2\beta^2\Lambda_{\parallel} - \frac{e^2}{8mc^2}\sum_N \langle x_N^2 + y_N^2\rangle\right) \tag{3-82}$$

Assuming that it is only the $(2S + 1)$ manifold of the ground state which is populated we get

$$\chi_{\parallel} = \frac{L_A g_{\parallel}^2 \beta^2}{k_BT} \frac{\sum\limits_{M=-S}^{S} M^2 \exp(-DM^2/k_BT)}{\sum\limits_{M=-S}^{S} \exp(-DM^2/k_BT)} + \alpha_{\parallel} \tag{3-83}$$

$$\alpha_{\parallel} = 2L_A k^2 \beta^2 \Lambda_{\parallel} - \frac{e^2 L_A}{4mc^2}\sum_N \langle x_N^2 + y_N^2\rangle \tag{3-84}$$

Expanding

$$\exp(-DM^2/k_BT) = 1 - \frac{DM^2}{k_BT} + \cdots \tag{3-85}$$

and using

$$\sum_{M=-S}^{S} M^2 = \frac{1}{3}S(S+1)(2S+1) \tag{3-86}$$

$$\sum_{M=-S}^{S} M^4 = \frac{1}{15} S(S+1)(2S+1)(3S^2+3S-1) \tag{3-87}$$

we get to first order in D:

$$\chi_{\parallel} = \frac{L_A g_{\parallel}^2 \beta^2}{3k_B T} S(S+1)\left[1 - \frac{D}{15k_B T}(2S+3)(2S-1)\right] + \alpha_{\parallel} \tag{3-88}$$

In order to calculate $\chi_{\perp}$ we notice from Eq. (3-66) that the operator giving us $(W^{(1)})^2$ is $g_{\perp}^2 \beta^2 \hat{S}_x^2$. With $\hat{S}^2 = \hat{S}_x^2 + \hat{S}_y^2 + \hat{S}_z^2$ and $\hat{S}_x^2 = \hat{S}_y^2$ the operator is equal to $g_{\perp}^2 \beta^2 \frac{1}{2}(\hat{S}^2 - \hat{S}_z^2)$. We get therefore

$$(W^{(1)})^2 = \tfrac{1}{2} g_{\perp}^2 \beta^2 (S(S+1) - M^2) \tag{3-89}$$

Insertion of Eq. (3-89) into Eq. (3-80), making use of Eqs. (3-86) and (3-87), leads to

$$\chi_{\perp} = \frac{L_A g_{\perp}^2 \beta^2}{3k_B T} S(S+1)\left[1 + \frac{D}{30k_B T}(2S+3)(2S-1)\right] + \alpha_{\perp} \tag{3-90}$$

$$\alpha_{\perp} = 2L_A k^2 \beta^2 \Lambda_{\perp} - \frac{e^2 L_A}{4mc^2} \sum_N \langle y_N^2 + z_N^2 \rangle \tag{3-91}$$

Notice that the susceptibility is made up of a temperature-dependent and a temperature-independent part. The positive part of the temperature-independent terms of Eqs. (3-84) and (3-91) is often referred to as the *Van Vleck* term. The negative part of Eqs. (3-84) and (3-91) is the so-called *diamagnetic* term.

Performing a susceptibility measurement on a powdered sample will yield an average value $\bar{\chi}$ equal to

$$\bar{\chi} = \tfrac{1}{3}(2\chi_{\perp} + \chi_{\parallel}) \tag{3-92}$$

Using Eqs. (3-88) and (3-90) we obtain the Curie–Weiss law

$$\bar{\chi} = \frac{C}{T - \Delta} + \alpha \tag{3-93}$$

where

$$C = \frac{2g_{\perp}^2 + g_{\parallel}^2}{3} \frac{L_A \beta^2 S(S+1)}{3k_B} \tag{3-94}$$

and

$$\Delta = \frac{g_{\perp}^2 - g_{\parallel}^2}{2g_{\perp}^2 + g_{\parallel}^2} \frac{D}{15k_B}(2S+3)(2S-1) \tag{3-95}$$

Normally the g factors are nearly isotropic, hence Δ is small, and the deviation from a simple Curie law is slight except at fairly low temperatures where our expansions become invalid.

As proven by Van Vleck[1] the temperature-independent term α in the magnetic susceptibility is invariant of the origin chosen for its calculation. This theorem can be used to estimate[8] the Van Vleck term in α. From the structure of Λ_{ab} we observe that this quantity vanishes for an atom. However, the vanishing of the Van Vleck term for an atom depends essentially on the absence of other centers of force. Consider a system consisting of two nuclei A and B and N electrons (see Fig. 3-2). Because for an atom $\langle x^2 \rangle = \langle y^2 \rangle = \langle z^2 \rangle = \frac{1}{3}\langle r^2 \rangle$, we get evaluating α with B as the center

$$\alpha = -\frac{e^2 L_A}{6mc^2} \sum_N \langle r_{NB}^2 \rangle \tag{3-96}$$

Provided no interaction takes place, due to the invariance of α with origin we must also have, choosing A as center

$$\alpha = -\frac{e^2 L_A}{6mc^2} \sum_N \langle r_{NA}^2 \rangle + \chi_{hf} \tag{3-97}$$

or

$$\chi_{hf} = \frac{e^2 L_A}{6mc^2} \sum_N (\langle r_{NA}^2 \rangle - \langle r_{NB}^2 \rangle) = \frac{e^2 L_A}{6mc^2} NR^2 \tag{3-98}$$

The Van Vleck term of a complex with j ligands each possessing N electrons and at a distance R from the center is therefore

$$\chi_{hf} \approx \frac{e^2 L_A}{6mc^2} NjR^2 \tag{3-99}$$

For a hexacoordinated complex with, say, 6 ammonia ligands at a distance $2a_0$, each ligand possessing 10 electrons, we get therefore an order of magnitude for χ_{hf} of 200×10^{-6}. A calculation of χ_{hf} using Eqs. (3-84) and (3-91) gives

$$\chi_{hf} = 2L_A \beta^2 \frac{k^2}{3} (\Lambda_{\parallel} + 2\Lambda_{\perp}) \tag{3-100}$$

For an octahedral Ni^{2+} complex $\Lambda_{\parallel} = \Lambda_{\perp} = 4/10Dq$. With $k \approx 1$ we get with $Dq = 1{,}100 \text{ cm}^{-1}$

$$\chi_{hf} = \frac{8L_A \beta^2}{10Dq} = 190 \times 10^{-6}$$

In case the orbital ground state of the complex is degenerate and can exhibit spin-orbit coupling, this feature will have to be considered before treating the

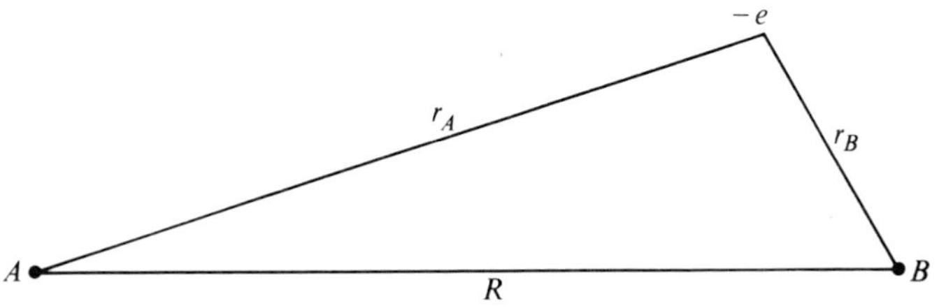

Figure 3-2 The two-center coordinate system for the evaluation of the high-frequency term χ_{hf}.

perturbations due to a magnetic field. When estimating the spin-orbit couplings of ${}^{2s+1}T_1$ and ${}^{2s+1}T_2$ states the use of a fictitious angular momentum $L' = 1$ greatly facilitates the calculations (see section 1-7). As an example we shall calculate the magnetic susceptibility $\bar{\chi}$ for a ${}^2T_{2g}(t_{2g})^1$ state.

Using the quantum number $M_J = M_L + M_S$ to classify the six components $|M_L, M_S\rangle$ of ${}^2T_{2g}$ we have

	M_J
$\|1, \frac{1}{2}\rangle$	$\frac{3}{2}$
$\|1, -\frac{1}{2}\rangle, \|0, \frac{1}{2}\rangle$	$\frac{1}{2}$
$\|-1, \frac{1}{2}\rangle, \|0, -\frac{1}{2}\rangle$	$-\frac{1}{2}$
$\|-1, -\frac{1}{2}\rangle$	$-\frac{3}{2}$

As our perturbation hamiltonian we take a modified operator of the type Eq. (3-48)

$$\mathscr{H}^{(1)} = -\zeta_M \mathbf{l}\cdot\mathbf{s} + \beta H(-k_{\pi,\pi}\hat{l}_z + 2\hat{s}_z) \tag{3-101}$$

where the effective Landé factor α for a ${}^2T_{2g}(t_{2g})^1$ state is taken to be -1 (compare Eqs. (1-128)–(1-130)). We get easily

$$W_{3/2} = -\tfrac{1}{2}\zeta_M + \beta H(-k_{\pi,\pi} + 1) \tag{3-102}$$

$$\begin{vmatrix} \frac{1}{2}\zeta_M - \beta H(k_{\pi,\pi} + 1) - W_{1/2} & \dfrac{-\sqrt{2}}{2}\zeta_M \\ \dfrac{-\sqrt{2}}{2}\zeta_M & \beta H - W_{1/2} \end{vmatrix} = 0 \tag{3-103}$$

$$\begin{vmatrix} \frac{1}{2}\zeta_M + \beta H(k_{\pi,\pi} + 1) - W_{-1/2} & \dfrac{-\sqrt{2}}{2}\zeta_M \\ \dfrac{-\sqrt{2}}{2}\zeta_M & -\beta H - W_{-1/2} \end{vmatrix} = 0 \tag{3-104}$$

$$W_{-3/2} = -\tfrac{1}{2}\zeta_M + \beta H(k_{\pi,\pi} - 1) \tag{3-105}$$

Expanding Eqs. (3-103) and (3-104) to second order in H we get

$$W_{1/2} = \begin{cases} -\dfrac{1}{2}\zeta_M + \left(\dfrac{1}{3} - \dfrac{1}{3}k_{\pi,\pi}\right)\beta H - \dfrac{4}{27}\dfrac{\beta^2(k_{\pi,\pi}+2)^2}{\zeta_M}H^2 & (3\text{-}106) \\ \zeta_M + \left(-\dfrac{1}{3} - \dfrac{2}{3}k_{\pi,\pi}\right)\beta H + \dfrac{4}{27}\dfrac{\beta^2(k_{\pi,\pi}+2)^2}{\zeta_M}H^2 & (3\text{-}107) \end{cases}$$

$$W_{-1/2} = \begin{cases} -\dfrac{1}{2}\zeta_M + \left(-\dfrac{1}{3} + \dfrac{1}{3}k_{\pi,\pi}\right)\beta H - \dfrac{4}{27}\dfrac{\beta^2(k_{\pi,\pi}+2)^2}{\zeta_M}H^2 & (3\text{-}108) \\ \zeta_M + \left(\dfrac{1}{3} + \dfrac{2}{3}k_{\pi,\pi}\right)\beta H + \dfrac{4}{27}\dfrac{\beta^2(k_{\pi,\pi}+2)}{\zeta_M}H^2 & (3\text{-}109) \end{cases}$$

In zero field the fourfold degenerate $G(W = -\zeta_M/2)$ state is the ground state, and placed at $W = \zeta_M$ we have a twofold degenerate $E_{5/2}$ state. Notice that for the G state we have a g factor $g = 2/3(1 - k_{\pi,\pi})$ and for the $E_{5/2}$ state $g = 2/3(1 + 2k_{\pi,\pi})$. The g factor for the G state is therefore in this approximation only different from zero by virtue of the deviations of the t_{2g} orbital from a d-orbital. For the $E_{5/2}$ state $g \approx 2$.

The magnetic susceptibility of a $(t_{2g})^1\ {}^2T_{2g}$ state is now calculated by the insertion of Eqs. (3-102) and (3-105) to (3-109) into Eq. (3-80). Calling $\zeta_M/k_BT = x$ this leads to

$$\bar{\chi} = \frac{L_A\beta^2}{3k_BT} \times \frac{30(k_{\pi,\pi} - 1)^2x + 8(k_{\pi,\pi} + 2)^2 + [3(2k_{\pi,\pi} + 1)^2x - 8(k_{\pi,\pi} + 2)^2]\exp(-3x/2)}{9x[2 + \exp(-3x/2)]} \tag{3-110}$$

The dimensionless quantity, the Bohr magneton number μ_{eff}, is defined in terms of the susceptibility by the relation

$$\bar{\chi} = \frac{L_A\beta^2\mu_{\text{eff}}^2}{3k_BT} \tag{3-111}$$

For $k_{\pi,\pi} \approx 1$ we get for $(t_{2g})^1\ {}^2T_{2g}$

$$\mu_{\text{eff}}^2 = \frac{8 + (3x - 8)\exp(-3x/2)}{x[2 + \exp(-3x/2)]} \tag{3-112}$$

For $T \to \infty$, μ_{eff} converges to $\sqrt{5}$, and for $T \to 0$ μ_{eff} goes to zero. As pointed out by Kotani[9] the Bohr-magneton number for $(t_{2g})^5\ {}^2T_{2g}$ can be obtained by changing the sign of x in Eq. (3-110). The formulae for $\bar{\chi}$ of the states ${}^3T_{1g}(t_{2g})^2$ and ${}^3T_{1g}(t_{2g})^4$ can also be found in Kotani's paper.

3-5 THE SPECIFIC HEAT TAIL

The electronic contribution to the specific heat of one mole is given from statistical mechanics[7] by

$$C_v = L_AT\frac{\partial^2(k_BT\ln Z)}{\partial T^2} \tag{3-113}$$

or expanding

$$C_v = 2L_Ak_BT\frac{1}{Z}\frac{\partial Z}{\partial T} + L_Ak_BT^2\frac{\partial}{\partial T}\left(\frac{1}{Z}\frac{\partial Z}{\partial T}\right) \tag{3-114}$$

L_A is Avogadro's number, and Z the partition function

$$Z = \sum \exp(-W/k_BT) \tag{3-115}$$

The sum extends over all occupied states of one molecule. For $T \to 0$, C_v goes to zero. However, when $k_B T$ is of the same order of magnitude as the molecular zero field splitting of the ground state, an electronic contribution to C_v emerges.

Expanding $\exp(-W/k_B T)$ assuming $W \ll k_B T$, we get

$$Z = n - \frac{\sum W}{k_B T} + \frac{1}{2}\frac{\sum (W)^2}{k_B^2 T^2} + \cdots \tag{3-116}$$

where n is the number of occupied states. Differentiating Eq. (3-116) and inserting the result in Eq. (3-114) gives the leading term

$$k_B T^2 C_v = \frac{L_A}{n}\sum W^2 - \frac{L_A}{n^2}\left(\sum W\right)^2 \tag{3-117}$$

For a Ni^{2+} complex for instance, where $S = 1$, the hamiltonian of Eq. (3-74) leads to a secular determinant

$$\begin{vmatrix} D - W & 0 & E \\ 0 & -W & 0 \\ E & 0 & D - W \end{vmatrix} = 0 \tag{3-118}$$

giving the solutions

$$W = 0 \qquad \text{and} \qquad W = D \pm E \tag{3-119}$$

With $n = 3$ we get from Eq. (3-117)

$$k_B T^2 C_v = \tfrac{2}{9} L_A (D^2 + 3E^2) \tag{3-120}$$

Measuring D and E in °K and introducing the gas constant R for one mole

$$C_v = \tfrac{2}{9} R (D^2 + 3E^2) T^{-2} \tag{3-121}$$

Equation (3-121) then gives the electronic contribution to C_v of a Ni^{2+} complex with $S = 1$ in the region where $k_B T$ is larger than the zero field splittings.

3-6 EXCHANGE INTERACTIONS

The magnetic electrons in a paramagnetic inorganic complex imbedded in a lattice can usually to good approximation be treated as being localized on a single unit. However, the wave functions for two units may overlap and a magnetic exchange interaction take place.[10] The coupled system can either have a ground state in which the spins are aligned or one in which the two units have antiparallel spins. In the first case we speak about ferromagnetism and in the latter about antiferromagnetism.

The coupling mechanism may be treated analogously to the formation of a chemical bond in the valence-bond approximation. Consider the two centers A and B of Fig. 3-3. With two electrons present the hamiltonian for the system is

$$\mathscr{H} = -\frac{1}{2}\nabla_1^2 - \frac{Z_A}{r_{A1}} - \frac{1}{2}\nabla_2^2 - \frac{Z_B}{r_{B2}} + \frac{1}{r_{12}} - \frac{Z_A}{r_{A2}} - \frac{Z_B}{r_{B1}} + \frac{Z_A Z_B}{R} \tag{3-122}$$

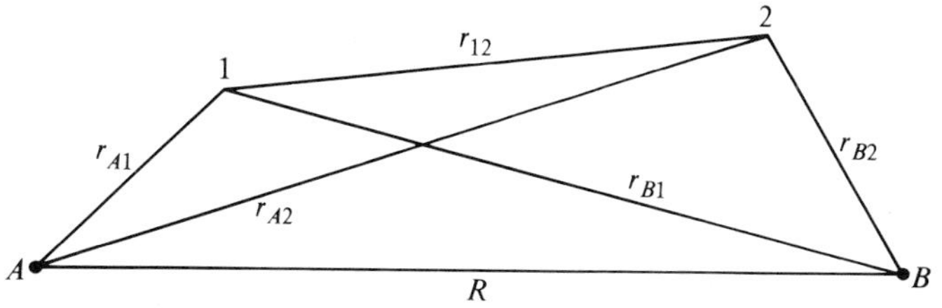

Figure 3-3 Two electrons in a two-center coordinate system.

With atomic orbitals χ_A and χ_B located respectively on centers A and B we have

$$\left(-\frac{1}{2}\nabla_1^2 - \frac{Z_A}{r_{A1}}\right)\chi_A(1) = \bar{h}_A\chi_A(1) \tag{3-123}$$

$$\left(-\frac{1}{2}\nabla_2^2 - \frac{Z_B}{r_{B2}}\right)\chi_B(2) = \bar{h}_B\chi_B(2) \tag{3-124}$$

and we take

$$\mathscr{H}^{(1)} = \frac{1}{r_{12}} - \frac{Z_A}{r_{A2}} - \frac{Z_B}{r_{B1}} \tag{3-125}$$

where we have ignored the last term in Eq. (3-122) since it will only give an additive constant to the energy.

A wave function for the spin triplet ($S = 1$) is in the valence-bond approximation

$$^3\Psi = \frac{1}{\sqrt{2 - 2S_{AB}^2}}[\chi_A(1)\chi_B(2) - \chi_A(2)\chi_B(1)] \tag{3-126}$$

where

$$S_{AB} = \langle\chi_A \mid \chi_B\rangle \tag{3-127}$$

is the overlap integral. For the spin singlet ($S = 0$) a wave function will be

$$^1\Psi = \frac{1}{\sqrt{2 + 2S_{AB}^2}}[\chi_A(1)\chi_B(2) + \chi_A(2)\chi_B(1)] \tag{3-128}$$

Hence

$$W(^3\Psi) = \bar{h}_A + \bar{h}_B + \frac{C - D}{1 - S_{AB}^2} \tag{3-129}$$

$$W(^1\Psi) = \bar{h}_A + \bar{h}_B + \frac{C + D}{1 + S_{AB}^2} \tag{3-130}$$

with

$$C = \iint \chi_A(1)\chi_B(2)\mathscr{H}^{(1)}\chi_A(1)\chi_B(2)\, d\tau_1\, d\tau_2 \tag{3-131}$$

$$D = \iint \chi_A(1)\chi_B(2)\mathscr{H}^{(1)}\chi_B(1)\chi_A(2)\, d\tau_1\, d\tau_2 \tag{3-132}$$

We get therefore

$$W(^1\Psi) - W(^3\Psi) = \frac{2D - 2CS_{AB}^2}{1 - S_{AB}^4} \approx 2(D - CS_{AB}^2) \tag{3-133}$$

Notice that in this direct coupling of the spin systems the energy difference Eq. (3-133) is of order S_{AB}^2. In the case where the two wave functions χ_A and χ_B are orthogonal to each other, $S_{AB} = 0$ and $D > 0$. In that case the spin triplet will always be the lowest state. The phenomenom of antiferromagnetism is therefore closely linked to the nonorthogonality of the wave functions centered on A and B.

It is also possible to have an indirect coupling of two spin systems through a ligand. This effect is called super-exchange. As an example we shall consider two magnetic ions A and B, each possessing one electron, each of which interacts with the electrons on a ligand L. We have

$$\langle \chi_A | \chi_B \rangle = 0 \tag{3-134}$$

and

$$\langle \chi_A | \chi_L \rangle = \langle \chi_B | \chi_L \rangle = S_{ML} \tag{3-135}$$

Calling

$$\langle \chi_A | \mathscr{H} | \chi_L \rangle = \langle \chi_B | \mathscr{H} | \chi_L \rangle = \beta \tag{3-136}$$

we get the secular equation

$$\begin{vmatrix} \bar{h}_M - w & \beta - wS_{ML} & 0 \\ \beta - wS_{ML} & \bar{h}_L - w & \beta - wS_{ML} \\ 0 & \beta - wS_{ML} & \bar{h}_M - w \end{vmatrix} = 0 \tag{3-137}$$

which with $\bar{h}_M \gg \bar{h}_L$ has the solutions

$$w = \bar{h}_M \tag{3-138}$$

$$w \approx \bar{h}_M + \frac{2(\beta - \bar{h}_M S_{ML})^2}{\bar{h}_M - \bar{h}_L} \tag{3-139}$$

$$w \approx \bar{h}_L - \frac{2(\beta - \bar{h}_L S_{ML})^2}{\bar{h}_M - \bar{h}_L} \tag{3-140}$$

The orbital level scheme will therefore be as pictured in Fig. 3-4. For small values of $\Delta = 2(\beta - \bar{h}_M S_{ML})^2/(\bar{h}_M - \bar{h}_L)$ a spin triplet will be energetically favored. For larger values of Δ, a spin-singlet ground state in which the electrons are paired in $1/\sqrt{2}(\chi_A - \chi_B)$ will have the lowest energy. The amplitude of the wave functions of the magnetic electrons in the overlap region L is proportional to S_{ML} and the interaction due to exchange between A and B therefore of order S_{ML}^4.

Formally we can write down a spin–spin coupling hamiltonian of the form[1,11]

$$\mathscr{H}^{(1)} = J\,\mathbf{S}_1 \cdot \mathbf{S}_2 \tag{3-141}$$

to account for the interaction of any two magnetic ions with spin S_1 and S_2,

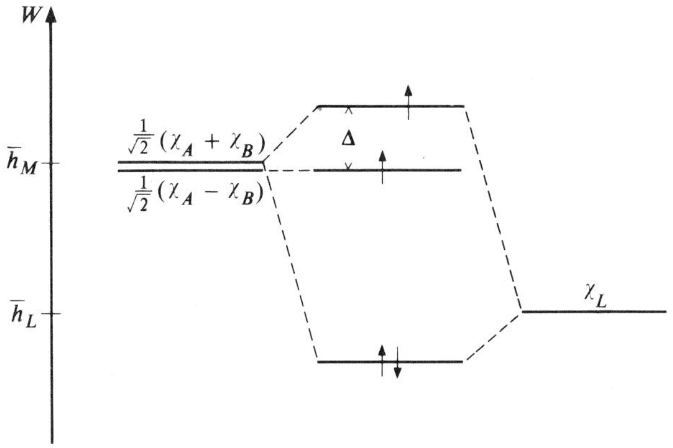

Figure 3-4 Molecular orbital level scheme for super-exchange.

respectively. With

$$\mathbf{S} = \mathbf{S}_1 + \mathbf{S}_2 \tag{3-142}$$

we get immediately for the eigenvalues of the $\mathbf{S}_1 \cdot \mathbf{S}_2$ operator in a $|\alpha SMS_1S_2\rangle$ representation

$$2\mathbf{S}_1 \cdot \mathbf{S}_2 = S(S+1) - S_1(S_1+1) - S_2(S_2+1) \tag{3-143}$$

and therefore

$$W^{(1)} = \tfrac{1}{2}J[S(S+1) - S_1(S_1+1) - S_2(S_2+1)] \tag{3-144}$$

with

$$S = S_1 + S_2, S_1 + S_2 - 1, \ldots, |S_1 - S_2|$$

With $S_1 = S_2 = \frac{1}{2}$, for example,

$$W^{(1)} = \begin{cases} -\frac{3}{4}J; & S = 0 \\ \frac{1}{4}J; & S = 1 \end{cases} \tag{3-145}$$

A comparison with Eq. (3-133) yields

$$-J = 2(D - CS^2) \tag{3-146}$$

The paramagnetic susceptibility for the antiferromagnetic system having $S_1 = S_2 = \frac{1}{2}$ can be calculated using Eqs. (3-80) and (3-145), neglecting the term in H^2. With the excited ^3X state being orbitally nondegenerate we get for $\bar{\chi}$ (see Fig. (3-5))

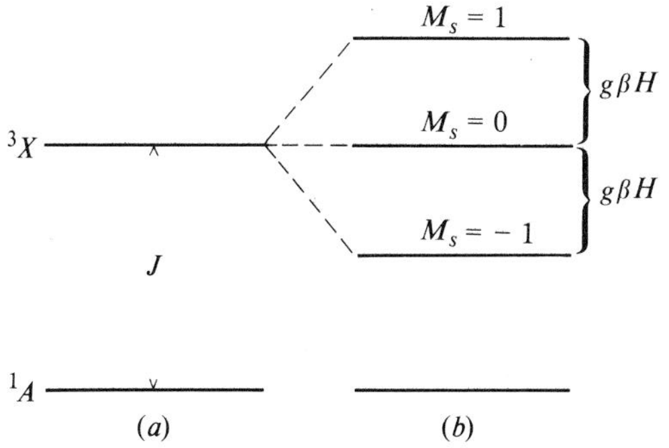

Figure 3-5 Energy-level diagram for an antiferromagnetic system having $S = 0$ and $S = 1$. (*a*) Without and (*b*) with a magnetic field present.

$$\bar{\chi} = L_A \frac{2\dfrac{g^2\beta^2}{k_BT}\exp(-J/k_BT)}{1 + 3\exp(-J/k_BT)} \tag{3-147}$$

or

$$\bar{\chi} = \frac{2g^2\beta^2 L_A}{k_BT} \frac{1}{3 + \exp(J/k_BT)} \tag{3-148}$$

The expression Eq. (3-148) is found to have a maximum for $J/k_BT = 1.604$.

The g factors for a coupled system consisting of two ions with respective spins $\mathbf{S}_1$ and $\mathbf{S}_2$ and no spatial degeneracy can be calculated in the following way. Let the total spin of the system be $\mathbf{S} = \mathbf{S}_1 + \mathbf{S}_2$. The wave functions for the coupled system are denoted $|\alpha SMS_1S_2\rangle$ with $\hat{S}_z|\alpha SMS_1S_2\rangle = M|\alpha SMS_1S_2\rangle$. We assume that the interaction between an external magnetic field $\mathbf{H}$ and the coupled system can be written $\boldsymbol{\mu}\cdot\beta\mathbf{H}$ where $\boldsymbol{\mu}$ is the magnetic-moment operator

$$\boldsymbol{\mu} = g_1\mathbf{S}_1 + g_2\mathbf{S}_2 \tag{3-149}$$

The diagonal matrix elements of $\hat{\mu}_z$ are evaluated using the relation given in Condon and Shortley[5], eq. (7) 8^3,

$$S(S+1)\langle\alpha SM|\hat{\mu}_z|\alpha SM'\rangle = \langle\alpha SM|\hat{S}_z|\alpha SM'\rangle\langle\alpha SM'|\mathbf{S}\cdot\boldsymbol{\mu}|\alpha SM\rangle \tag{3-150}$$

We get

$$\langle\alpha SMS_1S_2|\hat{\mu}_z|\alpha SMS_1S_2\rangle = \frac{M}{S(S+1)}\langle\alpha SMS_1S_2|g_1\mathbf{S}\cdot\mathbf{S}_1 + g_2\mathbf{S}\cdot\mathbf{S}_2|\alpha SMS_1S_2\rangle \tag{3-151}$$

Using $\mathbf{S}_1 = \mathbf{S} - \mathbf{S}_2$ and $\mathbf{S}_2 = \mathbf{S} - \mathbf{S}_1$ we find that $|\alpha SMS_1S_2\rangle$ are eigenfunctions of $\mathbf{S}\cdot\mathbf{S}_1$ and $\mathbf{S}\cdot\mathbf{S}_2$ with the eigenvalues

$$2\mathbf{S}\cdot\mathbf{S}_1 = S(S+1) + S_1(S_1+1) - S_2(S_2+1)$$

$$2\mathbf{S}\cdot\mathbf{S}_2 = S(S+1) - S_1(S_1+1) + S_2(S_2+1)$$

Equation (3-151) can therefore be written

$$\langle\alpha SMS_1S_2|\hat{\mu}_z|\alpha SMS_1S_2\rangle = M\left[\frac{1}{2}(g_1+g_2) + \frac{1}{2}(g_1-g_2)\frac{S_1(S_1+1) - S_2(S_2+1)}{S(S+1)}\right] \tag{3-152}$$

The factor in the square bracket is seen to be the g factor for the $|\alpha SMS_1S_2\rangle$ state of the coupled system.

As shown by Griffith[11] the above form of the g factor is quite general, and it is easily generalized to the cases where we have no isotropicity about the principal axes of the two g-tensors. The general problem of determining the relative energies and the magnetic susceptibilities in polynuclear transition metal complexes containing two, three, and four weakly interacting centers has further been considered by Griffith.[11,12]

REFERENCES

1. J. H. Van Vleck, *The Theory of Electric and Magnetic Susceptibilities*, Oxford University Press, 1932.
2. K. W. H. Stevens, *Proc. Roy. Soc.*, **A219**: 542, 1953.
3. M. Tinkham, *Proc. Roy. Soc.*, **A236**: 549, 1956.
4. M. H. L. Pryce, *Proc. Phys. Soc.*, **A63**: 25, 1950.
5. E. U. Condon and G. H. Shortley, *The Theory of Atomic Spectra*, Cambridge University Press, 1951.
6. H. A. Kramers, *Proc. Acad. Sci. Amst.*, **33**: 959, 1930.
7. J. C. Slater, *Introduction to Chemical Physics*, McGraw-Hill, New York, 1939.
8. C. Carter, Proc. Roy. Soc., **A235**: 321, 1956.
9. M. Kotani, *J. Phys. Soc.* (Japan), **4**: 293, 1949.
10. P. W. Anderson, *Solid State Physics*, **14**: 99, 1969.
11. J. S. Griffith, *Structure and Bonding*, **10**: 87, 1972.
12. J. S. Griffith, *Mol. Phys.*, **24**: 833, 1972.

CHAPTER

FOUR

THE ELECTRONIC TRANSITIONS

4-1 THE FERMI "GOLDEN RULE" NUMBER TWO

Suppose that at a time $t = 0$ a molecule is in the initial state Ψ_0 with energy W_0. In the next moment it is exposed to an electromagnetic radiation field. Due to the influence of the radiation, the molecule may then make a transition to another state Ψ_n with energy W_n. In order to calculate the probability of such a transition we must solve the time-dependent Schrödinger equation.[1]

The wave functions for the unperturbed molecule are determined by

$$i\hbar \frac{\partial \Psi_n(q, t)}{\partial t} = \mathscr{H}^0 \Psi_n(q, t) \tag{4-1}$$

$\mathscr{H}^0$ is independent of the time t. In addition to t, $\Psi_n(q, t)$ is a function of the nuclear and electronic coordinates q. The solutions to Eq. (4-1) are given by

$$\Psi_n(q, t) = \Psi_n(q) \exp(-iW_n t/\hbar) \tag{4-2}$$

where W_n is the energy of the system.

We now introduce the perturbing field represented by $\mathscr{H}^{(1)}(q, t)$. The wave equation is then determined by

$$(\mathscr{H}^0 + \mathscr{H}^{(1)})\Psi(q, t) = i\hbar \frac{\partial \Psi(q, t)}{\partial t} \tag{4-3}$$

In order to solve Eq. (4-3) we expand $\Psi(q, t)$ on the complete set of solutions for the unperturbed system

$$\Psi(q, t) = \sum_n \mathscr{A}_n(t) \Psi_n(q, t) \tag{4-4}$$

The quantity $|\mathscr{A}_n(t)|^2$ is the probability that the system is in the state Ψ_n at a time t. Substituting Eq. (4-4) into Eq. (4-3) leads to

$$i\hbar \sum_n \frac{d\mathscr{A}_n(t)}{dt} \Psi_n(q, t) = \mathscr{H}^{(1)} \sum_n \mathscr{A}_n(t) \Psi_n(q, t) \tag{4-5}$$

To obtain a solution of Eq. (4-5) we now assume that the perturbation is small, which means that during the perturbation the change in $\Psi(q, t)$ can be neglected. Using this assumption we replace $\sum_n \mathscr{A}_n(t) \Psi_n(q, t)$ on the right-hand side of Eq. (4-5) by the initial wave function $\Psi_0(q, t)$. Hence integrating over q after having multiplied Eq. (4-5) with $\Psi_m^*(q, t)$

$$i\hbar \frac{d\mathscr{A}_m(t)}{dt} = \int \Psi_m^*(q, t) \mathscr{H}^{(1)} \Psi_0(q, t) \, dq \tag{4-6}$$

Let $\mathscr{H}^{(1)}$ represent the action of an incoming light wave. The perturbation operator is then periodic in time, and with ν being the frequency of the light we can write

$$\mathscr{H}^{(1)} = \hat{F} \exp(i2\pi\nu t) + \hat{G} \exp(-i2\pi\nu t)$$

where $\hat{F}$ and $\hat{G}$ are functions of $(q, \partial/\partial q)$ but not of the time t. Since $\mathscr{H}^{(1)}$ is hermitian

$$\hat{F} \exp(2\pi i\nu t) + \hat{G} \exp(-2\pi i\nu t) = \hat{F}^* \exp(-2\pi i\nu t) + \hat{G}^* \exp(2\pi i\nu t)$$

or

$$\mathscr{H}^{(1)} = \hat{G} \exp(-2\pi i\nu t) + \hat{G}^* \exp(2\pi i\nu t) \tag{4-7}$$

Writing

$$\int \Psi_m^*(q) \hat{G} \Psi_0(q) \, dq = \langle m | \hat{G} | 0 \rangle$$

Eq. (4-6) is transformed into

$$i\hbar \frac{d\mathscr{A}_m(t)}{dt} = \langle m | \hat{G} | 0 \rangle \exp[2\pi i(\nu_{m0} - \nu)t] + \langle m | \hat{G}^* | 0 \rangle \exp[2\pi i(\nu_{m0} + \nu)t] \tag{4-8}$$

where we have written

$$W_m - W_0 = h\nu_{m0} \tag{4-9}$$

Because $\mathscr{A}_m(0) = 0$ we get from Eq. (4-8)

$$i\hbar \mathscr{A}_m(t) = \int_0^t \langle m | \hat{G} | 0 \rangle \exp[2\pi i(\nu_{m0} - \nu)t] \, dt + \int_0^t \langle m | \hat{G}^* | 0 \rangle \exp[2\pi i(\nu_{m0} + \nu)t] \, dt$$

Carrying out the integrations on the right-hand side of Eq. (4-8) gives

$$i\hbar \mathscr{A}_m(t) = \langle m | \hat{G} | 0 \rangle \frac{\exp[2\pi i(\nu_{m0} - \nu)t] - 1}{2\pi i(\nu_{m0} - \nu)} + \langle m | \hat{G}^* | 0 \rangle \frac{\exp[2\pi i(\nu_{m0} + \nu)t] - 1}{2\pi i(\nu_{m0} + \nu)} \tag{4-10}$$

This expression does not increase with time unless either $\nu_{m0} - \nu \approx 0$ or $\nu_{m0} + \nu \approx 0$. The first situation corresponds to absorption, the second to emission. Concentrating upon the absorption process we get for the probability that a transition $\Psi_0 \to \Psi_m$ has occurred.

$$|\mathscr{A}_m(t)|^2 = \frac{1}{\hbar^2} |\langle m | \hat{G} | 0 \rangle|^2 \frac{2 - \exp[2\pi i(\nu_{m0} - \nu)t] - \exp[-2\pi i(\nu_{m0} - \nu)t]}{4\pi^2(\nu_{m0} - \nu)^2}$$

$$= \frac{1}{\hbar^2} |\langle m | \hat{G} | 0 \rangle|^2 \frac{1 - \cos 2\pi(\nu_{m0} - \nu)t}{2\pi^2(\nu_{m0} - \nu)^2}$$

$$= \frac{1}{\hbar^2} |\langle m | \hat{G} | 0 \rangle|^2 \frac{\sin^2 \pi(\nu_{m0} - \nu)t}{\pi^2(\nu_{m0} - \nu)^2} \tag{4-11}$$

Using l'Hospital's rule, Eq. (4-11) is seen to go as t^2 for $\nu \to \nu_{m0}$. This is physically unreasonable. The absorption line ν_{m0} does, however, have a natural width. Integrating Eq. (4-11) "across the line," and introducing the energy density of the final states $\rho_W(= 1/W)$ the total probability of absorption is therefore

$$|a_m(t)|^2 = \frac{1}{\hbar^2} |\langle m | \hat{G} | 0 \rangle|^2 \int_{-\delta}^{+\delta} \frac{\sin^2 \pi(\nu_{m0} - \nu)t}{\pi^2(\nu_{m0} - \nu)^2} \rho_W \, d(h\nu)$$

Since by far the largest contributions to the integral occurs for $\nu_{m0} \approx \nu$ we may safely take the integration limits to be $+\infty$ and $-\infty$. Furthermore, at the energy $h\nu_{m0}$ we may treat ρ_W as approximately constant. Therefore with $x = \pi(\nu_{m0} - \nu)t$

$$|a_m(t)|^2 = \frac{1}{\hbar^2} |\langle m | \hat{G} | 0 \rangle|^2 \rho_W \frac{ht}{\pi} \int_{-\infty}^{\infty} \frac{\sin^2 x}{x^2} \, dx$$

$$= \frac{2\pi}{\hbar} |\langle m | \hat{G} | 0 \rangle|^2 \rho_W \cdot t$$

The total probability p_{0m} of a transition from Ψ_0 to the state Ψ_m per unit time is therefore given by the so-called Fermi "golden rule" number two:

$$p_{0m} = \frac{2\pi}{\hbar} |\langle m | \hat{G} | 0 \rangle|^2 \rho_W \tag{4-12}$$

4-2 THE ABSORPTION PROBABILITY

The vector potential **A** of radiation in the form of a plane-polarized wave moving in the $\boldsymbol{\kappa}$ direction and with **u** a unit vector perpendicular to $\boldsymbol{\kappa}$ and parallel to the electric field vector **E** is given by

$$\mathbf{A} = 2A^0(\omega)\mathbf{u} \cos\left(\frac{\omega}{c} \boldsymbol{\kappa} \cdot \mathbf{r} - \omega t\right) \tag{4-13}$$

or

$$\mathbf{A} = A^0(\omega)\mathbf{u} \exp\left(i\frac{\omega}{c} \boldsymbol{\kappa} \cdot \mathbf{r}\right) \exp(-i\omega t) + A^0(\omega)\mathbf{u} \exp\left(-i\frac{\omega}{c} \boldsymbol{\kappa} \cdot \mathbf{r}\right) \exp(i\omega t) \tag{4-14}$$

Here ω is the angular light frequency $2\pi\nu$, c the velocity of light, t the time, and $A^0(\omega)$ the real amplitude of the electromagnetic field.

Let us for simplicity take $\boldsymbol{\kappa}$ to be a unit vector in the z-direction with $\mathbf{u}$ in the x-direction. With $\mathbf{H} = \nabla \wedge \mathbf{A}$, where $\mathbf{H}$ is the magnetic field and $\mathbf{E} = -1/c\,\partial\mathbf{A}/\partial t$, we get from Eq. (4-13)

$$H_y = E_x = -\frac{2\omega}{c} A^0(\omega) \sin \omega \left(\frac{z}{c} - t\right) \tag{4-15}$$

The other components of $\mathbf{H}$ and $\mathbf{E}$ are zero. The energy density, $\Xi(\omega)$, of the light wave is classically given by

$$\Xi(\omega) = \frac{1}{8\pi} \overline{(|\mathbf{E}|^2 + |\mathbf{H}|^2)}$$

where the bar indicates an averaging over the period of the wave.

The use of Eq. (4-15) leads to

$$\Xi(\omega) = \frac{\omega^2}{2\pi c^2} [A^0(\omega)]^2 \tag{4-16}$$

With P photons of energy $\hbar\omega$ in a volume V we have also $\Xi(\omega) = P\hbar\omega/V$ or

$$A^0(\omega) = c\sqrt{\frac{2\pi P\hbar}{\omega V}} \tag{4-17}$$

We define the intensity $I(\omega)$ of the light beam as

$$I(\omega) = c\Xi(\omega) = \frac{\omega^2}{2\pi c} [A^0(\omega)]^2 \tag{4-18}$$

Furthermore, changing from the angular frequency scale ω to a linear frequency scale ν we must have

$$I(\omega)\,\mathrm{d}\omega = I(\nu)\,\mathrm{d}\nu$$

With $\mathrm{d}\omega = 2\pi\,\mathrm{d}\nu$

$$2\pi I(\omega) = I(\nu) = \frac{\omega^2}{c} [A^0(\omega)]^2 \tag{4-19}$$

Consider now a transition from the ground state $\Psi_0(\mathbf{r}, \xi)$ to an excited state $\Psi_j(\mathbf{r}, \xi)$. Here $\mathbf{r}$ stands for all the electronic coordinates, while ξ stands for the $3N - 6$ internal coordinates of the nuclear modes. The two states are orthogonal to each other in the electronic coordinates:

$$\int \Psi_0^*(\mathbf{r}, \xi)\Psi_j(\mathbf{r}, \xi)\,\mathrm{d}\mathbf{r} = 0 \qquad \text{for all } \xi \tag{4-20}$$

The transition probability per unit time of $\Psi_0 \to \Psi_j$ is given by the "golden rule", Eq. (4-12),

$$p_{0j} = \frac{2\pi}{\hbar^2} |\langle j|\,\hat{G}\,|0\rangle|^2 \rho_{0j} \tag{4-21}$$

The bracket indicates integration over the electronic coordinates as well as the $3N - 6$ nuclear coordinates. ρ_{0j} is the density of final states in ω-space.

The operator for the interaction of light and matter is to first order in the vector potential given by the first term of Eq. (3-8).

$$\mathscr{H}^{(1)} = \sum_{\eta} \frac{i\hbar Q_{\eta}}{M_{\eta} c} \mathbf{A}_{\eta} \cdot \nabla_{\eta} \tag{4-22}$$

The summation over η includes all the nuclei and electrons in the molecule. For a light wave polarized as in Eq. (4-15) the expression Eq. (4-22) is transformed into

$$\mathscr{H}^{(1)} = \sum_{\eta} \frac{i\hbar Q_{\eta}}{M_{\eta} c} A^{0}(\omega) \left[\exp\left(\frac{i\omega}{c} z_{\eta}\right) \exp(-i\omega t) + \exp\left(-\frac{i\omega}{c} z_{\eta}\right) \exp(i\omega t)\right] \frac{\partial}{\partial x_{\eta}} \tag{4-23}$$

A comparison of Eq. (4-23) with Eqs. (4-7) and (4-10) reveals that it is the first term inside the brackets which is responsible for absorption. For this process we may therefore take for the operator $\hat{G}$ in Eq. (4-21)

$$\hat{G}_{x} = \sum_{\eta} \frac{i\hbar Q_{\eta}}{M_{\eta} c} A^{0}(\omega) \exp\left(\frac{i\omega}{c} z_{\eta}\right) \frac{\partial}{\partial x_{\eta}} \tag{4-24}$$

$\hat{G}_x$ can be split into one part which depends upon the nuclear coordinates and one part which depends upon the electronic coordinates of the molecule. In the matrix element $\langle j | \hat{G} | 0 \rangle$ where the integrations are to be carried out over both electronic and nuclear coordinates, the interaction of the electromagnetic field with the nuclei drops out due to the orthogonality conditions of Eq. (4-20). Hence we may take

$$\hat{G}_{x} = \frac{-i\hbar e}{mc} A^{0}(\omega) \sum_{n} \exp\left(\frac{i\omega}{c} z_{n}\right) \frac{\partial}{\partial x_{n}} \tag{4-25}$$

where the summation now runs over the n electrons in the molecule.

Insertion of Eqs. (4-25) and (4-18) in (4-21) leads to

$$p_{0j}^{x}(\omega) = \frac{4\pi^2 e^2}{m^2 \omega^2} \rho_{0j} \Xi(\omega) \left| \langle j | \sum_{n} \exp\left(i\omega \frac{z_n}{c}\right) \frac{\partial}{\partial x_n} |0\rangle \right|^2 \tag{4-26}$$

Provided the molecular dimensions are comparable to the wavelength of the light we may expand

$$\exp(i\omega z/c) \approx 1 + \frac{i\omega z}{c} \tag{4-27}$$

The matrix element in Eq. (4-26) is then

$$\langle j | \sum_{n} \left(1 + \frac{i\omega}{c} z_n\right) \frac{\partial}{\partial x_n} |0\rangle$$

Using the commutator relation

$$[\mathscr{H}, x] = -\frac{\hbar^2}{m} \frac{\partial}{\partial x} \tag{4-28}$$

we have

$$\langle j|\sum_n \frac{\partial}{\partial x_n}|0\rangle = -\frac{m\omega_{j0}}{\hbar}\langle j|\sum_n x_n|0\rangle \tag{4-29}$$

with $W_j - W_0 = \hbar\omega_{j0}$. Further

$$\langle j|z\frac{\partial}{\partial x}|0\rangle = \frac{1}{2}\langle j|z\frac{\partial}{\partial x} - x\frac{\partial}{\partial z} + x\frac{\partial}{\partial z} + z\frac{\partial}{\partial x}|0\rangle$$

$$= -\frac{1}{2i\hbar}\langle j|\hat{l}_y|0\rangle + \frac{1}{2}\langle j|x\frac{\partial}{\partial z} + z\frac{\partial}{\partial x}|0\rangle$$

The commutator relation

$$[\mathscr{H}, xz] = -\frac{\hbar^2}{m}\left(x\frac{\partial}{\partial z} + z\frac{\partial}{\partial x}\right)$$

therefore gives us

$$\langle j|z\frac{\partial}{\partial x}|0\rangle = -\frac{1}{2i\hbar}\langle j|\hat{l}_y|0\rangle - \frac{m\omega_{j0}}{2\hbar}\langle j|xz|0\rangle \tag{4-30}$$

Inserting Eqs. (4-27), (4-29), and (4-30) in Eq. (4-26), and assuming that the line shape function ρ_{0j} peaks strongly at the angular frequency ω_{j0}, whereby all factors of ω can be approximated by ω_{j0}, leads to

$$p_{0j}^x = \frac{4\pi^2 e^2}{\hbar^2}\,\Xi(\omega_{j0})\left|\langle j|\sum_n x_n|0\rangle + \frac{\hbar}{2mc}\langle j|\sum_n \hat{l}_{yn}|0\rangle + i\frac{\omega_{j0}}{2c}\langle j|x_n z_n|0\rangle\right|^2$$

Changing to a ν scale, and introducing the Bohr magneton $\beta = e\hbar/2mc$, we get

$$p_{0j}^x = \frac{2\pi}{\hbar^2}\,\Xi(\nu_{j0})\,|M_x|^2 \tag{4-31}$$

with the factor M_x given by

$$M_x = \langle j|\sum_n ex_n|0\rangle + \langle j|\sum_n \beta\hat{l}_{yn}|0\rangle + \frac{i\pi\nu_{j0}}{c}\langle j|\sum_n ex_n z_n|0\rangle \tag{4-32}$$

The so-called Einstein coefficient B_{0j} for absorption[2] is now defined as follows. Let the molecule be immersed in a radiation field with density $\Xi(\nu_{j0})$. The chance that in the time interval dt it will make a transition from Ψ_0 to the state Ψ_j with the absorption of a quantum of radiation is taken to be $B_{0j}\Xi(\nu_{j0})\,dt$. A comparison with Eq. (4-31) then gives

$$B_{0j}^x = \frac{2\pi}{\hbar^2}|M_x|^2 \tag{4-33}$$

We now assume that the medium is isotropic. An averaging of Eq. (4-33) over all molecular orientations is then equivalent to a transformation from a space-fixed coordinate system to a molecular coordinate system. Having an optically inactive molecule the three terms in Eq. (4-32) represent components of different tensor operators. No cross terms will therefore appear after the averaging process and we get

$$B_{0j} = \frac{2\pi}{3\hbar^2}|\langle j|\sum_n e\mathbf{r}_n|0\rangle|^2 + \frac{2\pi}{3\hbar^2}|\langle j|\sum_n \beta\hat{\mathbf{l}}_n|0\rangle|^2 + \frac{2\pi^3\nu_{j0}^2}{5\hbar^2c^2}|\langle j|\sum_n e\tilde{q}_n|0\rangle|^2 \qquad (4\text{-}34)$$

where the quadrupole tensor $\tilde{q}_n$ has the active components xy, yz, xz, $\frac{1}{2}(x^2 - y^2)$, and $1/2\sqrt{3}(2z^2 - x^2 - y^2)$.

The electric dipole term, the magnetic dipole term, and the electric quadrupole term in Eq. (4-34) have the orders of magnitude

$$|\langle j|e\mathbf{r}|0\rangle|^2 \approx e^2a_0^2 = 6.5 \times 10^{-36} \text{ c.g.s.}$$

$$|\langle j|\beta\hat{\mathbf{l}}|0\rangle|^2 \approx \beta^2 = 8.7 \times 10^{-41} \text{ c.g.s.}$$

$$\frac{3\pi^2\nu_{j0}^2}{5c^2}|\langle j|e\tilde{q}_n|0\rangle|^2 \approx \frac{3\pi^2\nu_{j0}^2e^2a_0^4}{5c^2} = 4 \times 10^{-42} \text{ c.g.s. for } \lambda = 5{,}000 \text{ Å}.$$

Hence, provided the electric dipole term is different from zero, it is by far the dominant term.

4-3 THE OSCILLATOR STRENGTH

By an allowed electronic transition is understood a transition for which the electric dipole term in Eq. (4-34) is different from zero. Neglecting the other terms in Eq. (4-34), the strength of the transition is then proportional to the square of the transition moment $\mathbf{D}_{0j} = \langle j|\sum_n \mathbf{r}_n|0\rangle$. To measure the intensity of the transition it is convenient to introduce the dimensionless quantity f_{0j}, the oscillator strength, as

$$f_{0j} = \frac{2m}{3\hbar^2}h\nu_{0j}|\mathbf{D}_{0j}|^2 \qquad (4\text{-}35)$$

or

$$f_{0j} = \frac{8\pi^2mc}{3h}\bar{\nu}_{0j}|\mathbf{D}_{0j}|^2 = 1.085 \times 10^{11} \times \bar{\nu}_{0j}|\mathbf{D}_{0j}|^2 \qquad (4\text{-}36)$$

when $\bar{\nu}_{0j}$ is measured in cm^{-1}. Relating Eq. (4-35) with the B coefficient of Eq. (4-34) yields

$$f_{0j} = \frac{mh\nu_{0j}}{\pi e^2}B_{0j} \qquad (4\text{-}37)$$

This relation is valid when we are dealing with an absorption line. In the case of a broad band (Fig. 4-1) we define the total Einstein coefficient B for the band by

$$B = \int\left(\frac{dB}{d\nu}\right)d\nu \qquad (4\text{-}38)$$

In order to relate the calculated number f to an experimentally determined

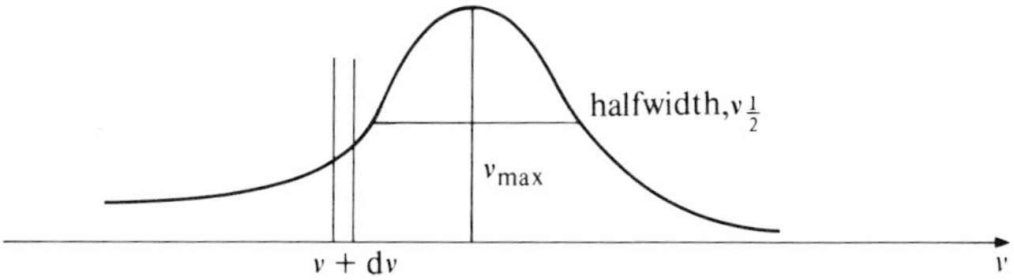

Figure 4-1 Broad-band absorption.

number we consider a container of cross section 1 cm^2 and length dl, with the number of molecules per cm^3 being L'. Each transition removes a quantum $h\nu$ of energy from the incident radiation, and the radiation encounters $L'\mathrm{d}l$ molecules. The density of the radiation is $\Xi(\nu)$ and from the definition of the B coefficient we get that the loss of energy per second of the radiation field is

$$-\mathrm{d}I(\nu) = \left(\frac{\mathrm{d}B}{\mathrm{d}\nu}\right)\Xi(\nu)h\nu L'\mathrm{d}l \qquad (4\text{-}39)$$

$I(\nu)$ is that energy which passes through a cross section of 1 cm^2 in 1 second. Using

$$c\Xi(\nu) = I(\nu) \qquad (4\text{-}40)$$

Eq. (4-39) can be written

$$-\frac{\mathrm{d}I(\nu)}{I(\nu)} = \left(\frac{\mathrm{d}B}{\mathrm{d}\nu}\right)\frac{h\nu}{c}L'\mathrm{d}l \qquad (4\text{-}41)$$

The Lambert–Beer law states[3]

$$-\frac{\mathrm{d}I(\nu)}{I(\nu)} = \ln 10\varepsilon(\nu)C\,\mathrm{d}l \qquad (4\text{-}42)$$

where $\varepsilon(\nu)$ is the molar extinction coefficient and C the molar concentration. With $L' \times 10^3 = C \cdot L_A$, where L_A is Avogadro's number, Eqs. (4-41) and (4-42) yield

$$\left(\frac{\mathrm{d}B}{\mathrm{d}\nu}\right) = \frac{10^3 \ln 10c}{hL_A}\frac{\varepsilon(\nu)}{\nu} \qquad (4\text{-}43)$$

From Eq. (4-37) we have

$$\left(\frac{\mathrm{d}f}{\mathrm{d}\nu}\right) = \frac{mh\nu}{\pi e^2}\left(\frac{\mathrm{d}B}{\mathrm{d}\nu}\right)$$

Therefore

$$\left(\frac{\mathrm{d}f}{\mathrm{d}\nu}\right) = \frac{10^3 mc \ln 10}{L_A\pi e^2}\varepsilon(\bar{\nu}) \qquad (4\text{-}44)$$

Integration over the broad band gives the total f number

$$f = \frac{10^3 mc^2 \ln 10}{L_A\pi e^2}\int \varepsilon(\bar{\nu})\,\mathrm{d}\bar{\nu} \qquad (4\text{-}45)$$

where we have changed to a wavenumber scale. Inserting numerical values

$$f = 4.332 \times 10^{-9} \int \varepsilon(\bar{\nu})\, d\bar{\nu} \tag{4-46}$$

Assuming a gaussian shape for the absorption band

$$\varepsilon(\bar{\nu}) = \varepsilon_{\max} \exp\left[-(\bar{\nu} - \bar{\nu}_{\max})^2/a^2\right] \tag{4-47}$$

we get easily for the experimental value of f

$$f = 4.61 \times 10^{-9}\, \varepsilon_{\max}\, \bar{\nu}_{1/2} \tag{4-48}$$

where $\bar{\nu}_{1/2}$ is the bandwidth at $\varepsilon = \frac{1}{2}\varepsilon_{\max}$ (see Fig. 4-1). It is therefore possible to compare the theoretically calculated f of Eq. (4-36) with the experimentally determined quantity given by Eqs. (4-46) or (4-48).

4-4 THE WIDTH OF MOLECULAR ABSORPTION LINES

A transition from one vibrational level associated with the molecular ground state to a vibrational level of an electronically excited state gives rise to an absorption line. The excited state will have a certain lifetime Δt, and this property of the state will determine the natural line width. The mean lifetime can be directly connected with the line width ΔW by means of the uncertainty principle

$$(\Delta W)(\Delta t) \geq \hbar \tag{4-49}$$

The observed radiative lifetime of an allowed transition is $\Delta t \approx 10^{-9}$ sec. Accordingly, $\Delta W \geq 5 \times 10^{-3}\,\text{cm}^{-1}$. However, in molecular spectroscopy we do not observe the natural line width. Instead we see a manifold of coupled unresolvable lines forming a structure which may show up as a single broad line.

With each vibrational level there is associated a dense envelope of either rotational levels and/or lattice vibrations, the so-called acoustical phonons. At 4.2 °K the thermal energy is about 3 cm^{-1}. Phonon broadening of about this order of magnitude is therefore expected. The band widths reflect in this case not only the hot-phonon broadening, but also the shape and nature of the potential energy surface for the molecule at its site in the medium. To account for the motions of the molecule in relation to its surroundings Moffitt and Moscowitz[4] introduced the "libration" quantum numbers. The density $\rho(W_j)$ of the libration states associated with the intramolecular vibration j is normalized as follows:

$$\int \rho(W_j)\, dW = 1 \tag{4-50}$$

$\rho(W_j)$ constitutes an effective line shape generated by the librations. This unresolved structure may give the molecular absorption lines a breadth of several wavenumbers.

Provided any sharp or medium-sharp lines (halfwidths less than 10 cm^{-1})

are observed, experience teaches us that they are only associated with the first few electronic transitions.[5] The reason is that a rapid intramolecular electronic relaxation takes place between the excited states and the underlying vibrational continuum built upon the lower-lying electronic states. The higher the excitation is in energy, the denser is the continuum, and the broader is the line.

The line shapes can be accounted for by using the model put forward by Bixon and Jortner.[6] Let ψ_s represent one vibronic state of energy W_s, belonging to the electronic state A. The set $\{\psi_l\}$ denotes vibronic levels associated with the state B. We shall assume that $\{\psi_l\}$ has nearly the same energy as W_s, and that it is uniformly spaced with an energy difference ε between the consecutive levels. The value of $l = 0$ is given to the nearest level from below W_s. The states lower than W_s have $l < 0$, while for states higher than W_s, $l > 0$. With $\alpha = W_s - W_0$ we have to zeroth order

$$\mathscr{H}\psi_l = (W_s - \alpha + l\varepsilon)\psi_l \tag{4-51}$$

and

$$\mathscr{H}\psi_s = W_s\psi_s \tag{4-52}$$

We now assume a perturbation $\mathscr{H}^{(1)}$ which will couple $\{\psi_l\}$ with ψ_s so that the matrix element

$$V = \langle\psi_s|\mathscr{H}^{(1)}|\psi_l\rangle \tag{4-53}$$

is assumed constant, independent of the index l.

The eigenstates of the molecular system are expressed as linear combinations of the zero-order wave functions in the form

$$\Psi_n = a_n\psi_s + \sum_l b_l^n\psi_l \tag{4-54}$$

The values of the coefficients a_n and b_l^n are given by the solutions to the secular determinant

$$\begin{vmatrix} W_s - W_n & V & V & & \\ & & & 0 & \\ V & W_l - W_n & & & \\ & & \cdot & & \\ V & & & \cdot & \\ \cdot & & & & \cdot \\ & 0 & & & \end{vmatrix} = 0$$

or

$$(W_s - W_n)\,a_n + V\sum_l b_l^n = 0 \qquad l = 0, \pm 1, \pm 2, \ldots \tag{4-55}$$

$$(W_l - W_n)b_l^n + Va_n = 0 \tag{4-56}$$

From the last equation

$$b_l^n = -\frac{Va_n}{W_s - \alpha + l\varepsilon - W_n} \tag{4-57}$$

Substituting Eq. (4-57) into Eq. (4-55) yields

$$\left[W_s - W_n + V^2 \sum_{l=-\infty}^{\infty} \frac{1}{W_n - W_s + \alpha - l\varepsilon}\right] a_n = 0 \tag{4-58}$$

Writing

$$\gamma_n = \frac{W_n - W_s + \alpha}{\varepsilon} \tag{4-59}$$

we get

$$W_s - W_n + \frac{V^2}{\varepsilon}\left[\frac{1}{\gamma_n} + \sum_{l=1}^{\infty}\left(\frac{1}{\gamma_n + l} + \frac{1}{\gamma_n - l}\right)\right] = 0$$

or

$$W_s - W_n + \frac{V^2}{\varepsilon}\frac{1}{\gamma_n} + \sum_{l=1}^{\infty} \frac{2\gamma_n}{\gamma_n^2 - l^2} = 0 \tag{4-60}$$

Using a formula from the *Handbook of Mathematical Functions*[7] we have

$$\sum_{l=1}^{\infty} \frac{2\gamma_n}{\gamma_n^2 - l^2} = -\frac{1}{\gamma_n} + \pi \cot \gamma_n \pi$$

or

$$W_s - W_n + \frac{V^2 \pi}{\varepsilon} \cot \gamma_n \pi = 0 \tag{4-61}$$

which determines $(W_n - W_s)$.

The value of the coefficient a_n is determined from the normalization condition

$$a_n^2 + \sum_l (b_l^n)^2 = 1 \tag{4-62}$$

Using Eqs. (4-57) and (4-59) in Eq. (4-62) leads to

$$a_n^2 + \frac{V^2 a_n^2}{\varepsilon^2} \sum_{l=-\infty}^{\infty} \frac{1}{(l - \gamma_n)^2} = 1 \tag{4-63}$$

Using[7]

$$\sum_{l=-\infty}^{\infty} \frac{1}{(\gamma_n - l)^2} = \pi^2 + \pi^2 \cot^2 \pi\gamma_n$$

we get from Eq. (4-63)

$$a_n^2 = \frac{1}{1 + (V^2\pi^2/\varepsilon^2)(1 + \cot^2 \pi\gamma_n)}$$

Substituting for $\cot \pi\gamma_n$ from Eq. (4-61) leads to

$$a_n^2 = \frac{V^2}{(W_s - W_n)^2 + V^2 + (\pi V^2/\varepsilon)^2} \tag{4-64}$$

This equation gives us a lorentzian envelope for the values of a_n^2 as a function of

W_n. The width Δ of the lorentzian amplitude distribution in a_n^2 is given by

$$\Delta^2 = 4\left[V^2 + \left(\frac{\pi V^2}{\varepsilon}\right)^2\right] \tag{4-65}$$

therefore

$$a_n^2 = \frac{V^2}{(W_s - W_n)^2 + \frac{1}{4}\Delta^2} \tag{4-66}$$

From Eqs. (4-55) and (4-66)

$$\left(\sum_l b_l^n\right)^2 = \frac{(W_s - W_n)^2}{(W_s - W_n)^2 + \frac{1}{4}\Delta^2} \tag{4-67}$$

Consider first the case where all the transition moments $\mathbf{D}_{0l}$ from the ground state ψ_0 to the continuum are zero. With $\mathbf{D}_{0s}$ the transition moment to the discrete state being different from zero, we get, using Eqs. (4-54) and (4-66), the line-shape function

$$G_{0n}(W_s - W_n, \Delta) = |\mathbf{D}_{0s}|^2 \frac{V^2}{(W_s - W_n)^2 + \frac{1}{4}\Delta^2}$$

The line shape will therefore be a lorentzian symmetric peak with a width of Δ.

Next we assume that the $\mathbf{D}_{0l}$ transition moments are all collinear and equal in magnitude. For a molecular system, the $\mathbf{D}_{0s}$ transition moment is, however, not necessarily collinear with $\mathbf{D}_{0l}$. In the general case[8] the line shape will be given by the expression $|\mathbf{D}_{0n} \cdot \mathbf{E}|^2$, where $\mathbf{E}$ denotes the electric vector of the electromagnetic field and $\mathbf{D}_{0n}$ is the transition moment to the scrambled state Ψ_n of Eq. (4-54).

For the normalized transition moment we have

$$\mathbf{D}_{0n} = \frac{a_n \mathbf{D}_{0s} + \sum b_l^n \mathbf{D}_{0l}}{|D_{0l}|}$$

The angle between $\mathbf{D}_{0s}$ and $\mathbf{D}_{0l}$ is called ϕ. With $\cos\phi = A$, and using the coordinate system defined in Fig. 4-2, we have

$$\mathbf{D}_{0n} = \frac{a_n |D_{0s}| \mathbf{k} + \sum b_l^n |D_{0l}| (A\mathbf{k} + \sqrt{1 - A^2}\,\mathbf{j})}{|D_{0l}|}$$

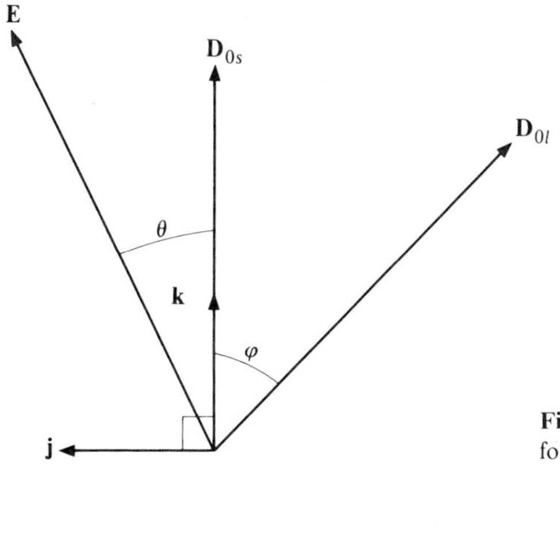

Figure 4-2 The transition moment dipole vectors for a discrete state imbedded in a continuum.

Introducing Eqs. (4-66) and (4-67) together with the reduced energy parameter w and the line-shape profile index q

$$w = \frac{W_s - W_n}{\frac{1}{2}\Delta} \qquad -\infty < w < \infty \tag{4-68a}$$

$$q = \frac{|D_{0s}|\,V}{|D_{0l}|\,\frac{1}{2}\Delta} \tag{4-68b}$$

we get

$$\mathbf{D}_{0n} = \frac{(q + Aw)\mathbf{k} + w\sqrt{1 - A^2}\,\mathbf{j}}{\sqrt{1 + w^2}}$$

Let θ be the angle which the electric vector $\mathbf{E}$ forms with $\mathbf{D}_{0s}$ in the plane defined by $\mathbf{D}_{0s}$ and $\mathbf{D}_{0l}$. Because the line shape is independent of the angle which the vector $\mathbf{E}$ forms with the $\mathbf{k}$, $\mathbf{j}$ plane we get for the line-shape function[8]

$$G_{0n}(\theta, w) = \frac{|(q + Aw)\cos\theta + w\sqrt{1 - A^2}\sin\theta|^2}{1 + w^2}$$

In a random phase experiment we have, averaging $G_{0n}(\theta, w)$ over θ,

$$\bar{G}_{0n}(w) \propto \frac{q^2 + 2q|A|w + w^2}{1 + w^2} \tag{4-69}$$

With $q = 0$, that is when the "background" has oscillator strength whereas the discrete state has none, the line-shape function is

$$\bar{G}_{0n}(w) \propto \frac{w^2}{1 + w^2}$$

The line will therefore show up as a symmetric dip with a width equal to Δ. Provided $|A| < 1$ the line will indeed be experimentally observed as a dip or "antiresonance" in the background absorption. In the case of $\mathbf{D}_{0s}$ and $\mathbf{D}_{0l}$ being colinear, $|A| = 1$, Eq. (4-69) reduces to the so-called Fano expression

$$\bar{G}_{0n}(w) \propto \frac{(q + w)^2}{1 + w^2}$$

In the region $0 < |q| < \infty$ we will therefore observe an asymmetric line with zero intensity at $w = -q$.

The order of magnitude of the theoretically expected line width Δ of an excited electronic state of a polyatomic molecule can be estimated to several hundred wavenumbers.[8] The spacing ε of the vibronic states can be related to the density ρ of vibrational energy levels by $\varepsilon^{-1} = \rho$. Classical or semiclassical considerations[9] give this quantity as a function of the vibrational frequencies of the molecule and of the energy of the excited state W. The interaction parameter V of Eq. (4-53) can be expanded in an electronic part v and in a vibrational overlap factor, $V = vS$. For high-lying states $V/\varepsilon \gg 1$ and we may write $\Delta \approx 2\pi v^2 S^2 \rho$.

With $\rho S^2 \approx 10^{-4} - 10^{-5}$ cm, which is a reasonable value when $W \approx 10{,}000\ \text{cm}^{-1}$, and estimating $v \approx 10^3\ \text{cm}^{-1}$ we get $\Delta \approx 10^2$–$10^3\ \text{cm}^{-1}$.

We may conclude that the principal reason for diffuseness, which is such a characteristic feature of nearly all electronic transitions in inorganic complexes, is the interaction of the excited vibronic lines with the background continuum. Other causes for band diffuseness, as for instance predissociation or overcrowding of absorption lines, may also be met with. We shall return to this last problem in section 4-7.

Furthermore, the scrambling of the discrete vibronic states with the surrounding continuum provides the mechanism for the intramolecular radiationless decay in an isolated molecule.[6] The condition for a nonradiative transition is that $V \gg \varepsilon$. This model may therefore also account for the rareness with which we observe light emission in inorganic complexes.

4-5 NATURAL OPTICAL ACTIVITY

It is well known that certain molecules have the ability to rotate a plane-polarized light wave, and such a molecule is said to be optically active. The phenomenom is closely linked to the feature that these molecules absorb left–circularly polarized light differently from right–circularly polarized light. In order to elucidate this effect we shall consider a transition from the molecular ground state $|0\rangle$ to an excited state $|j\rangle$. For absorption the relevant operator $\hat{G}$ can be written[10] (compare Eq. (4-25))

$$\hat{G}_\lambda = \frac{-\mathrm{i}\hbar e}{mc} A^0(\omega)\varepsilon_\lambda \sum_n \exp\left(\frac{\mathrm{i}\omega}{c}\varepsilon_3 \cdot \mathbf{r}_n\right)\nabla_n \tag{4-70}$$

where ε_λ is the polarization vector and ε_3 the propagation vector. The polarization vectors for left– and right–circularly polarized light are given by

$$\varepsilon_l = \frac{1}{\sqrt{2}}(\varepsilon_1 + \mathrm{i}\varepsilon_2), \qquad \varepsilon_r = \frac{1}{\sqrt{2}}(\varepsilon_1 - \mathrm{i}\varepsilon_2) \tag{4-71}$$

where ε_1, ε_2, and ε_3 form a right-handed system of mutually orthogonal unit vectors.

Inserting Eq. (4-70) in the "golden rule" Eq. (4-12) and making use of Eq. (4-18) leads to the transition probabilities $p_{0j}^{(l)}(\omega)$ and $p_{0j}^{(r)}(\omega)$ for left– and right–circular polarized light

$$p_{0j}^{(l),(r)}(\omega) = \frac{2\pi^2 e^2}{m^2 c\omega^2} I(\omega)\rho_j(\omega)\left|\langle j|(\varepsilon_1 \pm \mathrm{i}\varepsilon_2)\sum_n \exp\left(\frac{\mathrm{i}\omega}{c}\varepsilon_3 \cdot \mathbf{r}_n\right)\nabla_n|0\rangle\right|^2 \tag{4-72}$$

where the upper and lower signs give us $p_{0j}^{(l)}$ and $p_{0j}^{(r)}$, respectively. Equation (4-72) can be rewritten

$$p_{0j}^{(l),(r)}(\omega) = \frac{2\pi^2 e^2}{m^2 c\omega^2} I(\omega)\rho_j(\omega)\left[|\varepsilon_1 \cdot \mathbf{m}_j|^2 + |\varepsilon_2 \cdot \mathbf{m}_j|^2 \pm \varepsilon_3 \cdot \mathrm{i}(\mathbf{m}_j^* \wedge \mathbf{m}_j)\right] \tag{4-73}$$

Here $\mathbf{m}_j$ is the three-component quantity

$$\mathbf{m}_j = \langle j | \sum_n \exp\left(\frac{\mathrm{i}\omega}{c}\varepsilon_3 \cdot \mathbf{r}_n\right) \nabla_n |0\rangle \tag{4-74}$$

$\mathbf{m}_j$ can be manipulated as a vector, but it does not transform like a vector under rotations.

Classically the difference $k^l - k^r$ between the absorption coefficients for left– and right–circularly polarized light gives a measure of the so-called ellipticity. The ellipticity per unit length associated with the jth transition may be taken as[28]

$$\theta_j(\omega) = \tfrac{1}{4}[k_j^l(\omega) - k_j^r(\omega)] \tag{4-75}$$

We have further

$$I(\omega)k_j(\omega) = L'\hbar\omega p_j(\omega) \tag{4-76}$$

where L' is the number of absorbing molecules per unit volume. Using Eqs. (4-73), (4-75), and (4-76)

$$\theta_j(\omega) = \frac{\pi^2 e^2 \hbar L'}{m^2 c\omega}\rho_j(\omega)\,\mathrm{i}\varepsilon_3 \cdot (\mathbf{m}_j^* \wedge \mathbf{m}_j) \tag{4-77}$$

For natural optical activity it turns out to be convenient to define the rotatory strength R_{0j} for a transition as

$$R_{0j} = \frac{3\hbar c}{4\pi^2 L'}\int \theta_j(\omega)\frac{\mathrm{d}\omega}{\omega} \tag{4-78}$$

Inserting Eq. (4-77) in Eq. (4-78) leads to

$$R_{0j} = \frac{3\hbar^2 e^2}{4m^2}\int \mathrm{i}\varepsilon_3 \cdot (\mathbf{m}_j^* \wedge \mathbf{m}_j)\frac{\rho_j(\omega)}{\omega^2}\,\mathrm{d}\omega \tag{4-79}$$

The line shape $\rho_j(\omega)$ is assumed to have a strong peak at ω_{0j}. Hence to good approximation

$$R_{0j} = \frac{3\hbar^2 e^2}{4m^2\omega_{j0}^2}\mathrm{i}\varepsilon_3 \cdot (\mathbf{m}_{0j}^* \wedge \mathbf{m}_{0j}) \tag{4-80}$$

with

$$\mathbf{m}_{0j} = \langle j | \sum_n \exp\left(\frac{\mathrm{i}\omega_{j0}}{c}\varepsilon_3 \cdot \mathbf{r}_n\right) \nabla_n |0\rangle \tag{4-81}$$

Identifying the light-propagation direction ε_3 with the z axis we get

$$R_{0j}^z = \frac{3\hbar^2 e^2}{4m^2\omega_{j0}^2}\,\mathrm{i}\,[m_{0j}^*(x)m_{0j}(y) - m_{0j}^*(y)m_{0j}(x)] \tag{4-82}$$

where

$$m_{0j}(x) = \langle j | \sum_n \exp\left(\frac{\mathrm{i}\omega_{j0}}{c} z_n\right)\frac{\partial}{\partial x_n}|0\rangle$$

and

$$m_{0j}(y) = \langle j|\sum_n \exp\left(\frac{i\omega_{j0}}{c} z_n\right)\frac{\partial}{\partial y_n}|0\rangle$$

When dealing with natural optical activity the stationary wave functions can be assumed to be real. Expanding the exponential in $m_{0j}(x)$ and $m_{0j}(y)$ and remembering that R_{0j} is real, we get

$$R^z_{0j} = \frac{3\hbar^2 e^2}{2m^2 c\omega_{j0}}\left[-\langle j|\sum_n \frac{\partial}{\partial x_n}|0\rangle\langle j|\sum_n z_n\frac{\partial}{\partial y_n}|0\rangle + \langle j|\sum_n \frac{\partial}{\partial y_n}|0\rangle\langle j|\sum_n z_n\frac{\partial}{\partial x_n}|0\rangle\right] \tag{4-83}$$

Averaging over all orientations of the molecule leads to

$$R_{0j} = \frac{\hbar^2 e^2}{2m^2 c\omega_{j0}}\langle j|\sum_n \nabla_n|0\rangle \cdot \langle j|\sum_n (\mathbf{r}_n \wedge \nabla_n)|0\rangle \tag{4-84}$$

Making use of Eq. (4-29) and introducing the angular momentum operator $\mathbf{l}_n$ gives the equivalent form of Eq. (4-84)

$$R_{0j} = \frac{e^2}{2mc}\,\mathrm{Im}\,\langle j|\sum_n \mathbf{r}_n|0\rangle \cdot \langle j|\sum_n \mathbf{l}_n|0\rangle \tag{4-85}$$

where Im stands for the imaginary part. This expression is the Rosenfeld–Condon equation for the "rotatory strength" of a transition.

The ellipticity, likewise averaged over all orientations of the molecule, is found to be

$$\theta_j(\omega) = \frac{2\pi^2 e^2 \hbar L'}{3m^2 c^2}\langle j|\sum_n \nabla_n|0\rangle \cdot \langle j|\sum_n (\mathbf{r}_n \wedge \nabla_n)|0\rangle \rho_j(\omega) \tag{4-86}$$

The shape of the ellipticity curve as a function of ω is seen to be given by the line shape generated by the density of the final states. The $\theta_j(\omega)$ curve is therefore concentrated around the absorption band, with which it is associated.

4-6 THE FARADAY EFFECT

When a magnetic field is applied parallel to the direction of propagation of incident light, an otherwise inactive substance will exhibit optical activity. This is the so-called Faraday effect. It is found that the rotation of the plane-polarized light is proportional to the field H and the molar concentration, and the proportionality factor is called the Verdet constant.

The effect of the magnetic field is to produce a Zeeman effect on the levels between which the electronic transition takes place. Also it may cause mixing between the levels. For simplicity we shall ignore this last effect in what follows and concentrate upon the consequences of the first.

We now take both the light-propagation axis and the magnetic field direction to be the z axis. The hamiltonian is then

$$\mathscr{H} = \mathscr{H}^0 + \mathscr{H}^{(1)}$$

where

$$\mathscr{H}^{(1)} = -|\beta| H_z \sum_n (\hat{l}_{zn} + 2\hat{s}_{zn}) \tag{4-87}$$

$|\beta|$ is the numerical value of the Bohr magneton. Defining

$$\hat{\mu}_{zn} = -|\beta|(\hat{l}_{zn} + 2\hat{s}_{zn}) \tag{4-88}$$

we can write

$$\mathscr{H}^{(1)} = H_z \sum_n \hat{\mu}_{zn} \tag{4-89}$$

$\mathscr{H}^{(1)}$ is a purely imaginary operator, and being hermitian $\mathscr{H}^{(1)} = -(\mathscr{H}^{(1)})^*$.

Let ψ be a member of a zero-order degenerate manifold of $\mathscr{H}^0$. We further take ψ to be diagonal in $\mathscr{H}^{(1)}$. Then

$$\begin{aligned} \langle\psi^*|\mathscr{H}^{(1)}|\psi^*\rangle &= -\langle\psi^*|(\mathscr{H}^{(1)})^*|\psi^*\rangle \\ &= -\langle\psi|\mathscr{H}^{(1)}|\psi\rangle^* = -\langle\psi|\mathscr{H}^{(1)}|\psi\rangle \end{aligned}$$

Evidently ψ and ψ^* have equal but opposite first-order perturbation energy. Furthermore, the zero-order wave functions which are diagonal in $\mathscr{H}^{(1)}$ are made up by a real and imaginary part. With two real wave functions, ψ_x and ψ_y say, a conjugate pair diagonal in $\mathscr{H}^{(1)}$ will be $\psi_{-1} = 1/\sqrt{2}(\psi_x - i\psi_y)$ and $\psi_{+1} = -1/\sqrt{2}(\psi_x + i\psi_y)$. To every transition $0 \to j$ with a Zeeman displacement $\hbar\omega_{0j}^{(1)} H_z$ there is therefore a conjugate transition $0^* \to j^*$ with equal but opposite shift $-\hbar\omega_{0j}^{(1)} H_z$.

The circular dichroism CD of a transition $0 \to j$ is defined by[11]

$$\Delta k_{0-j} = [k_{0j}^l(\omega) - k_{0j}^r(\omega)]\,\rho_{0j}(\omega)$$

where $k^l - k^r$ is the difference between the absorption coefficients for left– and right–circularly polarized light and $\rho(\omega)$ is the normalized line-density function

$$\int \rho(\omega)\, d\omega = 1$$

We have, therefore, using Eqs. (4-73), (4-74), and (4-76);

$$\Delta k_{0j} = \frac{4\pi^2 e^2 \hbar L'}{m^2 c\omega} \rho_{0j}(\omega) i\varepsilon_3 \cdot (\mathbf{m}_j^* \wedge \mathbf{m}_j)$$

In contrast to the case of natural optical activity where the wave functions can be assumed to be real, the present wave functions contain an imaginary part. Hence in the calculation of the magnetic circular dichroism MCD, the expansion of the exponential factor in Eq. (4-74) can be broken off after the first term. We get therefore

$$\Delta k_{0j} = \frac{4\pi^2 e^2 \hbar L'}{m^2 c\omega} \rho_{0j}(\omega) \mathrm{i} \left[\langle j | \sum_n \frac{\partial}{\partial x_n} | 0 \rangle^* \langle j | \sum_n \frac{\partial}{\partial y_n} | 0 \rangle - \langle j | \sum_n \frac{\partial}{\partial y_n} | 0 \rangle^* \langle j | \sum_n \frac{\partial}{\partial x_n} | 0 \rangle \right] \tag{4-90}$$

Strictly speaking $\partial/\partial x$ should be $\partial/\partial x - \mathrm{i}e/\hbar c\, a_x$ and similarly for $\partial/\partial y$, where a_x is the x-component of the static outer-vector potential. However, changing from a dipole velocity to a dipole length expression using Eq. (4-29) corrects this omission. Hence, utilizing the fact that the line-shape factor peaks strongly at ω_{0j}, we get

$$\Delta k_{0j} = \frac{4\pi^2 e^2 \omega_{0j} L'}{\hbar c} \rho_{0j}(\omega) \mathrm{i} \left[\langle j | \sum_n x_n | 0 \rangle^* \langle j | \sum_n y_n | 0 \rangle - \langle j | \sum_n y_n | 0 \rangle^* \langle j | \sum_n x_n | 0 \rangle \right] \tag{4-91}$$

or

$$\Delta k_{0j} = \frac{8\pi^2 e^2 \omega_{0j} L'}{\hbar c} \rho_{0j}(\omega) \operatorname{Im} \langle j | \sum_n x_n | 0 \rangle^* \langle j | \sum_n y_n | 0 \rangle \tag{4-92}$$

where Im stands for the imaginary part.

The population of a Zeeman component M of the ground state $|0\rangle$ will with $\mathscr{H}^{(1)} = H_z \sum_n \hat{\mu}_{zn}$ be given by

$$\frac{L'_M}{L'} = \frac{\exp\left(- \dfrac{\langle 0, M | \sum_n \hat{\mu}_{zn} | 0, M \rangle H_z}{k_B T} \right)}{\exp \sum_M - \dfrac{\langle 0, M | \sum_n \hat{\mu}_{zn} | 0, M \rangle H_z}{k_B T}} \tag{4-93}$$

Expanding this expression, assuming the Zeeman splittings to be small compared to $k_B T$, we get, with d_0 as the degeneracy of the ground state,

$$\frac{L'_M}{L'} = \frac{1}{d_0} \left(1 - \frac{\langle 0, M | \sum_n \hat{\mu}_{zn} | 0, M \rangle H_z}{k_B T} \right) \tag{4-94}$$

Starting at the lowest Zeeman levels of the ground state we get, summing over all ground and excited state levels

$$\Delta k_{0j} = \frac{8\pi^2 e^2 \omega_{0j} L'}{\hbar c d_0} \sum_M \left(1 + \frac{\langle 0M | \sum_n \hat{\mu}_{zn} | 0M \rangle}{k_B T} H_z \right)$$
$$\times \rho \left(\omega - (\omega_{0j} + \langle 0M | \sum_n \hat{\mu}_{zn} | 0M \rangle H_z - \langle jM | \sum_n \hat{\mu}_{zn} | jM \rangle H_z) \right)$$
$$\times \operatorname{Im} \langle jM | \sum_n x_n | 0M \rangle^* \langle jM | \sum_n y_n | 0M \rangle \tag{4-95}$$

Expanding the density function appearing in Eq. (4-95) in a Taylor series, and retaining only terms which are linear in H_z, leads to

$$\Delta k_{0j} = \frac{8\pi^2 e^2 \omega_{0j} L'}{\hbar c d_0} H_z \sum_M \mathrm{Im}\,\langle jM|\sum_n x_n|0M\rangle^* \langle jM|\sum_n y_n|0M\rangle$$

$$\times\left\{\frac{\langle 0M|\sum_n \hat{\mu}_{zn}|0M\rangle}{k_B T}\rho_{0j}(\omega) + \left[\langle jM|\sum_n \hat{\mu}_{zn}|jM\rangle - \langle 0M|\sum_n \hat{\mu}_{zn}|0M\rangle\right]\left(\frac{\partial\rho}{\partial\omega}\right)_{0j}\right\} \tag{4-96}$$

The above derivation has assumed the molecules to be fixed in space. If the systems are randomly oriented we must perform an average over all possible orientations. Provided the symmetry of the molecule is sufficiently high, e.g., octahedral or tetrahedral, the x, y, and z directions are equivalent. Defining the so-called $\mathfrak{A}$ and $\mathfrak{C}$ terms

$$\mathfrak{A} = \frac{3}{d_0}\sum\left[\langle j|\sum_n \hat{\mu}_{zn}|j\rangle - \langle 0|\sum_n \hat{\mu}_{zn}|0\rangle\right]\mathrm{Im}\,\langle j|\sum_n x_n|0\rangle^* \langle j|\sum_n y_n|0\rangle \tag{4-97}$$

$$\mathfrak{C} = \frac{3}{d_0}\sum\langle 0|\sum_n \hat{\mu}_{zn}|0\rangle\,\mathrm{Im}\,\langle j|\sum_n x_n|0\rangle^* \langle j|\sum_n y_n|0\rangle \tag{4-98}$$

we can write for the averaged expression[11] in which the Zeeman splitting is small compared to the line width

$$\Delta\bar{k}_{0j} = -\frac{4\pi^2 \gamma L_A}{3\hbar}\left\{\mathfrak{A}\rho' + \frac{\mathfrak{C}}{k_B T}\rho\right\} H \tag{4-99}$$

where γ is a frequency-independent correction factor allowing for the effect of the medium on the electric field of the light wave.

Apart from the $\mathfrak{A}$ and $\mathfrak{C}$ terms appearing in Eq. (4-99) a more refined calculation would also have introduced a so-called $\mathfrak{B}$ term. This small term which we have neglected here is due to the mixing of the zero field states by $\mathscr{H}^{(1)}$. The $\mathfrak{B}$ term is to be multiplied with ρ and therefore has the same shape as the $\mathfrak{C}$ term.

The $\mathfrak{C}$ term is known as the *paramagnetic* term because it has the same temperature dependence as the paramagnetic susceptibility. Its shape is seen to be given by the density function associated with the excited state. The $\mathfrak{A}$ term is temperature-independent. Its shape is given by an S curve, viz. the derivative of the density function $\rho(\omega)$ of the excited state.

The appearance and sign convention of MCD curves have been given in Fig. 4-3. We observe that the presence of an $\mathfrak{A}$ term in the measured MCD curve is a certain sign of a degeneracy in the terminating electronic state.

We may define the dipole strength of the absorption as

$$\mathfrak{D} = \frac{3}{d_0}\sum\left|\langle j|\sum_n x_n|0\rangle\right|^2 \tag{4-100}$$

where the summation is again over all of the degenerate components. It follows that

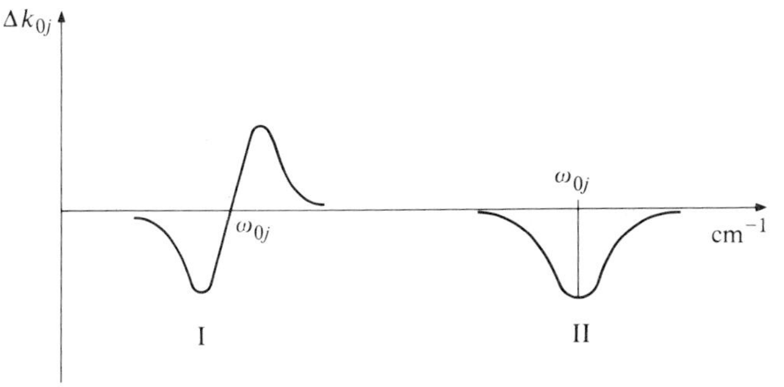

Figure 4-3 MCD curves. (I) Positive $\mathfrak{A}$ term; (II) Positive $\mathfrak{B}$ and $\mathfrak{C}$ terms.

the ratios $\mathfrak{A}/\mathfrak{D}$ and $\mathfrak{C}/\mathfrak{D}$ depend only upon the symmetries of the electronic states and the magnitude of the matrix element $\langle i|\sum_n \hat{\mu}_{zn}|i\rangle$. The use of these ratios removes to a large extent solvent effects. It also eliminates the calculation of transition-moment elements. Measurements of the magnetic circular dichroism can therefore provide us with the g factors of the excited states.

4-7 THE INTENSITY DISTRIBUTION IN AN ALLOWED TRANSITION

For an allowed electric dipole transition $|0\rangle \rightarrow |j\rangle$ the Einstein B coefficient, giving the transition probability, is equal to

$$B_{0j} = \frac{2\pi e^2}{3\hbar^2}|\mathbf{D}_{0j}|^2 \tag{4-101}$$

where

$$\mathbf{D}_{0j} = \langle j|\sum_n \mathbf{r}_n|0\rangle \tag{4-102}$$

In the adiabatic approximation the molecular wave function for the ground state is given by $|0\rangle = \psi_0^0(\mathbf{r})\chi''_{v''}(\xi)$. The electronic part of the wave function is a solution to the "electronic" Schrödinger Eq. (1-4) at the nuclear equilibrium conformation $\mathbf{Q}^0$. The nuclear part $\chi''(\xi)$ is a function of the $3N-6$ internal molecular vibrations described by the symmetry-adapted vibrational coordinates ξ. In the excited states $|j\rangle = \psi_j^0(\mathbf{r})\chi'_{v'}(\xi)$ the electronic functions are solutions to the same "electronic" hamiltonian as the one used for the ground state. The electronic transition moment $\mathbf{D}_{0j}$ of Eq. (4-102) is therefore evaluated at the equilibrium nuclear configuration of the ground state. This approximation constitutes the Franck–Condon principle. The physical basis for this approximation is that the time of the electronic transition, about 10^{-16} sec, is much smaller than the time of a nuclear vibration, about 10^{-14} sec. The internuclear distances do not therefore change during the electronic transition. Such a transition is called a *vertical electronic transition*, because it is seen to occur vertically, upwards or downwards, between two potential-energy surfaces (Fig. 4-4).

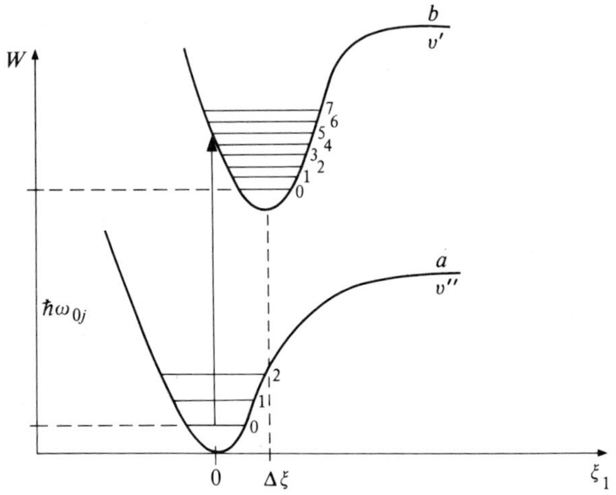

Figure 4-4 Vertical transition from the lowest vibrational level $v'' = 0$ in the ground state to $v' = 5$ in the electronically excited state.

The electronic transitions take place between the various vibrational levels, labelled $v'' = 0, 1, 2, \ldots$, of the lower electronic state and the various vibrational levels, labelled $v' = 0, 1, 2, \ldots$, of the upper electronic state. The total electronic intensity given by Eq. (4-101) is therefore distributed over the vibrational line structure of the absorption band. Using the adiabatic molecular wave functions we get for the transition moment of an allowed transition

$$\mathbf{D}_{0j} = \langle \psi_j^0(\mathbf{r}) | \sum_n \mathbf{r}_n | \psi_0^0(\mathbf{r}) \rangle \langle \chi''_{v''}(\xi) | \chi'_{v'}(\xi) \rangle \tag{4-103}$$

With

$$\chi(\xi) = \prod_{t=1}^{3N-6} \chi_v(\xi_t) \tag{4-104}$$

we find

$$\mathbf{D}_{0j} = \langle \psi_j^0(\mathbf{r}) | \sum_n \mathbf{r}_n | \psi_0^0(\mathbf{r}) \rangle \langle \chi_{v'}(\xi_1) | \chi_{v''}(\xi_1) \rangle \cdots \langle \chi_{v'}(\xi_{3N-6}) | \chi_{v''}(\xi_{3N-6}) \rangle \tag{4-105}$$

We observe that $\mathbf{D}_{0j}$ is made up of one factor depending upon the electronic motions, and $3N - 6$ factors having the forms of vibrational overlap integrals. The squares of the vibrational overlap integrals which appear in the expression Eq. (4-101) for the band intensity are often referred to as the Franck–Condon factors.

Plotting the adiabatic potential curves, states with different electronic configurations will have potential minima at different nuclear configurations. Figure 4-4 pictures a situation in which we have cut the potential hypersurface $W(\xi)$ with a plane containing a totally symmetric vibrational coordinate ξ_1. The vertical transition starts out from the lowest vibrational level, $v'' = 0$, and terminates on the excited state, $v' = 5$. The frequency of this transition is, of course, simply given by the energy difference $\hbar\omega_{0j}$ of the pure electronic transition plus five quanta of the vibrational frequency $\hbar\omega'_1$ associated with the excited state.

In practice there are various reasons why the observation of the individual lines may be prevented. A limitation on resolution is of course set by the finite slit width. More important in our case is the inherent line broadening brought about by the electronic relaxation phenomena dealt with in section 4-4. The absorption spectrum is also usually observed with the molecule or ion imbedded in a host lattice. The interaction of the guest with the host causes each molecular line to be associated with a manifold of low fundamental lattice vibrations. This so-called phonon spectrum may give each line a considerable width. Finally, a simple overcrowding of lines in a certain spectral range may wipe out the vibrational structure.

Dealing first with the trivial diffuseness due to the overcrowding of lines[5] we observe that the number of vibrational states associated with the ground potential surface rises steeply as a function of W. Indeed to good approximation the total number of states $f(W)$ goes as $\exp(\alpha W)$. The vibrational state density $df(W)/dW = \rho(W)$ is then $\alpha \exp(\alpha W)$. The number of transitions observed in the interval $\hbar\omega_{0j}$ to $\hbar\omega_{0j} + d(\hbar\omega_{0j})$ is therefore proportional to

$$\alpha \exp(\alpha W)\, d(\hbar\omega_{0j}) \exp\left(-\frac{W}{k_B T}\right) = \alpha\, d(\hbar\omega_{0j}) \exp\left(\alpha - \frac{1}{k_B T}\right) W \qquad (4\text{-}106)$$

Clearly this number becomes independent of W for $T_s = (\alpha k_B)^{-1}$. The implications are that for $T > T_s$ each line group will spread out over the entire spectrum. By going to low temperature one may, on the other hand, avoid the overcrowding.

The electric dipole moment operator in Eq. (4-102) transforms as (X, Y, Z) in the point group of the molecule. The selection rule for an allowed electric dipole transition is then that the direct product $\Gamma(\psi_0^0) \times \Gamma(X, Y, Z) \times \Gamma(\psi_j^0)$ should contain at least one component which transforms as a totally symmetric representation in the point group of the molecule. In the case where the two electronic states have different equilibrium conformation, only those symmetry elements which are common to both point groups should be considered.

In order to calculate the electronic band shape, we shall first deal with the case of an allowed electric dipole transition. We assume that the molecule retains its shape in the excited state. It follows from this assumption that excitations in a non-totally symmetric vibrational frequency can only be associated with vibrational quantum number changes of 0, 2, 4, ... The proof makes use of the properties of the Hermite polynomials, Eq. (1-36), since by symmetry $\langle \chi_{v'}(\xi_i) \,|\, \chi_{v''}(\xi_i) \rangle$ equals zero if the integral contains an odd power in a non-totally symmetric coordinate ξ_i.

Consider now the Franck–Condon factors for a transition when only a totally symmetric vibration with the coordinate ξ_1 is excited. For such a vibration all of the vibrational wave functions are totally symmetric and the matrix element $\langle \chi_{v'}(\xi_1) \,|\, \chi_{v''}(\xi_1) \rangle$ can therefore not be zero due to symmetry. Hence the vibrational quantum number may change by any number of quanta during the electronic transition.

Let us assume that the transition starts out as in Fig. 4-4 from the zero vibrational level of the ground state. The difference between the minima of the two potential curves a and b is called $\Delta\xi$, and we shall assume that both potentials are harmonic. The Franck–Condon factors are then given by the squares of the integral

$$\langle \chi_0 \,|\, \chi_v \rangle = \int \chi_0(\xi, \omega'')\chi_v(\xi + \Delta\xi, \omega')\, d\xi$$

The two values of the angular vibrational frequency, ω'' and ω', reflect that the vibrational frequencies need not be the same for the two potential surfaces a and b.

The wave function for a harmonic oscillator is for instance given in Schiff[1]

$$\chi_v = \sqrt{\left(\sqrt{\frac{\alpha}{\pi}}\frac{1}{2^v v!}\right)}\exp\left(-\tfrac{1}{2}\alpha\xi^2\right) H_v(\sqrt{\alpha}\,\xi)$$

where H_v is a Hermite polynomial, $\alpha = M\omega/\hbar$, and M the effective mass of the vibration. We can write after a little manipulation

$$H_v[\sqrt{\alpha'}(x + \Delta x)] = H_v\left[\sqrt{\frac{2\alpha'}{\alpha' + \alpha''}}\sqrt{\frac{\alpha' + \alpha''}{2}}\left(x + \frac{\alpha'}{\alpha' + \alpha''}\Delta x\right) + \sqrt{\frac{\alpha'' - \alpha'}{\alpha'' + \alpha'}}\frac{\alpha''}{\alpha' + \alpha''}\sqrt{\frac{\alpha'(\alpha'' + \alpha')}{\alpha'' - \alpha'}}\Delta x\right]$$

Using the general addition formula for the Hermite polynomials[12] we have

$$H_v[\sqrt{\alpha'}(x + \Delta x)] = \left(\frac{\alpha'' - \alpha'}{\alpha'' + \alpha'}\right)^{v/2} H_0\left[\sqrt{\frac{\alpha' + \alpha''}{2}}\left(x + \frac{\alpha'}{\alpha' + \alpha''}\Delta x\right)\right] \times H_v\left[\frac{\alpha''}{\alpha' + \alpha''}\sqrt{\frac{\alpha'(\alpha' + \alpha'')}{\alpha'' - \alpha'}}\Delta x\right] + \text{higher terms } H_j H_{v-j},\, j \neq 0$$

We find easily by direct evaluation

$$\langle \chi_0 \,|\, \chi_0 \rangle = \exp\left[-\frac{\alpha'\alpha''}{2(\alpha' + \alpha'')}(\Delta\xi)^2\right]\sqrt{\frac{\sqrt{\alpha'\alpha''}}{\frac{1}{2}(\alpha' + \alpha'')}} \tag{4-107}$$

and using the orthogonality properties of χ

$$\langle \chi_0 \,|\, \chi_{v'} \rangle = \langle \chi_0 \,|\, \chi_0 \rangle \sqrt{\frac{1}{2^{v'} v'!}}\left(\frac{\alpha'' - \alpha'}{\alpha'' + \alpha'}\right)^{v'/2} H_{v'}\left[\frac{\alpha''}{\alpha'' + \alpha'}\sqrt{\frac{\alpha'(\alpha'' + \alpha')}{\alpha'' - \alpha'}}\Delta\xi\right] \tag{4-108}$$

In the case of $\alpha'' = \alpha'$, Eq. (4-108) reduces to

$$\langle \chi_0 \,|\, \chi_{v'} \rangle = \langle \chi_0 \,|\, \chi_0 \rangle \sqrt{\frac{(\alpha')^{v'}}{2^{v'} v'!}}(\Delta\xi)^{v'}$$

The intensity of a line $|a0\rangle \to |bv''\rangle$ is proportional to $|\langle \chi_0 \,|\, \chi_{v'} \rangle|^2$. Defining

$$S_1 = \tfrac{1}{2}\alpha_1'(\Delta\xi_1)^2 \tag{4-109}$$

we find for the intensity $\mathfrak{I}(a0'' \to bv')$

$$\mathfrak{I}(a0'' \to bv') \propto \exp(-S_1) \frac{S_1^{v'}}{v'!} \tag{4-110}$$

In the electronic ground state, the totally symmetric zero-point vibration is governed by the wave function $\chi_0(\xi, \omega'')$. The magnitude of displacement from the equilibrium position is given by

$$\bar{\xi} = [\langle \chi_0(\xi, \omega'') | \, \xi^2 \, | \chi_0(\xi, \omega'') \rangle]^{1/2}$$

or,

$$\bar{\xi} = \frac{1}{\sqrt{2\alpha''}}$$

Substituting for α'' in $S = \frac{1}{2}\alpha''(\Delta\xi)^2$ leads to $S = (\Delta\xi/2\bar{\xi})^2$.

The intensity distributions for S equal to 1 and 4 have been pictured in Fig. 4-5. Evidently, if $\Delta\xi$ is large compared with $2\bar{\xi}$, the maximum line intensity will fall on a high vibrational quantum number in the excited state. For reasonably large values of S this can easily be found to occur at $v'_{\max} \approx S$.

If we expand $\mathfrak{I}(a0'' \to bv')$ around $\mathfrak{I}_{\max}$ in a Taylor series, we get

$$\mathfrak{I}(a0'' \to bv') = \mathfrak{I}_{\max} \exp\left[-(v_{\max} - v')^2/2v_{\max}\right]$$

Notice that the absorption band will have the shape of a gaussian curve.

The distribution of intensity over the vibrational levels can also be calculated in the case where the transitions originate from a set of vibrational levels populated according to the Boltzmann distribution. The calculation can be carried through in closed form provided $\omega' = \omega''$. Let the ground state vibrational wave functions $\chi_{v''}(\xi_1 - \frac{1}{2}\Delta\xi_1)$ be centered at $+\frac{1}{2}\Delta\xi_1$ and the excited state vibrational wave functions $\chi_{v'}(\xi_1 + \frac{1}{2}\Delta\xi_1)$ at $-\frac{1}{2}\Delta\xi_1$. The intensity distribution $\mathfrak{I}(av'' \to bv')$ with $\omega = \omega' = \omega''$ is determined by

$$|\langle \chi_{v''}(\xi_1 - \tfrac{1}{2}\Delta\xi_1) \,|\, \chi_{v'}(\xi_1 + \tfrac{1}{2}\Delta\xi_1) \rangle|^2$$

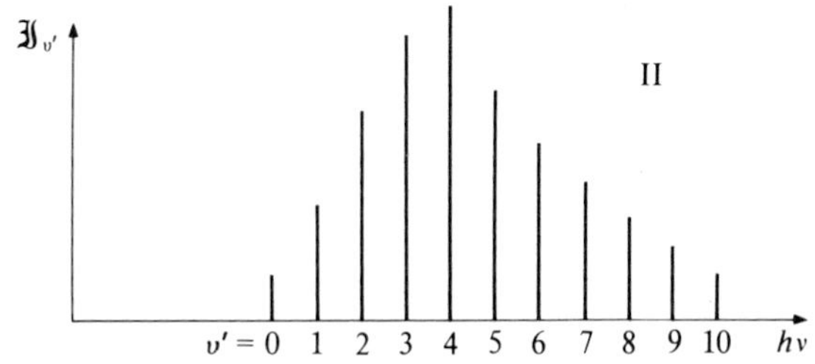

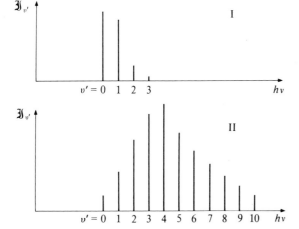

Figure 4-5 The intensity distributions for a transition $v'' = 0 \to v'$ for (I) $\Delta\xi/2\bar{\xi} = 1$ and (II) $\Delta\xi/2\bar{\xi} = 2$.

Using the relation[12]

$$H_n(x+y) = \sum_{t=0}^{n} (2y)^{n-t} \binom{n}{t} H_t(x)$$

we get easily by utilizing the orthogonality of the Hermite polynomials with $v' \geq v''$

$$\langle \chi_{v''}(\xi_1 - \tfrac{1}{2}\Delta\xi_1, \omega) \,|\, \chi_{v'}(\xi_1 + \tfrac{1}{2}\Delta\xi_1, \omega)\rangle = \exp(-\tfrac{1}{2}S)\, S^{(v'+v'')/2}$$

$$\times \sqrt{\frac{1}{v''!\,v'!}} \sum_{u=0}^{v''} (-1)^{(v''-u)} S^{-u} u! \binom{v'}{u}\binom{v''}{u} \qquad (4\text{-}111)$$

where S as before equals $\frac{1}{2}\alpha(\Delta\xi_1)^2$ and $\alpha = M\omega/\hbar$.

A slight rearrangement turns Eq. (4-111) into

$$\langle \chi_{v''} \,|\, \chi_{v'}\rangle = \exp(-\tfrac{1}{2}S)\, S^{(v'+v'')/2} \sqrt{\frac{v''!}{v'!}} \sum_{u=0}^{v''} \frac{1}{u!} \binom{v'}{v''-u} (-S)^u$$

The collection of the mathematical formulae known as the "Bateman manuscript"[12] gives us the relation

$$L_n^{(\alpha)}(x) = \sum_{k=0}^{n} \binom{n+\alpha}{n-k} \frac{(-x)^k}{k!}$$

where $L_n^{(\alpha)}(x)$ is a Laguerre polynomial. Using Eq. (4-111) we get then for the Franck–Condon factors

$$|\langle \chi_{v''} \,|\, \chi_{v'}\rangle|^2 = \exp(-S)\, S^{v'-v''} \frac{v''!}{v'!} [L_{v''}^{(v'-v'')}(S)]^2 \qquad (4\text{-}112)$$

The initial vibrational states are now each weighted with the Boltzmann factor $\exp[-\hbar\omega v''/k_B T]$. The total intensity at the energy difference $h\nu_{ab} + \hbar\omega(v' - v'')$, where $h\nu_{ab}$ is the energy difference of the pure electronic states, is then given by

$$|\mathbf{D}_{ba}^0|^2 \frac{\displaystyle\sum_{v''=0}^{\infty} \exp\left(-\frac{\hbar\omega}{k_B T} v''\right) \exp(-S) S^p \frac{v''!}{(p+v'')!} [L_{v''}^{(p)}(S)]^2}{\displaystyle\sum_{v''=0}^{\infty} \exp\left(-\frac{\hbar\omega}{k_B T} v''\right)} \qquad (4\text{-}113)$$

where $p = v' - v''$ and $\mathbf{D}_{ba}^0$ is the electronic transition moment of Eq. (4-102) evaluated at the equilibrium configuration of the ground state. Turning once more to the Bateman manuscript[12] we find the formula

$$\sum_{n=0}^{\infty} \frac{n!}{(n+\alpha)!} z^n [L_n^{(\alpha)}(x)]^2 = \frac{1}{1-z} \exp\left(-\frac{2xz}{1-z}\right) (zx^2)^{-\alpha/2} I_\alpha\left(\frac{2x\sqrt{z}}{1-z}\right)$$

where I_α is a modified Bessel function. We find, identifying $z = \exp[-\hbar\omega/k_B T]$, that Eq. (4-113) equals

$$|\mathbf{D}_{ba}^0|^2 z^{-p/2} \exp\left(-S\frac{1+z}{1-z}\right) I_p\left(2S\frac{\sqrt{z}}{1-z}\right) \qquad (4\text{-}114)$$

or for the individual absorption lines

$$\mathfrak{I}(av'' \to bv') = |\mathbf{D}^0_{ba}|^2 \exp\left(-S\coth\frac{\hbar\omega}{2k_BT}\right)\exp\left(\frac{p\hbar\omega}{2k_BT}\right)I_p\left(S\operatorname{csch}\frac{\hbar\omega}{2k_BT}\right)$$
$$\times\, \delta(h\nu_{ab} + p\hbar\omega - h\nu) \qquad (4\text{-}115)$$

where $\operatorname{csch}\hbar\omega/2k_BT = (\sinh\hbar\omega/2k_BT)^{-1}$.

It is sometimes convenient to introduce the average quantum number $\langle\bar{v}\rangle_T$ of a mode giving the number of phonons in a certain mode at temperature T. We have

$$\langle\bar{v}\rangle_T = \frac{\sum_{v=0}^{\infty} v\exp\left(-\frac{(v+\frac{1}{2})\hbar\omega}{k_BT}\right)}{\sum_{v=0}^{\infty}\exp\left(-\frac{(v+\frac{1}{2})\hbar\omega}{k_BT}\right)} = \frac{1}{\exp\left(\frac{\hbar\omega}{k_BT}\right)-1}$$

For $T \to 0$ we find $\langle\bar{v}\rangle_T \to 0$. Introducing this quantity, Eq. (4-115) is turned into

$$\mathfrak{I}(h\nu) = |\mathbf{D}^0_{ba}|^2\exp[-S(1+2\langle\bar{v}\rangle_T)]\left(\frac{\langle\bar{v}\rangle_T+1}{\langle\bar{v}\rangle_T}\right)^{p/2}$$
$$\times\, I_p\left[2S\sqrt{\langle\bar{v}\rangle_T(\langle\bar{v}\rangle_T+1)}\right]\delta(h\nu_{ab}+p\hbar\omega-h\nu) \qquad (4\text{-}116)$$

In the high-temperature limit for which

$$y = S\operatorname{csch}\frac{\hbar\omega}{2k_BT} \gg 1$$

we may use the following expansion for $I_p(y)$

$$I_p(y) \approx \exp(y)\frac{1}{\sqrt{2\pi y}}\exp\left(-\frac{p^2}{2y}\right)$$

The maximum intensity will occur for $d\mathfrak{I}(h\nu)/dp = 0$ or $p_{\max} = y\hbar\omega/2kT$. This result enables us to write Eq. (4-115) as

$$\mathfrak{I}(av'' \to bv') = \mathfrak{I}_{\max}\exp[-(h\nu - h\nu_{\max})^2/2y(\hbar\omega)^2] \qquad (4\text{-}117)$$

The intensity distribution is seen to be gaussian. In the low-temperature limit $\langle\bar{v}\rangle_T \to 0$ and a valid expansion of $I_p(y)$ for small y is

$$I_p(y) \approx \frac{(y/2)^p}{p!} \qquad (4\text{-}118)$$

Insertion of this expression in (4-116) gives us back the Poisson distribution of Eq. (4-109), $\exp(-S)S^p/p!$. In the limit where $\langle\bar{v}\rangle_T = 0$ we observe that p can be identified with v'.

The above derivations of the band shape are valid for a band profile consisting of excitations of one totally symmetric mode in an electronically allowed transition. However, the adiabatic potential depends on all the normal modes. These will also contribute to the band width; however, their influence will generally be much smaller than that of the totally symmetric modes. From our assumption that the symmetry of the molecule is preserved in the excited state, the non-totally symmetric frequencies are, as we have already observed, associated with vibrational quantum number changes of 0, 2, 4, ... Hence, only provided the non-totally symmetric vibrational frequencies associated with the two potential surfaces differ, can a 0–2 line acquire intensity. Using Eq. (4-107) with $(\Delta\xi) = 0$ it is easy to show that the intensity of the $0'' \to 0'$ line to the sum of all the line intensities $0'' \to 2v'$ is given by

$$\frac{\Im(0'' \to 0')}{\sum_{v'=0}^{\infty} \Im(0'' \to 2v')} = \frac{\sqrt{\omega'\omega''}}{\frac{1}{2}(\omega'' + \omega')} \tag{4-119}$$

which, with ω' being not too different from ω'', is seen to collect virtually all of the intensity in the 0–0 line.

In general, the intensity of a $v \to v + 2$ transition is proportional to

$$|\langle \chi_v(\sqrt{\alpha''}\,\xi)\,|\,\chi_{v+2}(\sqrt{\alpha'}\,\xi)\rangle|^2$$

Provided ω'' is not too different from ω', we can expand $\chi_v(\sqrt{\alpha'}\,\xi)$ on the set $\chi_v(\sqrt{\alpha''}\,\xi)$. Retaining only first orders in the expansion parameter $(\omega'' - \omega')/\omega'$ the intensity of a $v \to v + 2$ transition is found to be proportional to $(v + 2)(v + 1)$. For low ($< 100\ \text{cm}^{-1}$) non-totally symmetric vibrations, the higher vibrational levels can be populated even at quite low temperatures. Additional lines may thus appear in the spectrum corresponding to excitations in quanta of two in a non-totally symmetric mode. Built upon such lines we may further find a progression in single quanta of a totally symmetric mode. This last situation is referred to as a *false vibronic origin*. However, in general all we expect to see, provided the excited electronic state of the molecule has the same molecular shape as the ground state, is a progression in single quanta of the totally symmetric mode(s).

A molecule which is electronically excited does, however, in many cases possess an equilibrium conformation with a shape different from that of the ground state. Evidently, there will be a distorting non-totally symmetric mode which will carry the molecule to the new conformation. This mode will therefore in the point group of the new equilibrium conformation transform as a totally symmetric vibration, and the "distorting" vibration will behave in the same way as a "normal" totally symmetric vibration. We expect, therefore, in this situation to see a progression in single quanta of the distorting mode.

Thus, the observation of a vibrational progression in an electronic spectrum provides information on the relative geometric shapes of the two electronic states involved. It requires only that we can identify the occurring vibration, and this is usually done by a comparison with the vibrational frequencies found by an infrared or Raman analysis of the ground state. In practice, this means that the vibrational frequencies in the ground and excited states must not be too different, since otherwise we may not be able to make an identification. If possible the measurements of isotopic ratios in the vibrational frequencies are always of great value for the correct identification of vibrations.[27]

4-8 GUEST-HOST COUPLING

For a molecule or ion imbedded in a matrix, two sets of local modes are coupled to the electronic states of the impurity; the vibrations of the guest and the lattice modes of the host. The vibrations of the guest are very much higher in frequency than the lattice modes. The gross shape of the absorption spectrum of the guest is therefore determined by the molecular vibrations, but the shape of the individual absorption lines is governed by the coupling to the lattice phonons.[13]

We shall assume that no symmetry restrictions are imposed upon the excitation of the lattice vibrations. Let the collective vibrations of the N nuclei of the host lattice be described by $3N - 6$ uncoupled harmonically oscillating modes, Q_J. The eigenvalue equation is

$$[T(Q) + W_j(Q)]\chi_{jv}(Q) = w_v \chi_{jv}(Q)$$

where $W_j(Q)$ is the adiabatic potential in which the lattice nuclei move when the guest molecule is in the electronic state j. We now take the harmonic modes of the ground state, described by $\chi_{0v''}(Q_J)$ and of the excited states, $\chi_{jv'}(Q_J)$, to have the same frequencies and to differ only in their equilibrium positions of the nuclei. The shape function of an absorption line $0 \to j$ is given by

$$G(\omega) = \mathrm{Av}_{v''} \sum_{v'} |\langle v' | v'' \rangle|^2 \, \delta\left(\omega - \omega_{0j} - \sum_J \omega_J (v' - v'')\right) \tag{4-120}$$

$\mathrm{Av}_{v''}$ is understood to be an average over the initial phonon states weighted with the Boltzmann factors. The phonon Franck–Condon factors for the transition $0 \to j$ are

$$|\langle v' | v'' \rangle|^2 = \left| \prod_J \langle v'_J | v''_J \rangle \right|^2$$

The shape function $G_p(\omega)$ as a function of the number of absorbed phonons, $p = v' - v'' \geq 0$, is given by Eq. (4-116)

$$G_p(\omega) = \prod_J \exp\left[-S_J(1 + 2\langle \bar{v}''_J \rangle_T)\right] \left(\frac{\langle \bar{v}''_J \rangle_T + 1}{\langle \bar{v}''_J \rangle_T} \right)^{p/2}$$
$$\times I_p\left(2S_J \sqrt{\langle \bar{v}''_J \rangle_T [\langle \bar{v}''_J \rangle_T + 1]}\right) \delta\left(\omega - \omega_{0j} - \sum_J \omega_J p\right) \tag{4-121}$$

For small values of $2S_J \sqrt{\langle v''_J \rangle_T (\langle \bar{v}''_J \rangle_T + 1)}$ we may expand Eq. (4-121) and get

$$G_p(\omega) = \prod_J \exp\left[-S_J(1 + 2\langle \bar{v}''_J \rangle_T)\right] \frac{(\langle \bar{v}''_J \rangle_T + 1)^p S_J^p}{p!} \delta\left(\omega - \omega_{0j} - \sum_J \omega_J p\right) \tag{4-122}$$

In case $v'' - v' = p \geq 0$ the formula (4-122) is

$$G_{-p}(\omega) = \prod_J \exp\left[-S_J(1 + 2\langle \bar{v}''_J \rangle_T)\right] \frac{(\langle \bar{v}''_J \rangle_T)^p S_J^p}{p!} \delta\left(\omega - \omega_{0j} + \sum_J \omega_J p\right) \tag{4-123}$$

For the no-phonon line, $p = 0$ for all ω_J. Then $G_0(\omega)$ assumes the form

$$G_0(\omega) = \prod_J \exp\left[-S_J(1 + 2\langle\bar{v}_J''\rangle_T)\right]\delta(\omega - \omega_{0j})$$

$$= \exp\left[-\sum_J S_J(1 + 2\langle\bar{v}_J''\rangle_T)\right]\delta(\omega - \omega_{0j}) \qquad (4\text{-}124)$$

Defining

$$S = \sum_J S_J(1 + 2\langle\bar{v}_J''\rangle_T) \qquad (4\text{-}125)$$

we get

$$G_0(\omega) = \exp(-S)\,\delta(\omega - \omega_{0j}) \qquad (4\text{-}126)$$

The no-phonon line is therefore a δ function, with an intensity which depends strongly on temperature; the lower the temperature, the more intense is the line.

Consider now the absorption of one quantum of one of the $3J - 6$ phonons. The absorption probability is proportional to

$$\exp\left[-S_J(1 + 2\langle\bar{v}_J''\rangle_T)\right](1 + \langle\bar{v}_J''\rangle_T)S_J\,\delta(\omega - \omega_{0j} - \omega_J)$$

Since any of the $3J - 6$ phonons may be absorbed we find

$$G_1(\omega) = \exp(-S)\sum_J (1 + \langle\bar{v}_J''\rangle_T)S_J\,\delta(\omega - \omega_{0j} - \omega_J) \qquad (4\text{-}127)$$

The density of phonon states in the host crystal is of course very great. We can therefore replace the summation in Eq. (4-127) by integration over the density function $\Xi(\omega)$. Hence

$$G_1(\omega) = \exp(-S)\int_0^{\omega_{\max}} \Xi(\omega)\left[1 + \langle\bar{v}_J''\rangle_T\right]S(\omega)\,d\omega \qquad (4\text{-}128)$$

Equation (4-128) indicates that the immediate high-frequency side of the no-phonon line should show an intensity distribution corresponding to a weighted phonon density. The intensity distribution on the low-frequency side is given by

$$G_{-1}(\omega) = \exp(-S)\int_0^{\omega_{\max}} \Xi(\omega)\left[\langle\bar{v}_J''\rangle_T\right]S(\omega)\,d\omega \qquad (4\text{-}129)$$

At very low temperatures the multiphonon processes will therefore contribute to the high-energy side of the no-phonon line. The absorption envelope will become broader and broader the more phonons are absorbed. As the temperature increases the absorption spectrum extends to the low-frequency side of the no-phonon line, and at the same time that line goes down in intensity. The stronger the guest–host coupling is, the less likely is it therefore to observe molecular-vibrational fine structure even at quite low temperatures.

4-9 INTENSITY DISTRIBUTION IN AN ORBITALLY "FORBIDDEN" TRANSITION

We want now to investigate the case where the electronic dipole transition moment

$$\mathbf{D}_{0j} = \langle \psi_0^0(\mathbf{r}) | \sum_n \mathbf{r}_n | \psi_j^0(\mathbf{r}) \rangle$$

as evaluated in the adiabatic approximation is equal to zero for symmetry reasons. This situation is referred to as an *orbitally forbidden* transition. In order to see how such electronic transitions may gain intensity, we expand the "electronic" hamiltonian Eq. (1-3) in a Taylor series around the equilibrium point of the molecular ground state $\mathbf{Q}^0$:

$$\mathscr{H}_E = \mathscr{H}_E^0 + \sum_i \left(\frac{\partial V}{\partial \xi_i} \right)_0 \xi_i + \cdots$$

In this expansion, ξ_i represents a vibrational symmetry coordinate of species Γ_i. Since the potential energy V transforms like the totally symmetric representation in the point group of the molecule, we observe that $(\partial V/\partial \xi_i)_0$ must transform like ξ_i in order for the product $(\partial V/\partial \xi_i)_0 \xi_i$ to transform like a totally symmetric representation Γ_1.

Using first-order perturbation theory we take

$$\mathscr{H}^{(1)} = \mathscr{H}_E - \mathscr{H}_E^0 = \sum_i \left(\frac{\partial V}{\partial \xi_i} \right)_0 \xi_i \tag{4-130}$$

As zero-order electronic wave functions we take the set $\psi_j^0(\mathbf{r})$ found by solving $\mathscr{H}_E^0 \psi_j^0(\mathbf{r}) = W_j^0 \psi_j^0(\mathbf{r})$. The perturbed electronic functions in a Herzberg–Teller scheme (section 1-3) are then

$$\psi_j(\mathbf{r}, \xi) = \psi_j^0(\mathbf{r}) + \sum_{j \neq m} a_{m,j}(\xi) \psi_m^0(\mathbf{r}) \tag{4-131}$$

with

$$a_{m,j}(\xi) = \frac{1}{W_m^0 - W_j^0} \sum_i \langle \psi_j^0(\mathbf{r}) | \left(\frac{\partial V}{\partial \xi_i} \right)_0 | \psi_m^0(\mathbf{r}) \rangle \, \xi_i \tag{4-132}$$

Let us assume that we have a molecule with three well-separated states, $\psi_0^0(\mathbf{r})$ with electronic energy W_0^0, $\psi_m^0(\mathbf{r})$ with energy W_m^0, and $\psi_j^0(\mathbf{r})$ with energy W_j^0. Further

$$\mathbf{D}_{0j} = \langle \psi_0^0(\mathbf{r}) | \sum_n \mathbf{r}_n | \psi_j^0(\mathbf{r}) \rangle = 0 \tag{4-133}$$

but

$$\mathbf{D}_{0m} = \langle \psi_0^0(\mathbf{r}) | \sum_n \mathbf{r}_n | \psi_m^0(\mathbf{r}) \rangle \neq 0 \tag{4-134}$$

Provided therefore that ψ_j and ψ_m can mix under the vibronic perturbation operator of Eq. (4-130), we find – to first order – a transition moment for the symmetry forbidden transition $|0\rangle \rightarrow |j\rangle$ equal to

$$\mathbf{D}_{0j}(\xi) = a_{m,j}(\xi)\mathbf{D}_{0m}$$

The intensity of the "forbidden" electronic transition $|0\rangle \rightarrow |j\rangle$ will evidently be scaled down from that of an allowed transition by the factor $|\langle \chi_{v''}| a_{m,j}(\xi) | \chi_{v'} \rangle|^2$. With reasonable orders of magnitude for the various molecular quantities appearing in the expression Eq. (4-132) for $a_{m,j}$, this will give a "symmetry-forbidden" transition an intensity of about 10^{-3} times that of a "symmetry-allowed" transition.

From a group-theoretical point of view, we must evidently have from Eq. (4-132)

$$\Gamma(\psi_m^0) \times \Gamma(\xi_i) \supset \Gamma(\psi_j^0) \tag{4-135}$$

where we have recalled that the electronic operator $(\partial V/\partial \xi_i)_0$ transforms like ξ_i. From Eq. (4-134) we get

$$\Gamma(\psi_0^0) \times \Gamma(X, Y, Z) \supset \Gamma(\psi_m^0) \tag{4-136}$$

Eliminating the intermediate state $\psi_m^0(\mathbf{r})$ from which the intensity is "stolen," Eqs. (4-135) and (4-136) can be combined leading to the general "vibronic" selection rule for a symmetry-forbidden transition $|0\rangle \rightarrow |j\rangle$

$$\Gamma(\psi_0^0) \times \Gamma(X, Y, Z) \times \Gamma(\xi_i) \times \Gamma(\psi_j^0) \supset \Gamma_1 \tag{4-137}$$

where Γ_1 is the totally symmetric representation in the molecular point group corresponding to the conformation $\mathbf{Q}^0$.

As an example we now assume that the above molecule possesses only two vibrational coordinates. The vibrational symmetry coordinate ξ_1 is totally symmetric, while ξ_2 is a onefold degenerate non-totally symmetric vibrational symmetry coordinate. The "vibronic" mixing coefficient is given by

$$c_{m,j} = \langle \psi_m^0(\mathbf{r}) | \left(\frac{\partial V}{\partial \xi_2} \right)_0 | \psi_j^0(\mathbf{r}) \rangle \neq 0$$

With these simplifying assumptions, which in no way cause our conclusions to be less general, the perturbed wave function $\Psi_j(\mathbf{r}, \xi)$ equals

$$\Psi_j = \left[\psi_j^0(\mathbf{r}) + \frac{c_{m,j}\xi_2}{W_m^0 - W_j^0} \psi_m^0(\mathbf{r}) \right] \chi_{v_1'}(\xi_1)\chi_{v_2'}(\xi_2)$$

The integrals over the vibrational coordinates for $|0\rangle \rightarrow |j\rangle$ are then

$$\langle \chi_{v_1''} | \chi_{v_1'} \rangle \langle \chi_{v_2''} | \xi_2 | \chi_{v_2'} \rangle$$

Assuming the vibrational wave functions in ξ_2 to be solutions to the same harmonic potential function, we get from the second of these matrix elements that $v_2'' = v_2' \pm 1$ if we are to get a value different from zero. We must therefore excite one quantum of a non-totally symmetric vibration in order to get a transition probability different from zero; hence the name "vibronic" transition. The totally symmetric vibration associated with the transition may, on the other hand, change by any number of quanta.

In an orbitally forbidden transition we will, therefore, not see the $|0, v_1'' = 0, v_2'' = 0, \ldots, v_{3N-6}'' = 0\rangle \rightarrow |j, v_1' = 0, v_2^1 = 0, \ldots, v_{3N-6}' = 0\rangle$ line; the first observed line will correspond to an excitation of one quantum of a non-totally symmetric vibration on which progressions in the totally symmetric vibration(s) can be built. If more than one non-totally symmetric vibration can introduce intensity into an otherwise forbidden transition, we will have as many "false" origins, followed by progressions in the totally symmetric vibration(s), as we have intensity-introducing vibrations.

It is a characteristic feature of a "vibronic" transition that the total integrated band intensity is temperature dependent. In this, a "vibronic" transition differs from an allowed transition. In the latter case the intensity may shift around inside the band when the temperature is altered, but the integrated band intensity remains constant under temperature variations. This can be seen as follows.

Consider the energy levels of a totally symmetric harmonic vibration, associated with the electronic ground state. For a particular vibrational level v'' the number of molecules present is proportional to the Boltzmann factor $\exp(-\hbar\omega v''/k_B T)$, where ω is the angular frequency of the vibration.

The intensity associated with a progression of lines starting out from v'' is then proportional to

$$|\mathbf{D}_{0j}|^2 \frac{\exp\left(-\dfrac{\hbar\omega v''}{k_B T}\right) \sum_{v'=0}^{\infty} \left[\int \chi_{v''}(\xi)\chi_{v'}(\xi)\,d\xi\right]^2}{\sum_{v''=0}^{\infty} \exp\left(-\dfrac{\hbar\omega v''}{k_B T}\right)}$$

and for the total of all lines we have

$$|\mathbf{D}_{0j}|^2 \frac{\sum_{v''=0}^{\infty} \exp\left(-\dfrac{\hbar\omega v''}{k_B T}\right) \sum_{v'=0}^{\infty} \left[\int \chi_{v''}(\xi)\chi_{v'}(\xi)\,d\xi\right]^2}{\sum_{v''=0}^{\infty} \exp\left(-\dfrac{\hbar\omega v''}{k_B T}\right)} \tag{4-138}$$

Expanding $\chi_{v''}(\xi)$ on the complete orthonormal set $\chi_{v'}(\xi)$

$$\chi_{v''}(\xi) = \sum_{v'=0}^{\infty} a_{v',v''}\,\chi_{v'}(\xi)$$

The coefficients $a_{v',v''}$ obey

$$1 = \sum_{v'=0}^{\infty} (a_{v',v''})^2$$

Hence

$$\sum_{v'=0}^{\infty} \left[\int \chi_{v''}(\xi)\chi_{v'}(\xi)\,d\xi\right]^2 = 1$$

and the total intensity is then, according to Eq. (4-138), given by $|\mathbf{D}_{0j}|^2$, and is independent of temperature.

On the other hand, for an orbitally forbidden band, the intensity to be distributed is governed by integrals of the type $\langle \chi_{v''}(\xi)|\,\xi\,|\chi_{v'}(\xi)\rangle$. Let us assume that the force constants and equilibrium positions are the same in the electronic ground state $|0\rangle$ and in the excited state $|j\rangle$. The integral is then only different from zero provided $v'' = v' \pm 1$ and evaluation shows it to be proportional to the square root of v'' or v', whichever is the greater.[1]

Assuming only one "active" vibration, and for simplicity dropping the index on the vibrational quantum number, we get for the total "stolen" intensity

$$\mathfrak{J} \propto \frac{\sum_{v=0}^{\infty} \exp\left(-\frac{\hbar\omega v}{k_B T}\right)\left\{\left[\int \chi_v(\xi)\xi\chi_{v+1}(\xi)\,\mathrm{d}\xi\right]^2 + \left[\int \chi_v(\xi)\xi\chi_{v-1}(\xi)\,\mathrm{d}\xi\right]^2\right\}}{\sum_{v=0}^{\infty} \exp\left(-\frac{\hbar\omega v}{k_B T}\right)}$$

At higher temperatures, therefore, the stronger $1 \rightarrow 2$, $2 \rightarrow 3, \ldots$ lines appear, corresponding to larger amplitudes of vibration and greater distortions. Writing $\exp(-\hbar\omega/k_B T) = X$, the above expression reduces to

$$\mathfrak{J} \propto \left[\int \chi_0(\xi)\xi\chi_1(\xi)\,\mathrm{d}\xi\right]^2 \frac{\sum_{v=0}^{\infty} X^v[v+1] + \sum_{v=1}^{\infty} X^v \cdot v}{\sum_{v=0}^{\infty} X^v} \tag{4-139}$$

For $X < 1$ we have

$$\sum_{v=0}^{\infty} X^v = \frac{1}{1-X} \quad \text{and} \quad \sum_{v=0}^{\infty} vX^v = \frac{X}{(1-X)^2}$$

For $T \rightarrow 0$, $X \rightarrow 0$. Introducing the total band intensity $\mathfrak{J}_0$ found when $T = 0$, we get

$$\mathfrak{J} = \mathfrak{J}_0 \frac{1+X}{1-X}$$

or substituting back for X

$$\mathfrak{J} = \mathfrak{J}_0 \coth \frac{\hbar\omega}{2k_B T} = \mathfrak{J}_0 \cdot (1 + 2\langle \bar{v} \rangle_T)$$

where $\langle \bar{v} \rangle_T$ as before is the averaged vibrational quantum number. If more than one perturbing vibration is active, we get simply[14]

$$\mathfrak{J} = \mathfrak{J}_0 \sum_i \coth \frac{\hbar\omega_i}{2k_B T} \tag{4-140}$$

The temperature variation of the total integrated intensity is seen to be particularly sensitive to the influence of low active vibrations.

We notice that in principle an investigation of the total band intensity as a function of temperature could be used to distinguish "vibronic" from "allowed" transitions. Due to experimental difficulties in estimating the integrated intensities, the results are unfortunately rarely conclusive.

A better method uses the emission properties of a molecule. Provided we can measure both the fluorescence and absorption spectrum for the same electronic transition, it is seen that for an orbitally allowed transition the $(0, 0\ldots0)'' \leftrightarrow (0, 0\ldots0)'$ line will coincide in absorption and emission. The two spectra will, to good approximation, be mirror images of each other.

For vibronically induced transitions, the $(0, 0\ldots0)'' \leftrightarrow (0, 0\ldots0)'$ line will be missing, and there will be a gap between absorption and emission spectra. The first line in the absorption spectrum will correspond to the excitation of one quantum of the "perturbing" vibration in the excited state, and the first line in the emission spectrum will have one quantum of the same vibration excited in the ground state. The gap will, therefore, be approximately equal to two quanta of this vibration, provided no "hot" bands are present.

In octahedral complexes for instance all of the ligand field bands arise from formally forbidden $g \rightarrow g$ transitions. Within the framework of a one-electron approximation Fenske[15] has, however, shown that the $t_{2g} \rightarrow e_g$ transitions cannot borrow intensity via a vibronic mechanism from a charge transfer transition ψ_u(ligand) $\rightarrow t_{2g}$(metal).

Let the "mixed" states, between which the ligand field transition takes place, be

$$|A'\rangle = |A_g\rangle + \sum_j |C_{ju}\rangle \frac{\langle C_{ju}|\mathscr{H}^{(1)}|A_g\rangle}{W_A - W_{C_j}}$$

$$|B'\rangle = |B_g\rangle + \sum_j |C_{ju}\rangle \frac{\langle C_{ju}|\mathscr{H}^{(1)}|B_g\rangle}{W_B - W_{C_j}}$$

In order to mix the even and odd electronic states we have used $\mathscr{H}^{(1)} = \sum_n \sum_i (\partial V_n/\partial \xi_{iu})_0 \, \xi_{iu}$ where the summation in n is over the electrons in the system. The dipole transition matrix element between $|A'\rangle$ and $|B'\rangle$ is

$$\langle A'|\sum_n \mathbf{r}_n|B'\rangle = \sum_j \frac{\langle A_g|\sum_n \mathbf{r}_n|C_{ju}\rangle\langle C_{ju}|\mathscr{H}^{(1)}|B_g\rangle}{W_B - W_{C_j}} + \sum_j \frac{\langle B_g|\sum_n \mathbf{r}_n|C_{ju}\rangle\langle C_{ju}|\mathscr{H}^{(1)}|A_g\rangle}{W_A - W_{C_j}} \qquad (4\text{-}141)$$

With $(\psi_u)^k$ being a filled ligand orbital, the state functions can be represented by

$$|A_g\rangle = |(\psi_u)^k (t_{2g})^l (e_g)^m|$$

$$|B_g\rangle = |(\psi_u)^k (t_{2g})^{l-1} (e_g)^{m+1}|$$

$$|C_{1u}\rangle = |(\psi_u)^{k-1} (t_{2g})^{l+1} (e_g)^m|$$

$$|C_{2u}\rangle = |(\psi_u)^{k-1} (t_{2g})^l (e_g)^{m+1}|$$

We notice that $|B_g\rangle$ and $|C_{ju}\rangle$ can only differ by the occupation of one orbital in order for Eq. (4-141) to be different from zero. Excited states of the type $|C_{1u}\rangle$ will therefore have to be ruled out, which proves Fenske's point.

4-10 SPIN-FORBIDDEN TRANSITIONS

Since the operator $\hat{G}$ of Eq. (4-25) for the electronic transition probability is independent of the electronic spin coordinates, no intensity should be associated with a transition in the cases where the two states do not have the same spin quantum number S. The presence of the spin-orbit coupling term in the hamiltonian will, however, to a greater or lesser degree break down the validity of the spin selection rule $\Delta S = 0$, and introduce intensity to a "spin-forbidden" electronic transition. The "stolen" or borrowed intensity is usually calculated by first-order perturbation theory.

Let us assume that we have a ground spin singlet state ^1X and an excited spin triplet state ^3T with energy W_T^0. At still higher energy, we have another spin singlet ^1B having the energy W_B^0. We assume further that the transition moment $\mathbf{D}_{XB}^0$ is different from zero. The "mixed" wave function for ^3T is then given by first-order perturbation theory as

$$\Psi(^3T) + \frac{\langle \Psi(^3T)|\mathscr{H}^{(1)}|\Psi(^1B)\rangle}{W_T^0 - W_B^0}\Psi(^1B)$$

where $\mathscr{H}^{(1)}$ is the spin-orbit coupling term given in Eq. (1-122). Therefore, for the "stolen" transition moment $\mathbf{D}_{XT}$

$$\mathbf{D}_{XT} = \frac{\langle \Psi(^3T)|\mathscr{H}^{(1)}|\Psi(^1B)\rangle}{W_T^0 - W_B^0}\mathbf{D}_{XB}^0$$

The intensity of a spin-forbidden transition is usually reduced by a factor of about 100 from the intensity of the band from which it is assumed to "steal" its intensity.

The band profiles of the spin-forbidden bands are, on the other hand, completely determined by the potential surfaces for the two states between which the transition takes place. We can look at the intensity stealing as providing a small "bump" in the potential surfaces, making the transitions "slightly allowed." However, the vibrational wave functions associated with the potential surfaces will remain unaltered. The vibrational pattern to be observed for a spin-forbidden transition is therefore governed by the electronic configuration of the "forbidden" state.

In the case where the electronic configuration is the same but S is different in the ground and excited states, we expect the two potential surfaces to be nearly identical. In particular they will both possess approximately the same equilibrium points $\mathbf{Q}^0$. For that reason a host lattice will be indifferent as to whether the guest is in its ground or excited state, and the change in the guest–host coupling will be very small. Due to the orthogonality of the vibrational wave functions the "band envelope" therefore appears as a sharp line. In inorganic spectroscopy the best known such example is the spin-forbidden ${}^4A_{2g}(t_{2g})^3 \rightarrow {}^2E_g(t_{2g})^3$ transitions seen in octahedral Cr^{3+} complexes.

4-11 BAND PROFILES TREATED SEMICLASSICALLY

For a transition $|av''\rangle \rightarrow |bv'\rangle$ (see Fig. 4-4) the quantum-mechanical expression for the band-intensity distribution as a function of the frequency ν of the incoming light is

$$G_{ba}(\nu) = \mathrm{Av}_{v''} \sum_{v'} |\langle \chi_{bv'}(\xi)|\, \mathbf{D}_{ba}(\xi)\, |\chi_{av''}(\xi)\rangle|^2\, \delta(W_{bv'} - W_{av''} - h\nu) \qquad (4\text{-}142)$$

where in general the electronic transition moment $\mathbf{D}_{ba}(\xi)$ is a function of the vibrational symmetry coordinates. The average $\mathrm{Av}_{v''}$ is understood to be an average over the initial vibrational states, weighting each with the Boltzmann factor $\exp(-W_{av''}/k_BT)$ and

$$W_{bv'} - W_{av''} = \hbar\omega_{ab} + \sum_{j=1}^{3N-6} [(v_{j'} + \tfrac{1}{2})\hbar\omega_{bj} - (v_{j''} + \tfrac{1}{2})\hbar\omega_{aj}]$$

where $\hbar\omega_{ab}$ is the energy difference $W_b^0 - W_a^0$ of the pure electronic states. The use of an integral representation of the δ function,

$$\delta(W_{bv'} - W_{av''} - h\nu) = \frac{1}{h}\int_{-\infty}^{\infty} \exp\left[\mathrm{i}(W_{bv'} - W_{av''} - h\nu)\frac{t}{\hbar}\right] \mathrm{d}t \qquad (4\text{-}143)$$

transforms Eq. (4-142) into

$$G_{ba}(\nu) = \frac{1}{h}\int_{-\infty}^{\infty} \exp(-2\pi\mathrm{i}\nu t)\, G_{ba}(t)\, \mathrm{d}t \qquad (4\text{-}144)$$

where

$$G_{ba}(t) = \mathrm{Av}_{v''} \sum_{v'} \langle av''|\, \mathbf{D}_{ba}^*\, |bv'\rangle \langle bv'|\, \mathbf{D}_{ba}\, |av''\rangle \exp\left[\frac{\mathrm{i}t}{\hbar}(W_{bv'} - W_{av''})\right] \qquad (4\text{-}145)$$

We can also write Eq. (4-145) as

$$G_{ba}(t) = \mathrm{Av}_{v''} \sum_{v'} \langle av''|\, \mathbf{D}_{ba}^*\, |bv'\rangle \langle bv'| \exp\left(\frac{\mathrm{i}t}{\hbar} W_{bv'}\right) \mathbf{D}_{ba}\, |av''\rangle \exp\left(-\frac{\mathrm{i}t}{\hbar} W_{av''}\right)$$

Now

$$\chi_{bv'} \exp\left(\frac{\mathrm{i}t}{\hbar} W_{bv'}\right) = \chi_{bv'}\left[1 + \frac{\mathrm{i}t}{\hbar}W_{bv'} + \frac{1}{2}\left(\frac{\mathrm{i}t}{\hbar}\right)^2 W_{bv'}^2 + \cdots\right]$$

$$= \left[1 + \frac{\mathrm{i}t}{\hbar}\mathscr{H}_b + \frac{1}{2}\left(\frac{\mathrm{i}t}{\hbar}\right)^2 \mathscr{H}_b^2 + \cdots\right]\chi_{bv'}$$

$$= \exp\left(\frac{\mathrm{i}t}{\hbar}\mathscr{H}_b\right)\chi_{bv'}$$

and similarly for $\chi_{av''} \exp(-\mathrm{i}t/\hbar\, W_{av''})$.

Hence Eq. (4-145) can be transformed into

$$G_{ba}(t) = \mathrm{Av}_{v''} \sum_{v'} \langle av'' | \mathbf{D}^*_{ba} | bv' \rangle \langle bv' | \exp\left(\frac{it}{\hbar}\mathscr{H}_b\right) \mathbf{D}_{ba} \exp\left(-\frac{it}{\hbar}\mathscr{H}_a\right) | av'' \rangle$$

We have replaced $W_{bv'}$ with the operator because whereas $W_{bv'}$ is dependent upon the quantum numbers v', $\mathscr{H}_b$ is not. Since $|bv'\rangle$ constitutes a complete set we can perform a closure over $|bv'\rangle$ and get for Eq. (4-145):

$$G_{ba}(t) = \mathrm{Av}_{v''} \langle av'' | \mathbf{D}^*_{ba} \exp\left(\frac{it}{\hbar}\mathscr{H}_b\right) \mathbf{D}_{ba} \exp\left(-\frac{it}{\hbar}\mathscr{H}_a\right) | av'' \rangle \qquad (4\text{-}146)$$

Assuming the Condon approximation we can take $\mathbf{D}^0_{ba}$ out

$$G_{ba}(t) = |\mathbf{D}^0_{ba}|^2 \, \mathrm{Av}_{v''} \langle av'' | \exp\left(\frac{it}{\hbar}\mathscr{H}_b\right) \exp\left(-\frac{it}{\hbar}\mathscr{H}_a\right) | av'' \rangle \qquad (4\text{-}147)$$

As can be seen by expansion $\exp[(it/\hbar)\mathscr{H}_b] \exp[(-it/\hbar)\mathscr{H}_a]$ is only equal to $\exp[(it/\hbar)(\mathscr{H}_b - \mathscr{H}_a)]$ provided $\mathscr{H}_a$ and $\mathscr{H}_b$ commute. Assuming this to be the case, that is neglecting some few commutators, we take

$$\exp\left(\frac{it}{\hbar}\mathscr{H}_b\right) \exp\left(-\frac{it}{\hbar}\mathscr{H}_a\right) = \exp\left[\frac{it}{\hbar}(\mathscr{H}_b - \mathscr{H}_a)\right] = \exp\left[\frac{it}{\hbar}(V_b - V_a)\right] \qquad (4\text{-}148)$$

where V_b and V_a are the potential energy terms in $\mathscr{H}_a$ and $\mathscr{H}_b$. The last equality sign in Eq. (4-148) follows because the kinetic energy term drops out in $\mathscr{H}_b - \mathscr{H}_a$. The semiclassical approximation to the band shape is indeed contained in the assumption $[\mathscr{H}_a, \mathscr{H}_b] = 0$, which holds true for $h \to 0$. Using Eq. (4-148) we get from Eq. (4-147)

$$G_{ba}(t) = |\mathbf{D}^0_{ba}|^2 \, \mathrm{Av}_{v''} \langle av'' | \exp\left[\frac{it}{\hbar}\Delta V(\xi)\right] | av'' \rangle \qquad (4\text{-}149)$$

where

$$\Delta V(\xi) = V_b(\xi) - V_a(\xi) \qquad (4\text{-}150)$$

Combining Eqs. (4-149) and (4-144)

$$G_{ba}(\nu) = \frac{1}{h}\int_{-\infty}^{\infty} dt \exp(-2\pi i \nu t) |\mathbf{D}^0_{ba}|^2 \, \mathrm{Av}_{v''} \int_{-\infty}^{\infty} \chi_{av''}(\xi)\chi_{av''}(\xi) \exp\left[\frac{it}{\hbar}\Delta V(\xi)\right] d\xi \qquad (4\text{-}151)$$

We again use the δ function definition Eq. (4-143) and get

$$G_{ba}(\nu) = |\mathbf{D}^0_{ba}|^2 \, \mathrm{Av}_{v''} \int_{-\infty}^{\infty} \chi_{av''}(\xi)\chi_{av''}(\xi)\delta[\Delta V(\xi) - h\nu] \, d\xi \qquad (4\text{-}152)$$

The so-called Mehlers formula is now utilized. For mass-normalized vibrations (force constants $k_j = \omega_j^2$) we have[16]

$$\mathrm{Av}_{v''}[\chi_{v''}(\xi)\chi_{v''}(\xi)] = \sqrt{\frac{\omega_j}{2\pi\hbar \sinh(\hbar\omega_j/2k_B T)}} \exp[-(\omega_j/\hbar)\xi^2 \tanh(\hbar\omega_j/2k_B T)] \qquad (4\text{-}153)$$

Ignoring the normalizing factor contained in the square root in Eq. (4-153) we get finally for the semiclassical expression for the shape function

$$G_{ba}(\nu) = |\mathbf{D}^0_{ba}|^2 \int_{-\infty}^{\infty} \exp\left(-\frac{\xi^2}{\gamma^2}\right) \delta[\Delta V(\xi) - h\nu]\, d\xi \qquad (4\text{-}154)$$

with

$$\gamma^2 = \frac{\hbar}{\omega_j} \coth \frac{\hbar\omega_j}{2k_B T} \qquad (4\text{-}155)$$

Consider now the band shape of a transition originating from the potential surfaces V_a and terminating on the potential surface V_b (see Fig. 4-6). With $V_a = \frac{1}{2}\omega_j^2 \xi^2$ and $V_b = \frac{1}{2}\omega_j^2(\xi + \Delta\xi)^2$, where ω_j is the angular frequency of the vibration, we get

$$\Delta V(\xi) = \hbar\omega_{ab} + \tfrac{1}{2}\omega_j^2(\Delta\xi)^2 + \omega_j^2 \xi \Delta\xi = w_0 + \omega_j^2 \xi \Delta\xi \qquad (4\text{-}156)$$

where

$$w_0 = \hbar\omega_{ab} + \tfrac{1}{2}\omega_j^2(\Delta\xi)^2 \qquad (4\text{-}157)$$

w_0 is seen to correspond to the energy at the "vertical" transition.

Using Eq. (4-156), the δ function in Eq. (4-154) demands

$$h\nu = w_0 + \omega_j^2 \xi \Delta\xi \qquad (4\text{-}158)$$

or

$$\xi = \frac{h\nu - w_0}{\omega_j^2 \Delta\xi} \qquad (4\text{-}159)$$

Carrying out the integration in Eq. (4-154) therefore leads to a gaussian shaped absorption band with the maximum intensity occurring at the "vertical" transition

$$G_{ba}(\nu) = |\mathbf{D}^0_{ba}|^2 \exp\left[-\frac{(h\nu - w_0)^2}{u^2}\right] \qquad (4\text{-}160)$$

with $u = \omega_j^2 \gamma \Delta\xi$.

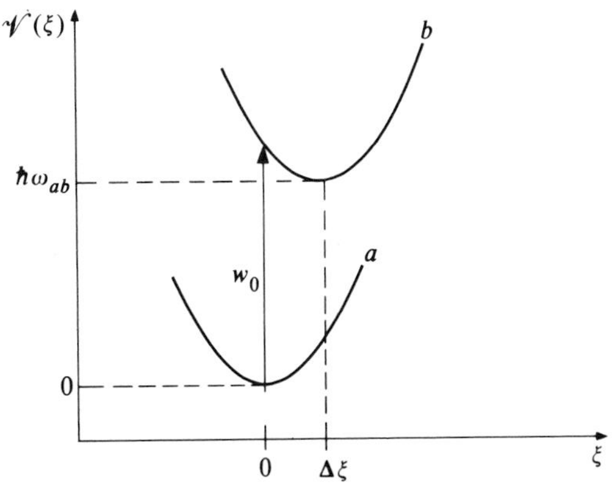

Figure 4-6 Two potential curves with different minima, but with identical force constants $k' = k'' = \omega^2$.

In the case where, for example, two vibrations are involved, Eq. (4-154) is replaced by

$$G_{ba}(\nu) = |\mathbf{D}_{ba}^{0}|^2 \int_{-\infty}^{\infty} d\xi_1 \exp\left(-\frac{\xi_1^2}{\gamma_1^2}\right) \int_{-\infty}^{\infty} d\xi_2 \exp\left(-\frac{\xi_2^2}{\gamma_2^2}\right) \delta[\Delta V(\xi_1, \xi_2) - h\nu] \tag{4-161}$$

With

$$V_a = \tfrac{1}{2}\omega_1^2\xi_1^2 + \tfrac{1}{2}\omega_2^2\xi_2^2 \tag{4-162}$$

and

$$V_b = \tfrac{1}{2}\omega_1^2(\xi_1 + \Delta\xi_1)^2 + \tfrac{1}{2}\omega_2^2(\xi_2 + \Delta\xi_2)^2 \tag{4-163}$$

we get

$$\Delta V(\xi_1, \xi_2) = \tilde{w}_0 + \omega_1^2\xi_1\Delta\xi_1 + \omega_2^2\xi_2\Delta\xi_2 \tag{4-164}$$

with

$$\tilde{w}_0 = \hbar\omega_{ab} + \tfrac{1}{2}\omega_1^2(\Delta\xi_1)^2 + \tfrac{1}{2}\omega_2^2(\Delta\xi_2)^2$$

$\tilde{w}_0$ is seen to be the energy of the "vertical" transition.

The δ function in Eq. (4-161) is used to fix ξ_2:

$$\xi_2 = \frac{h\nu - \tilde{w}_0 - \omega_1^2\xi_1\Delta\xi_1}{\omega_2^2\Delta\xi_2} \tag{4-165}$$

Inserting Eq. (4-165) in Eq. (4-161) yields

$$I_{ba}(\nu) = |\mathbf{D}_{ba}^{0}|^2 \int_{-\infty}^{\infty} d\xi_1 \exp\left(-\frac{\xi_1^2}{\gamma_1^2}\right) \exp\left[-\frac{(h\nu - \tilde{w}_0 - \omega_1^2\xi_1\Delta\xi_1)^2}{(\gamma_2\omega_2^2\Delta\xi_2)^2}\right]$$

which by integration leads to

$$I_{ba}(\nu) = |\mathbf{D}_{ba}^{0}|^2 \exp\left[-\frac{(h\nu - \tilde{w}_0)^2}{(u_1^2 + u_2^2)}\right] \tag{4-166}$$

with

$$u_1^2 + u_2^2 = \gamma_1^2\omega_1^4(\Delta\xi_1)^2 + \gamma_2^2\omega_2^4(\Delta\xi_2)^2 \tag{4-167}$$

Again the absorption curve is gaussian with its maximum at the vertical transition.

To show how powerful the semiclassical method is we shall further give as an example the calculation of the band shape of an electronically allowed transition $\psi(A) \rightarrow \psi(E)$ where the excited state $\psi(E)$ is twofold degenerate and subjected to a Jahn–Teller configurational instability.[17]

Let the molecule have C_3 symmetry with two vibrational symmetry coordinates ξ_{2a} and ξ_{2b} spanning the vibrational symmetry coordinate ε. From the development in section 1-6, the potential surfaces of the excited states will then be given (compare Eq. (1-88)) as

$$V_b = \tfrac{1}{2}\omega_\varepsilon^2(\xi_{2a}^2 + \xi_{2b}^2) \pm a_\varepsilon\sqrt{\xi_{2a}^2 + \xi_{2b}^2} + \hbar\omega_{ab} \tag{4-168}$$

For V_a, the surface of the ground state, we have

$$V_a = \tfrac{1}{2}\omega_\varepsilon^2(\xi_{2a}^2 + \xi_{2b}^2) \tag{4-169}$$

and therefore

$$\Delta V = \hbar\omega_{ab} \pm a_\varepsilon\sqrt{\xi_{2a}^2 + \xi_{2b}^2} \tag{4-170}$$

According to Eq. (4-161) we can write

$$G_{ba}(\omega) = |\mathbf{D}_{ba}^0|^2 \frac{1}{2}\sum \iint d\xi_1\, d\xi_2 \exp\left(-\frac{\xi_1^2 + \xi_2^2}{\gamma^2}\right)\delta(\hbar\omega_{ab} \pm a_\varepsilon\sqrt{\xi_{2a}^2 + \xi_{2b}^2} - \hbar\omega) \tag{4-171}$$

where the summation is over the two potential curves V_b^+ and V_b^- (see Fig. 4-7). Substituting

$$\xi_{2a} = \rho\cos\phi$$

$$\xi_{2b} = \rho\sin\phi$$

we transform Eq. (4-171) into

$$G_{ba}(\omega) = |\mathbf{D}_{ba}^0|^2 \frac{1}{2}\sum \int_0^{2\pi} d\phi \int_0^\infty \rho\, d\rho \exp\left(-\frac{\rho^2}{\gamma^2}\right)\delta(\hbar\omega_{ab} \pm a_\varepsilon\rho - \hbar\omega) \tag{4-172}$$

From the δ function

$$\rho = \frac{|\hbar\omega - \hbar\omega_{ab}|}{a_\varepsilon} \tag{4-173}$$

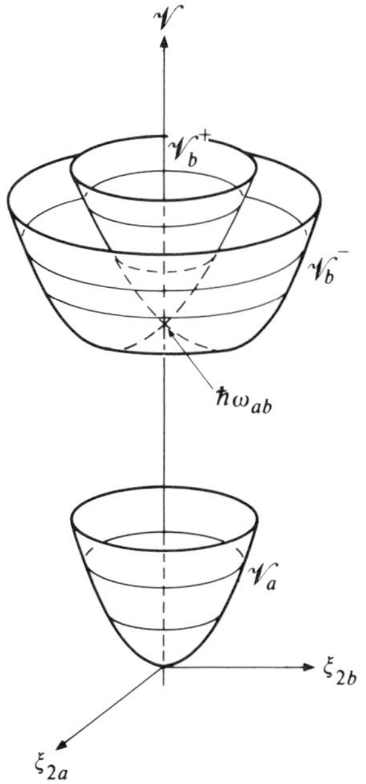

Figure 4-7 The three potential surfaces of $\psi_a(A)$ and $\psi_b(E_+$ and $E_-)$.

and therefore

$$G_{ba}(\omega) = |\mathbf{D}_{ba}^{0}|^2 \pi \frac{|\hbar\omega - \hbar\omega_{ab}|}{a_\varepsilon} \exp\left(-\frac{|\hbar\omega - \hbar\omega_{ab}|^2}{\gamma^2 a_\varepsilon^2}\right) \tag{4-174}$$

The absorption band shape is pictured in Fig. 4-8. $G_{ba}(\omega)$ is seen to be zero at $\omega = \omega_{ab}$. By differentiation of G_{ba} the separation of the two peaks is found to be proportional to γ. For $k_B T \gg \hbar\omega_\varepsilon$ expansion of Eq. (4-155) shows therefore that the peak separation is proportional to $\sqrt{T}$.

The physical explanation for the vibronic band splitting is as follows. Let the vibrational wave function for the ground state be $\chi_{n,m}(\rho, \phi)$. The probability distribution over the ρ coordinate is then $2\pi\rho|\chi_{n,m}|^2\,d\rho$. This is zero at $\rho = 0$. In the semiclassical Franck–Condon approximation the most probable transition is therefore not at $\rho = 0$. The upward transition will therefore hit the two excited potential surfaces (Fig. 4-7) where these are separated, leading to two band maxima. Note, however, that a strong coupling to a totally symmetric mode may wipe out the Jahn–Teller splitting.

Consider finally the band shape of an allowed transition originating from a onefold degenerate state ψ_0 and terminating on a threefold degenerate electronic state, in which a Jahn–Teller type interaction in a twofold degenerate vibration is active. For a doubly degenerate Jahn–Teller active vibration, with vibrational coordinates ξ_{2a} and ξ_{2b}, we write as usual

$$\mathscr{H}^{(1)} = \left(\frac{\partial V}{\partial \xi_{2a}}\right)_0 \xi_{2a} + \left(\frac{\partial V}{\partial \xi_{2b}}\right)_0 \xi_{2b} \tag{4-175}$$

The symmetric product of, say, a T_{2g} representation in O_h contains an E representation. Using the symmetry elements of O_h we get easily the diagonal perturbation matrix

$$\begin{array}{cccc} & T_{2g,1} & T_{2g,2} & T_{2g,3} \\ c_\varepsilon & \left(\begin{array}{c} -\frac{1}{2}\xi_{2a} - \frac{\sqrt{3}}{2}\xi_{2b} \\ 0 \\ 0 \end{array}\right. & \begin{array}{c} 0 \\ -\frac{1}{2}\xi_{2a} + \frac{\sqrt{3}}{2}\xi_{2b} \\ 0 \end{array} & \left.\begin{array}{c} 0 \\ 0 \\ \xi_{2a} \end{array}\right) \end{array} \tag{4-176}$$

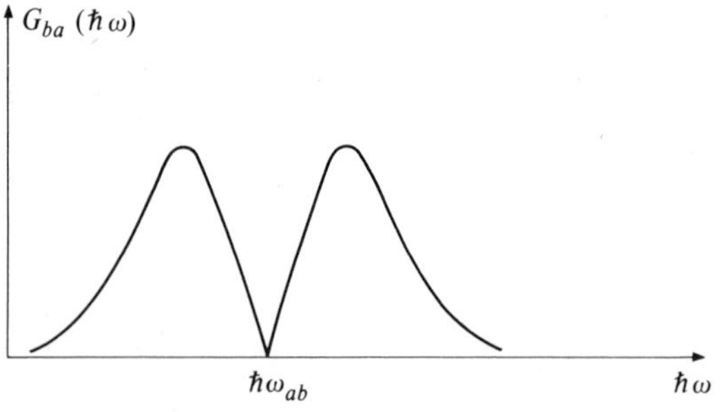

Figure 4-8 Absorption band shape for a transition $\Psi(A) \rightarrow \Psi(E)$.

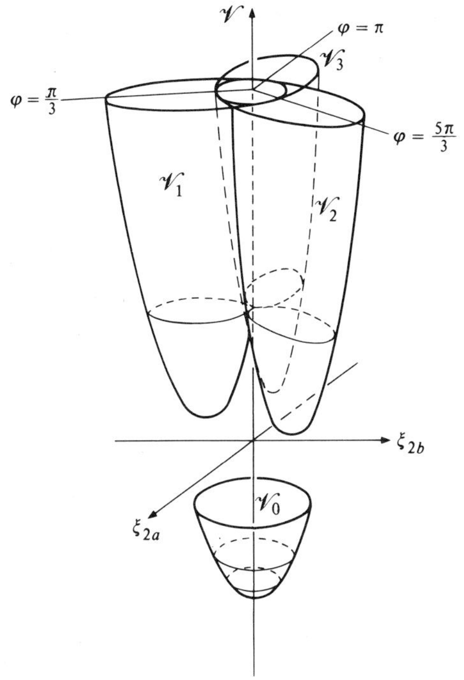

Figure 4-9 A threefold degenerate electronic state split under a Jahn–Teller effect in an ε vibration.

The addition of the harmonic term $\frac{1}{2}k_\varepsilon(\xi_{2a}^2 + \xi_{2b}^2)$ to Eq. (4-176) leads to the three potential surfaces

$$V_1 = \frac{1}{2}k_\varepsilon(\xi_{2a}^2 + \xi_{2b}^2) + c_\varepsilon\left(-\frac{1}{2}\xi_{2a} - \frac{\sqrt{3}}{2}\xi_{2b}\right) \tag{4-177}$$

$$V_2 = \frac{1}{2}k_\varepsilon(\xi_{2a}^2 + \xi_{2b}^2) + c_\varepsilon\left(-\frac{1}{2}\xi_{2a} + \frac{\sqrt{3}}{2}\xi_{2b}\right) \tag{4-178}$$

$$V_3 = \frac{1}{2}k_\varepsilon(\xi_{2a}^2 + \xi_{2b}^2) + c_\varepsilon\,\xi_{2a} \tag{4-179}$$

Minimizing these three potential functions and changing the coordinate systems to the position of the minima gives us three potential functions of the form

$$V_i = \frac{1}{2}k_\varepsilon(q_{ia}^2 + q_{ib}^2) - \frac{c_\varepsilon^2}{2k_\varepsilon} \qquad i = 1, 2, 3 \tag{4-180}$$

The original threefold degenerate surface therefore splits into three displaced Jahn–Teller stabilized surfaces (see Fig. 4-9).

It is intuitively clear that the band shape of an electronic transition originating on V_0 and terminating on (V_1, V_2, V_3) will not show any splittings. Any one of the $V_i(i = 1, 2, 3)$ potential surfaces can be obtained from the other by a rotation of $2\pi/3$. The resulting band shape is therefore identical to what we have considered previously, a transition between two states having their minima displaced.

An interesting situation occurs when we have an additional perturbation which is off-diagonal between the three distorted Jahn–Teller states.[18] A spin-orbit coupling or a stress along a threefold axis in the octahedron would be a case in point. In the last situation, group theory predicts that the T_2 state will split into an E and A state. The perturbation is entirely off-diagonal between the T_2 components. Hence for the vibronic wave functions we have to evaluate matrix elements of the type

$$H_{1,2} = \langle \psi_1(\mathbf{r}) | \mathscr{H}^{(1)}(\mathbf{r}) | \psi_2(\mathbf{r}) \rangle \langle \chi^1_{v''}(\xi) | \chi^2_{v''}(\xi) \rangle \tag{4-181}$$

Calling the electronic matrix element $D\sigma$, and the vibrational overlap integral

$$S_{00} = \langle \chi^1_{v''=0} | \chi^2_{v''=0} \rangle = \langle \chi^1_{v''=0} | \chi^3_{v''=0} \rangle = \langle \chi^2_{v''=0} | \chi^3_{v''=0} \rangle$$

we calculate the splitting of the original threefold degenerate level $\psi_i(\mathbf{r})\chi^i_{v''=0}$, $i = 1, 2, 3$, to be

$$W = W_0 - D\sigma \times S_{00} \qquad \text{(twofold)} \tag{4-182}$$

$$W = W_0 + 2D\sigma \times S_{00} \qquad \text{(onefold)} \tag{4-183}$$

The overlap integral S_{00} is easily calculated to be

$$S_{00} = \exp\left(-\frac{3}{2} \frac{\Delta V_{J-T}}{\hbar\omega_\varepsilon} \right) \tag{4-184}$$

In other words, the larger the Jahn–Teller stabilization ΔV_{J-T} is compared with an active vibrational quantum $\hbar\omega_\varepsilon$, the smaller the splitting of the no-phonon line. The electronic splitting is quenched by a vibrational overlap factor. We refer to this phenomenon as the *Ham effect*. The Ham effect is indeed quite general, since all nondiagonal electronic matrix elements are reduced in size by the action of a vibrational overlap integral.[18]

4-12 BAND MOMENTS

The band moments $M^{(n)}$ of an electronic transition $|av''\rangle \rightarrow |bv'\rangle$ are defined from the shape function $G_{ba}(\nu)$ of the band by

$$M^{(n)} = \int (h\nu)^n G_{ba}(\nu)\, \mathrm{d}(h\nu) \tag{4-185}$$

From an experimental point of view we notice that

$$M^{(0)} = \int G_{ba}(\nu)\, \mathrm{d}(h\nu) \tag{4-186}$$

is the area of the absorption band, giving a measure of the intensity of the electronic transition. Further

$$M^{(1)} = \int (h\nu)^1 G_{ba}(\nu)\, \mathrm{d}(h\nu) \equiv (h\nu_0) M^{(0)} \tag{4-187}$$

gives us the centroid $(h\nu_0)$ of the shape function. Provided $G_{ba}(\nu)$ has a pronounced maximum $(h\nu_0)$ fall close to this, and if $G_{ba}(\nu)$ is symmetric around the maximum $(h\nu_0)$ will be the energy at the maximum.

The shape width $(h\nu_{1/2})$ of a band is defined by

$$\int (h\nu - h\nu_0)^2 \, G_{ba}(\nu) \, \mathrm{d}(h\nu) \equiv (h\nu_{1/2})^2 M^{(0)} \tag{4-188}$$

The identity

$$\int (h\nu)^2 \, G_{ba}(\nu) \, \mathrm{d}(h\nu) = \int [(h\nu - h\nu_0)^2 + 2(h\nu - h\nu_0)h\nu_0 + (h\nu_0)^2] \, G_{ba}(\nu) \, \mathrm{d}(h\nu)$$

gives us immediately

$$M^{(2)} = (h\nu_{1/2})^2 \, M^{(0)} + (h\nu_0)^2 \, M^{(0)} \tag{4-189}$$

Normalizing $M^{(0)} = 1$ we have therefore

$$(h\nu_0) = M^{(1)} \tag{4-190}$$

and

$$(h\nu_{1/2}) = \sqrt{M^{(2)} - (M^{(1)})^2} \tag{4-191}$$

Provided the band shape is gaussian, $2(h\nu_{1/2})$ equals the half-width of the band, that is the width at $\varepsilon = \frac{1}{2}\varepsilon_{\max}$.

The quantum-mechanical calculation of $M^{(n)}$ proceeds as follows. From Eq. (4-144) we have

$$G_{ba}(\nu) = \frac{1}{h} \int_{-\infty}^{\infty} \exp(-2\pi i \nu t) \, G_{ba}(t) \, \mathrm{d}t \tag{4-192}$$

with (see Eq. (4-146))

$$G_{ba}(t) = \mathrm{Av}_{v''} \langle av'' | \, \mathbf{D}_{ba}^* \exp\left(\frac{it}{\hbar} \mathscr{H}_b\right) \mathbf{D}_{ba} \exp\left(-\frac{it}{\hbar} \mathscr{H}_a\right) | av'' \rangle \tag{4-193}$$

A Fourier inversion of Eq. (4-192) gives

$$G_{ba}(t) = \int G_{ba}(\nu) \exp(2\pi i \nu t) \, \mathrm{d}(h\nu) \tag{4-194}$$

By differentiating Eq. (4-194) with respect to t, and putting $t = 0$, we obtain

$$\left(\frac{\mathrm{d}^n G_{ba}(t)}{\mathrm{d}t^n}\right)_{t=0} = \int (2\pi i \nu)^n \, G_{ba}(\nu) \mathrm{d}(h\nu) \tag{4-195}$$

Differentiating Eq. (4-193) with respect to t, setting $t = 0$, and comparing with Eq. (4-195) leads to the following expressions[19] for $M^{(n)}$

$$M^{(0)} = \mathrm{Av}_{v''} \langle av'' | \, \mathbf{D}_{ba}^* \mathbf{D}_{ba} \, | av'' \rangle \tag{4-196}$$

$$M^{(1)} = \mathrm{Av}_{v''} \langle av'' | \, \mathbf{D}_{ba}^* \mathscr{H}_b \mathbf{D}_{ba} - \mathbf{D}_{ba}^* \mathbf{D}_{ba} \mathscr{H}_a \, | av'' \rangle \tag{4-197}$$

$$M^{(2)} = \mathrm{Av}_{v''}\langle av''|\, \mathbf{D}^*_{ba}\mathscr{H}_b^2\, \mathbf{D}_{ba} - 2\mathbf{D}^*_{ba}\mathscr{H}_b\, \mathbf{D}_{ba}\mathscr{H}_a + \mathbf{D}^*_{ba}\mathbf{D}_{ba}\mathscr{H}_a^2\,|av''\rangle \tag{4-198}$$

Introducing the Condon approximation, which assumes $\mathbf{D}_{ba}$ to be independent of the nuclear coordinates ξ, turns the above expressions into

$$M^{(0)} = |\mathbf{D}^0_{ba}|^2 \tag{4-199}$$

$$M^{(1)} = |\mathbf{D}^0_{ba}|^2\, \mathrm{Av}_{v''}\langle av''|\, V_b - V_a\,|av''\rangle \tag{4-200}$$

$$M^{(2)} = |\mathbf{D}^0_{ba}|^2\, \mathrm{Av}_{v''}\langle av''|\,(V_b - V_a)^2\,|av''\rangle \tag{4-201}$$

where the kinetic energy in the hamiltonian has cancelled out. We have for future reference

$$\mathrm{Av}_{v''}\langle av''|\,1\,|av''\rangle = 1 \tag{4-202}$$

$$\mathrm{Av}_{v''}\langle av''|\,\xi_t\,|av''\rangle = 0 \tag{4-203}$$

$$\mathrm{Av}_{v''}\langle av''|\,\xi_t^2\,|av''\rangle = \frac{\hbar}{\omega_t}\frac{\sum\limits_{v''=0}^{\infty}(v''+\tfrac{1}{2})\exp\left(-\dfrac{v''\hbar\omega_t}{k_BT}\right)}{\sum\limits_{v''=0}^{\infty}\exp\left(-\dfrac{v''\hbar\omega_t}{k_BT}\right)} = \frac{\hbar}{2\omega_t}\coth\frac{\hbar\omega_t}{2k_BT} \tag{4-204}$$

Let us take

$$V_a = \tfrac{1}{2}\sum_t k_a^{tt}\xi_t^2 \tag{4-205}$$

and for the excited state

$$V_b = \hbar\omega_{ab} + \tfrac{1}{2}\sum_t k_b^{tt}(\xi_t - \Delta\xi_t)^2 + \sum_{s\neq t} k_b^{st}\xi_s\xi_t \tag{4-206}$$

The excited state potential is seen to be translated and rotated relative to the ground state. For an allowed transition, Eq. (4-199) yields

$$M^{(0)} = |\mathbf{D}^0_{ba}|^2 \tag{4-207}$$

As seen previously, the total band intensity is independent of the vibrational potentials. We have further

$$V_b - V_a = w_0 - \sum_t k_b^{tt}(\Delta\xi_t)\xi_t + \tfrac{1}{2}\sum_t (k_b^{tt} - k_a^{tt})\xi_t^2 + \sum_{s\neq t} k_b^{st}\xi_s\xi_t \tag{4-208}$$

with

$$w_0 = \hbar\omega_{ab} + \tfrac{1}{2}\sum k_b^{tt}(\Delta\xi_t)^2$$

From Eq. (4-200) we get then for the first moment of the transition $|av''\rangle \to |bv'\rangle$ using Eqs. (4-202), (4-203), and (4-204), and putting $M^{(0)} = 1$

$$M^{(1)} = w_0 + \tfrac{1}{2}\sum_t (k_b^{tt} - k_a^{tt})\frac{\hbar}{2\omega_t}\coth\frac{\hbar\omega_t}{2k_BT} \tag{4-209}$$

Provided $\hbar\omega_t < 2k_BT$ we may expand Eq. (4-209), and with $\omega_t^2 = k_a^{tt}$ the result is

$$M^{(1)} \approx w_0 + \tfrac{1}{2}\sum_t \frac{(k_b^{tt} - k_a^{tt})}{k_a^{tt}} k_BT \tag{4-210}$$

In the case where $\sum_t (k_b^{tt} - k_a^{tt}) < 0$, that is, where the force constants in the excited state are less than the force constants in the ground state, the "center of gravity" for the absorption band will move to the blue when the temperature is lowered.

The shape width of the band will be largely determined by the linear terms $\sum_t k_b^{tt}(\Delta\xi_t)\xi_t$ in Eq. (4-208) since ξ_t^2 and $\xi_t\xi_s$ contributions will depend upon the fourth power of the zero point amplitudes. Retaining therefore only the linear terms, we get easily using Eq. (4-201) with $M^{(0)} = 1$

$$M^{(2)} = w_0^2 + \sum_t (k_b^{tt}\Delta\xi_t)^2 \frac{\hbar}{2\omega_t} \coth \frac{\hbar\omega_t}{2k_B T} \tag{4-211}$$

With the shape width of the band given in Eq. (4-191) and assuming $\hbar\omega_t/2k_B T < 1$, $h\nu_{1/2}$ is found to be proportional to $\sqrt{T}$.

For an orbitally forbidden transition we may expand the transition moment $\mathbf{D}_{ba}$ around the equilibrium position

$$\mathbf{D}_{ba} = \sum_t \left(\frac{\partial \mathbf{D}_{ba}}{\partial \xi_t}\right)_0 \xi_t$$

Using Eq. (4-196) we find

$$M^{(0)} = \sum_t \left[\left(\frac{\partial D_{ba}}{\partial \xi_t}\right)_0\right]^2 \frac{\hbar}{2\omega_t} \coth \frac{\hbar\omega_t}{2k_B T}$$

As previously found the band intensity is in this situation seen to be temperature dependent. For $\hbar\omega_t/2k_B T < 1$ the intensity is indeed proportional to T.

4-13 COUPLED CHROMOPHORES

Molecules or ions which from a theoretical point of view can be considered as being made up of two interacting chromophores are not uncommon. In section 3-6 we considered the magnetic properties of two strongly coupled units in which the wave functions of the groups overlapped. In this section we shall look at coupled systems which exhibit a weak interaction. The essential feature in a weakly coupled chromophore model is the neglect of all direct bonding between the two chromophores. Consequently the overlap of the wave functions of the two chromophores A and B is put equal to zero. The advantage of such a model is that it makes full use of our knowledge of the transformation properties and energies of the self-consistent field orbitals for the single chromophore. The method is therefore equivalent to a molecular orbital treatment with a restricted configuration interaction included.[20,21]

We divide the electrons of the system up into two groups, the electrons on chromophore A and the electrons on chromophore B. The ground state of the system is

$$\Psi_A^0 \Psi_B^0$$

where both of the molecular fragments are in their lowest state. Excited states in which one electron has been excited are, with an obvious nomenclature, represented by

$$\Psi^1_{A,-1}\Psi^0_B,\ \Psi^0_A\Psi^1_{B,-1},\ \Psi^0_{A,-1}\Psi^1_B,\ \Psi^1_A\Psi^0_{B,-1}$$

Both fragments may also be excited as for instance exemplified by a state

$$\Psi^1_{A,-1},\ \Psi^1_{B,-1}$$

Configurations in which a single electron has been excited are conveniently divided into locally excited configurations, referring respectively to an excitation of an electron within orbitals belonging to the same chromophore and to a transfer of an electron from one chromophore to the other. Doubly excited configurations can correspond to mixed configurations.

The hamiltonian for the combined system AB may be written

$$\mathscr{H} = \mathscr{H}_A + \mathscr{H}_B + V \tag{4-212}$$

With s electrons on A and t electrons on B we have

$$V = -\sum_{j=1}^{s}\frac{Z_A}{r_{Aj}} - \sum_{i=1}^{t}\frac{Z_B}{r_{Bi}} + \frac{Z_A Z_B}{R} + \sum_{i=1}^{s}\sum_{j=1}^{t}\frac{1}{r_{ij}} \tag{4-213}$$

Using Fig. 4-10 we can write $\mathbf{r}_{ij} - \mathbf{r}_{Bj} - \mathbf{R} + \mathbf{r}_{Ai} = 0$. Then

$$\frac{1}{r_{ij}} = \frac{1}{\sqrt{R^2 + |\mathbf{r}_{Bj} - \mathbf{r}_{Ai}|^2 + 2\mathbf{R}\cdot(\mathbf{r}_{Bj} - \mathbf{r}_{Ai})}} \tag{4-214}$$

$$\frac{1}{r_{Bi}} = \frac{1}{\sqrt{R^2 + r_{Ai}^2 - 2\mathbf{r}_{Ai}\cdot\mathbf{R}}} \tag{4-215}$$

$$\frac{1}{r_{Aj}} = \frac{1}{\sqrt{R^2 + r_{Bj}^2 + 2\mathbf{r}_{Bj}\cdot\mathbf{R}}} \tag{4-216}$$

Assuming $|R| \gg |r_{Ai}|$ and $|r_{Bj}|$ we can expand Eqs. (4-214) to (4-216) using

$$\frac{1}{\sqrt{1+x}} = 1 - \frac{1}{2}x + \frac{3}{8}x^2 - \cdots \tag{4-217}$$

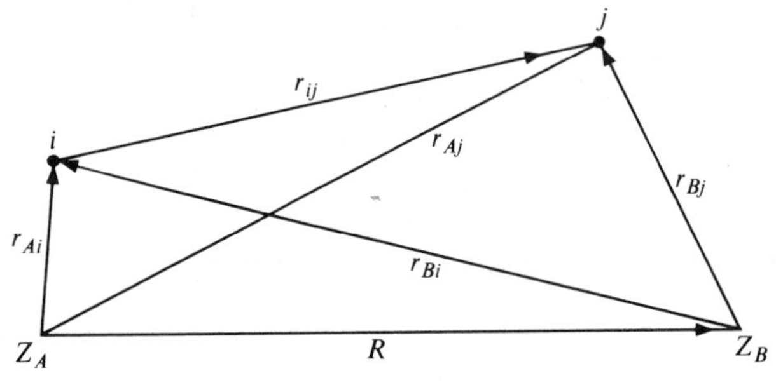

Figure 4-10 Electronic coordinates for a coupled chromophore system.

Defining $\mathbf{n} = \mathbf{R}/R$ and

$$q_A = Z_A - s \qquad q_B = Z_B - t \tag{4-218}$$

we see, after some manipulation, that Eq. (4-213) is transformed into

$$\begin{aligned} V = \frac{q_A \cdot q_B}{R} &+ \frac{1}{R^2}\left[q_A \sum_{j=1}^{s} (\mathbf{n} \cdot \mathbf{r}_{Bj}) - q_B \sum_{i=1}^{t} (\mathbf{n} \cdot \mathbf{r}_{Ai}) \right] \\ &+ \frac{1}{R^3}\left[\sum_{i=1}^{s} \mathbf{r}_{Ai} \cdot \sum_{j=1}^{t} \mathbf{r}_{Bj} - 3 \sum_{i=1}^{s} (\mathbf{n} \cdot \mathbf{r}_{Ai}) \sum_{j=1}^{t} (\mathbf{n} \cdot \mathbf{r}_{Bj}) \right] \\ &+ \frac{1}{2R^3}\left[q_A \sum_{i=1}^{t} \left(r_{Bj}^2 - 3(\mathbf{n} \cdot \mathbf{r}_{Bj})(\mathbf{n} \cdot \mathbf{r}_{Bj}) \right) + q_B \sum_{i=1}^{s} \left(r_{Ai}^2 - 3(\mathbf{n} \cdot \mathbf{r}_{Ai})(\mathbf{n} \cdot \mathbf{r}_{Ai}) \right) \right] \end{aligned} \tag{4-219}$$

The four terms correspond respectively to (charge–charge), (charge–dipole), (dipole–dipole), and (charge–quadrupole) interactions. This form of V is valid as long as we do not want to consider the mixing of charge transfer states with locally excited states.

In the case where the two chromophores are electrically neutral groups $q_A = q_B = 0$. With three states $|0, 0\rangle = \Psi_A^0 \Psi_B^0$, $|0, 1\rangle = \Psi_A^0 \Psi_{B,-1}^1$, and $|1, 0\rangle = \Psi_{A,-1}^1 \Psi_B^0$ the interaction under V is given by

V	$\|0, 0\rangle$	$\|0, 1\rangle$	$\|1, 0\rangle$
$\|0, 0\rangle$	$\mathbf{D}_{00}^A \cdot \mathbf{D}_{00}^B$	$\mathbf{D}_{00}^A \cdot \mathbf{D}_{01}^B$	$\mathbf{D}_{01}^A \cdot \mathbf{D}_{00}^B$
$\|0, 1\rangle$		$\mathbf{D}_{00}^A \cdot \mathbf{D}_{11}^B$	$\mathbf{D}_{01}^A \cdot \mathbf{D}_{10}^B$
$\|1, 0\rangle$			$\mathbf{D}_{11}^A \cdot \mathbf{D}_{00}^B$

Here

$$\begin{aligned} \mathbf{D}_{pv} \cdot \mathbf{D}_{kl} = \frac{1}{R^3}\Bigg[&\langle \Psi_A^p \,|\, \sum_{i=1}^{s} \mathbf{r}_{Ai} \,|\, \Psi_A^v \rangle \langle \Psi_B^k | \sum_{j=1}^{t} \mathbf{r}_{Bj} \,|\, \Psi_B^l \rangle \\ &- 3\langle \Psi_A^p | \sum_{i=1}^{s} (\mathbf{n} \cdot \mathbf{r}_{Ai}) \,|\, \Psi_A^v \rangle \langle \Psi_B^k | \sum_{j=1}^{t} (\mathbf{n} \cdot \mathbf{r}_{Bj}) \,|\, \Psi_B^l \rangle \Bigg] \end{aligned} \tag{4-220}$$

If, in addition to being electrically neutral, the groups A and B do not possess a static dipole moment, the only term which is left under the interaction operator V is $\mathbf{D}_{01}^A \cdot \mathbf{D}_{10}^B$. Provided the two chromophores are identical the degenerate states $|0, 1\rangle$ and $|1, 0\rangle$ will therefore be split by twice this quantity. We may in this case write for the two excited states

$$|+\rangle = \frac{1}{\sqrt{2}}(|0, 1\rangle + |1, 0\rangle) \qquad W^{(1)} = \mathbf{D}_{01}^A \cdot \mathbf{D}_{10}^B \tag{4-221}$$

and

$$|-\rangle = \frac{1}{\sqrt{2}}(|0, 1\rangle - |1, 0\rangle) \qquad W^{(1)} = -\mathbf{D}_{01}^A \cdot \mathbf{D}_{10}^B \tag{4-222}$$

The electric transition dipole moment from the ground state to $|+\rangle$ or $|-\rangle$ is given by

$$\left\langle 0,0 \right| \sum_{i=1}^{s} \mathbf{r}_{Ai} + \sum_{j=1}^{s} \mathbf{r}_{Bj} \left| \frac{1}{\sqrt{2}}(|0,1\rangle \pm |1,0\rangle) \right. = \frac{1}{\sqrt{2}}(\mathbf{D}_{01}^{B} \pm \mathbf{D}_{10}^{A}) \qquad (4\text{-}223)$$

The intensities for the two transitions are therefore proportional to

$$|\mathbf{D}_{01}^{A}|^2 \pm \mathbf{D}_{10}^{A} \cdot \mathbf{D}_{01}^{B} = \begin{cases} 2|\mathbf{D}_{01}^{A}|^2 \\ 0 \end{cases}$$

The minus combination is found to carry no intensity while the intensity of the plus combination is seen to be proportional to the band splitting.

Notice that in the cases where we are dealing with charged groups (q_A and $q_B \neq 0$) and/or groups with a permanent dipole, both the charge–dipole term and/or the dipole–dipole term couple the excited states to the ground state. This may lead to additional splittings of the excited states. However, in these cases there will not be any simple relation between the band intensities and splittings.

The mechanisms which give intensity to the transitions terminating on doubly excited configurations are of particular interest. With the transition moment operator being given as a sum of one-electron operators, transitions from the ground state $|AB\rangle$ to an excited state $|A^*B^*\rangle$ where both A and B are excited, are not allowed. A possible intensity-giving mechanism can, however, be found in terms of a configuration interaction between singly and doubly excited states.[22]

Let the wave function for the ground state be $\Psi_A^0\Psi_B^0$. Using perturbation theory we may write a general expression for the doubly excited states

$$|A^*B^*\rangle = |\Psi_A^1\Psi_B^1\rangle + \sum_{a,b} \frac{\langle \Psi_A^a\Psi_B^b | V | \Psi_A^1\Psi_B^1\rangle}{W_A^a + W_B^b - W_A^1 - W_B^1} |\Psi_A^a\Psi_B^b\rangle$$

In particular Ψ_A^a or Ψ_B^b may be Ψ_A^0 or Ψ_B^0. Then

$$|A^*B^*\rangle = \Psi_A^1\Psi_B^1 + \sum_{a} \frac{\langle \Psi_A^a\Psi_B^0 | V | \Psi_A^1\Psi_B^1\rangle}{W_A^a + W_B^0 - W_A^1 - W_B^1} |\Psi_A^a\Psi_B^0\rangle + \sum_{b} \frac{\langle \Psi_A^0\Psi_B^b | V | \Psi_A^1\Psi_B^1\rangle}{W_A^0 + W_B^b - W_A^1 - W_B^1} |\Psi_A^0\Psi_B^b\rangle \qquad (4\text{-}224)$$

The electric dipole transition moment from the ground state to a doubly excited configuration is therefore given by

$$\langle AB | \sum_{i=1}^{s+t} \mathbf{r}_i | A^*B^*\rangle = \sum_{a} \frac{\langle \Psi_A^a\Psi_B^0 | V | \Psi_A^1\Psi_B^1\rangle}{W_A^a + W_B^0 - W_A^1 - W_B^1} \langle \Psi_A^0 | \sum_{i=1}^{s} \mathbf{r}_{Ai} | \Psi_A^a\rangle + \sum_{b} \frac{\langle \Psi_A^0\Psi_B^b | V | \Psi_A^1\Psi_B^1\rangle}{W_A^0 + W_B^b - W_A^1 - W_B^1} \langle \Psi_B^0 | \sum_{j=1}^{t} \mathbf{r}_{Bj} | \Psi_B^b\rangle \qquad (4\text{-}225)$$

The transition moment therefore depends, in this mechanism, on transitions located on either A or B, but not on matrix elements directly connecting the ground and excited states of interest.

Consider finally the situation in which the intermediate state is a charge-transfer state. In the framework of the model we can take the orbitals on chromophore A and chromophore B to be symmetrically orthogonalized, that is

$$|a\rangle = \psi_a - \tfrac{1}{2}S_{ab}\psi_b \qquad (4\text{-}226)$$

$$|b\rangle = \psi_b - \tfrac{1}{2}S_{ab}\psi_a \qquad (4\text{-}227)$$

Inspection of Eq. (4-225) reveals that in this case both the transition moments $\langle\Psi_A^0|\sum \mathbf{r}|\Psi_A^a\rangle$, $\langle\psi_B^0|\sum \mathbf{r}|\psi_B^b\rangle$ and the mixing coefficient will be scaled by a factor S_{ab}. Hence the calculated intensity will be smaller by a factor S_{ab}^4. Therefore unless the charge transfer state and the doubly excited state have comparable energies (in which case our perturbation approach breaks down) doubly excited states cannot profitably "steal" intensity from charge-transfer states.

In some interesting cases absorption of exchange-coupled pairs of metal ions may be seen.[23,24] In a system $Mn^{2+} - F^- - Ni^{2+}$ the ground state of Mn^{2+} has $S = \frac{5}{2}$ and the ground state of Ni^{2+} has $S = 1$. The excited states of Mn^{2+} have $S = \frac{3}{2}$ and those of Ni^{2+} have $S = 0$. The total spin of the coupled system is, in the ground state Mn–F–Ni, $S = \frac{3}{2}, \frac{5}{2}, \frac{7}{2}$. For the excited states Mn*–F–Ni, $S = \frac{1}{2}, \frac{3}{2}, \frac{5}{2}$; and for Mn–F–Ni*, $S = \frac{5}{2}$. The total spin S is taken to be a good quantum number, and the intensity of the transitions is accounted for by using an "effective" perturbation hamiltonian coupling the radiation field and the pair of paramagnetic ions. The intensity of the low-lying transitions is in these cases explained as being "stolen" from charge-transfer configurations. However, we are dealing here with strongly coupled systems, and the foregoing developments are therefore not valid. On the other hand for the bridged system Cu_2(acetate)$_4 \cdot 2H_2O$ a weakly coupled chromophore model seems to work admirably.[21]

4-14 THE RAMAN EFFECT. TWO-PHOTON ABSORPTION

If a system in a state Ψ_a absorbs a photon of energy $\hbar\omega$ and then emits a photon of energy $\hbar\omega'$ thereby ending up in a state Ψ_b, this process is called scattering of light. If Ψ_a is identical to Ψ_b, $\hbar\omega = \hbar\omega'$, we have *coherent* or *Rayleigh* scattering. If $\Psi_a \neq \Psi_b$ we have *incoherent* or *Raman* scattering.

The scattering of photons is best treated as a two-stage process, as given in the diagram.

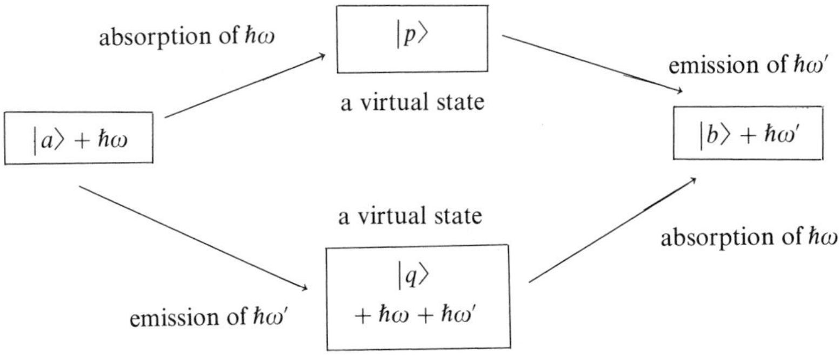

The system in moving from Ψ_a to Ψ_b can either first absorb a photon of energy $\hbar\omega$ and then emit a photon of energy $\hbar\omega'$, or it can perform the processes in reverse order.

For the transitions to the intermediate states $|p\rangle$ and $|q\rangle$ we have according to Eq. (4-8)

$$i\hbar \frac{\partial \mathscr{A}_p}{\partial t} = \langle p|\,\hat{G}\,|a\rangle \exp\left[i(\omega_{pa} - \omega)t\right] \tag{4-228}$$

$$i\hbar \frac{\partial \mathscr{A}_q}{\partial t} = \langle q|\,\hat{G}^*\,|a\rangle \exp\left[i(\omega_{qa} + \omega')t\right] \tag{4-229}$$

because according to Eq. (4-10) the first part of Eq. (4-8) corresponds to absorption, the second to emission.

Here $\hbar\omega_{pa} = W_p - W_a$, $\hbar\omega_{qa} = W_q - W_a$ and (see Eq. (4-24))

$$\hat{G} = \sum_N \frac{i\hbar Q_N}{M_N c} A^0(\omega)\mathbf{u} \exp\left(\frac{i\omega}{c}\boldsymbol{\kappa}\cdot\mathbf{r}_N\right)\cdot\nabla_N \tag{4-230}$$

where the summation over N includes both nuclei and electrons. Looking apart from arbitrary constants of integration which do not contribute to the scattering process, we get from Eqs. (4-228) and (4-229)

$$\mathscr{A}_p = \langle p|\,\hat{G}\,|a\rangle \frac{\exp\left[i(\omega_{pa} - \omega)t\right]}{\hbar(\omega - \omega_{pa})} \tag{4-231}$$

$$\mathscr{A}_q = -\langle q|\,\hat{G}^*\,|a\rangle \frac{\exp\left[i(\omega_{qa} + \omega')t\right]}{\hbar(\omega' + \omega_{qa})} \tag{4-232}$$

No physical reality should be attributed to the intermediate states. This is clear from the fact that the energy conservation law is not obeyed. They are thus only a computational convenience.

The transitions from the intermediate states to the final states are governed by

$$\begin{aligned} i\hbar \frac{\partial \mathscr{A}_b}{\partial t} &= \sum_p \mathscr{A}_p\langle b|\,\hat{G}^*|p\rangle \exp\left[i(\omega_{bp} + \omega')t\right] \\ &\quad + \sum_q \mathscr{A}_q\langle b|\,\hat{G}\,|q\rangle \exp\left[i(\omega_{bq} - \omega)t\right] \end{aligned} \tag{4-233}$$

Substituting for $\mathscr{A}_p$ and $\mathscr{A}_q$ we get, changing the summation index to s,

$$\begin{aligned} i\hbar \frac{\partial(\mathscr{A}_b)}{\partial t} &= \sum_s \left[\frac{\langle b|\,\hat{G}^*\,|s\rangle\langle s|\,\hat{G}\,|a\rangle}{\hbar(\omega - \omega_{sa})} - \frac{\langle b|\,\hat{G}\,|s\rangle\langle s|\,\hat{G}^*\,|a\rangle}{\hbar(\omega' + \omega_{sa})}\right] \\ &\quad \times \exp\left[\frac{i}{\hbar}(W_b + h\nu' - (W_a + h\nu))\right] \end{aligned} \tag{4-234}$$

In the case where $c/\omega \ll r_N$ we may expand $\hat{G}$ and get

$$\hat{G} = \sum_N \frac{i\hbar Q_N}{M_N c} A^0(\omega)\mathbf{u}\cdot\nabla_N \tag{4-235}$$

Using Eqs. (4-17) and (4-29) we obtain easily, normalizing the electric field strength A^0 to one photon per unit volume,

$$i\hbar\frac{\partial \mathscr{A}_b}{\partial t} = M_{ba}\exp\left[\frac{i}{\hbar}(W_b + h\nu' - (W_a + h\nu))\right] \tag{4-236}$$

with

$$M_{ba} = 2\pi\sqrt{\omega_{bs}\omega_{sa}}\sum_s\left[\frac{(\mathbf{u}'\cdot\mathbf{R}_{bs})(\mathbf{u}\cdot\mathbf{R}_{sa})}{\omega-\omega_{sa}} - \frac{(\mathbf{u}\cdot\mathbf{R}_{bs})(\mathbf{u}'\cdot\mathbf{R}_{sa})}{\omega'+\omega_{sa}}\right] \tag{4-237}$$

where

$$\mathbf{R}_{bs} = \langle b|\sum_N Q_N\mathbf{r}_N|s\rangle \tag{4-238}$$

and corresponding expressions. $\mathbf{u}$ and $\mathbf{u}'$ are, respectively, the polarization unit vectors of the incident and emitted photon.

Exactly as when we derived Fermi's "golden rule" (Eq. (4-12)) the integration of Eq. (4-236) leads to a transition probability per unit time equal to

$$p_{Ob} = \frac{2\pi}{\hbar}|M_{ba}|^2\delta[W_b + h\nu' - (W_a + h\nu)] \tag{4-239}$$

The energy of the system is seen to be conserved.

It follows from the form of M_{ba} that in order to get incoherent scattering $(\mathbf{u}'\cdot\mathbf{R}_{bs})(\mathbf{u}\cdot\mathbf{R}_{sa}) \neq 0$. In the pertinent cases this is the same as demanding $\langle a|\mathbf{RR}|b\rangle \neq 0$, where $\mathbf{RR}$ includes all bilinear forms XX, $XY, \ldots, ZZ$. The scattering is therefore governed by the transformation properties of the bilinear terms.

In a two-photon absorption process the system proceeds, after having reached the intermediate state $|p\rangle$ by absorbing one photon, to absorb one more photon, thereby reaching a final state $|c\rangle$.

Using Eq. (4-228) we get immediately

$$i\hbar\frac{\partial \mathscr{A}_c}{\partial t} = \sum_p\langle p|\hat{G}|a\rangle\frac{\exp[i(\omega_{pa}-\omega)t]}{\hbar(\omega-\omega_{pa})}\langle c|\hat{G}|p\rangle\exp[i(\omega_{cp}-\omega)t] \tag{4-240}$$

$$\frac{\partial \mathscr{A}_c}{\partial t} = \sum_p\frac{\langle c|\hat{G}|p\rangle\langle p|\hat{G}|a\rangle}{i\hbar^2(\omega-\omega_{pa})}\exp[i(\omega_{ca}-2\omega)t] \tag{4-241}$$

A resonance evidently appears at $\omega_{ca} = 2\omega$ corresponding to a two-photon absorption process. The probability of the process occurring is proportional to

$$\sum_p\frac{|\langle c|\hat{G}|p\rangle\langle p|\hat{G}|a\rangle|^2}{\hbar^4(\omega-\omega_{pa})^2} \tag{4-242}$$

As for the Raman process the selection rules for the two-photon transition are given by the transformation properties of the bilinear terms XX, $XY, \ldots, ZZ$.

For electric dipole transitions we have in a system possessing a center of symmetry the usual $u \rightleftarrows g$ selection rule. Notice that both for Raman scattering and two-photon absorption the selection rules will be of the form $g \rightleftarrows g$ and $u \rightleftarrows u$. Hence states can be reached which are inaccessible to single-photon electric dipole transitions.

In the adiabatic approximation we take $\Psi_a = \psi_a \chi_{av''}$ and $\Psi_b = \psi_b \chi_{bv'}$. In the case $\psi_a = \psi_b$ the summation over the N particles in Eq. (4-235) only leaves the terms $\langle \chi_{av''} | \mathbf{RR} | \chi_{av'} \rangle$. A Raman transition between two vibrational states belonging to the same potential surface is therefore allowed, provided the product $\Gamma(\chi_{v''})\, \Gamma(\chi_{v'})$ has the same species as at least one of the bilinear forms.

For the electronic Raman effect, the summation over N leaves only terms of the type

$$e^2 \langle \psi_a | \sum_n \mathbf{r}_n \sum_n \mathbf{r}_n | \psi_b \rangle \langle \chi_{av''} | \chi_{bv'} \rangle$$

In the pure electronic Raman effect $v'' = v'$. A progression in a totally symmetric vibration may also be seen, scaled in intensity by the Franck–Condon factors $|\langle \chi_{v''} | \chi_{v'} \rangle|^2$. Further complications may be encountered by using a Herzberg–Teller expansion of Ψ_a and Ψ_b, leading to the appearance of "vibronic" electronic Raman transitions. In that case, one quantum of a non-totally symmetric vibration is emitted simultaneously with the change in electronic energy.

Under conditions in which a molecule is excited with a laser line whose frequency is close to the band maximum of an allowed electronic transition, a so-called resonance Raman spectrum may be obtained. Such spectra are characterized by a large increase in the intensity of a band arising from a totally symmetric fundamental of the molecule. A good example is found in the $Mo_2Cl_8^{4-}$ ion.[26] Irradiation within the contour of the lowest allowed transition leads to long progressions in the axial metal–metal stretching fundamental. This demonstrates that the resonant electronic transition is also axially polarized, and it can be assigned as $\delta \rightarrow \delta^*(^1A_{1g} \rightarrow {}^1A_{2u})$.

REFERENCES

1. L. I. Schiff, *Quantum Mechanics*, 3d ed., McGraw-Hill, 1968.
2. A. Einstein, *Z. Physik*, **18**: 121, 1917. English translation in B. L. Van der Waerden (ed.), *Sources of Quantum Mechanics*, North-Holland, 1967.
3. H. H. Jaffé and M. Orchin, *Theory and Applications of Ultraviolet Spectroscopy*, John Wiley & Sons, 1962.
4. W. Moffitt and A. Moscowitz, *J. Chem. Phys.*, **30**: 648, 1959.
5. J. P. Byrne and I. G. Ross, *Aust. J. Chem.*, **24**: 1107, 1971.
6. M. Bixon and J. Jortner, *J. Chem. Phys.*, **48**: 715, 1968.
7. *Handbook of Mathematical Functions*, National Bureau of Standards, Applied Math. Series 55, Washington D.C., 1964, p. 75.
8. B. Sharf, *Chem. Phys. Lett.*, **5**: 456, 459, 1970.

9. P. C. Haarhoff, *Mol. Phys.*, **7**: 101, 1963.
10. Aa. E. Hansen and J. Avery, *Chem. Phys. Lett.*, **13**: 396, 1972.
11. P. J. Stephens, *J. Chem. Phys.*, **52**: 3489, 1970, and *Ann. Rev. Phys. Chem.*, **25**: 201, 1974.
12. H. Bateman, *Higher Transcendental Functions*. Vol. II. McGraw-Hill, vol. II, p. 192, 1953.
13. T. H. Keil, *Phys. Rev.*, **140A**: 601, 1965.
14. A. D. Liehr and C. J. Ballhausen, *Phys. Rev.*, **106**: 1161, 1957.
15. R. F. Fenske, *J. Am. Chem. Soc.*, **89**: 252, 1967.
16. J. J. Markham, *Rev. Mod. Phys.*, **31**: 956, 1959.
17. Y. Toyozawa and M. Inoue, *J. Phys. Soc.* (Japan), **21**: 1663, 1966.
18. F. S. Ham, *Phys. Rev.*, **138**: A1727, 1965.
19. M. Lax, *J. Chem. Phys.*, **20**: 1752, 1952.
20. C. Longuet-Higgins and J. N. Murrell, *Proc. Phys. Soc.*, **A68**: 601, 1955.
21. Aa. E. Hansen and C. J. Ballhausen, *Trans. Far. Soc.*, **61**: 631, 1965.
22. D. L. Dexter, *Phys. Rev.*, **126**: 1962, 1962.
23. J. Ferguson, H. J. Guggenheim, and Y. Tanabe, *J. Phys. Soc.* (Japan), **21**: 692, 1966.
24. J. Ferguson, H. J. Guggenheim, and Y. Tanabe, *J. Chem. Phys.*, **45**: 1134, 1966.
25. M. Göppert-Mayer, *Ann. Physik.*, **9**: 273, 1931.
26. R. J. H. Clark, M. L. Franks, and P. C. Turtle, *J. Am. Chem. Soc.*, **29**: 2473, 1977.
27. R. J. Tacon, P. Day, and R. G. Denning, *J. Chem. Phys.*, **61**: 751, 1974.
28. A. Moscowitz, *Adv. Chem. Phys.*, **4**: 67, 1964.

CHAPTER

FIVE

THE CHARACTERIZATION OF GROUND STATES. SPECIFIC EXAMPLES

5-1 HIGH-SPIN AND LOW-SPIN STATES

Knowing the molecular structure of a certain inorganic transition metal complex, one may classify the various molecular orbitals after their transformation properties in an effective point group characteristic of the system. A schematic molecular-orbital diagram can be constructed, and the molecular orbitals filled by the appropriate number of electrons. As a rough-and-ready guide to the energetic placement of the molecular orbitals one may use the size of the group overlap integral between the metal ion and the ligands. The larger the group overlap integral, the deeper the bonding molecular orbital is placed energetically and the higher the corresponding antibonding orbital. An antibonding pi orbital is therefore normally lower in energy than an antibonding sigma orbital.

Consider the case of an octahedral molecule. In Table 5-1 we see the various states arising from the electronic configurations $(t_2)^m(e)^n$. The core energies of the states can be written as $m(-4Dq) + n(6Dq)$. We notice, however, from Table 5-1 that for $m + n = 4$, 5, 6, and 7 the highest spin-multiplicities are not associated with the states having the lowest core energy. In d^4, for instance, the high-spin configuration $(t_2\alpha)^3(e\alpha)^1$ has $6[= (4 \cdot 3)/1 \cdot 2)]$ pairs of electrons with parallel spin. The low-spin configuration $(t_2\alpha)^3(t_2\beta)^1$ have only $3[= (3 \cdot 2)/(1 \cdot 2)]$ such pairs. The exchange energy which occurs with negative sign in the expression for the total energy is therefore larger for the high-spin than for the low-spin configurations. A loss of exchange energy may, however, be compensated by a gain in core energy, and a competition takes place as to which ground state is energetically most favorable.[1,2]

The above principle is well illustrated by the octahedral $[Co^{3+}L_6]$ complexes.

Table 5-1 Electronic configurations and states for octahedral and tetrahedral symmetry

(The states for 6 to 9 valence electrons can be obtained from the table by looking at the distribution of the holes in the e and t_2 shells.)

$(t_2)^2: {}^1A_1 + {}^1E + {}^1T_2 + {}^3T_1$
$(t_2)^1(e)^1: {}^1T_1 + {}^1T_2 + {}^3T_1 + {}^3T_2$
$(e)^2: {}^1A_1 + {}^1E + {}^3A_2$
$(t_2)^3: {}^2E + {}^2T_1 + {}^2T_2 + {}^4A_2$
$(t_2)^2(e)^1: {}^2A_1 + {}^2A_2 + 2{}^2E + 2{}^2T_1 + 2{}^2T_2 + {}^4T_1 + {}^4T_2$
$(t_2)^1(e)^2: 2{}^2T_1 + 2{}^2T_2 + {}^4T_1$
$(e)^3: {}^2E$
$(t_2)^4: {}^1A_1 + {}^1E + {}^1T_2 + {}^3T_1$
$(t_2)^3(e)^1: {}^1A_1 + {}^1A_2 + {}^1E + 2{}^1T_1 + 2{}^1T_2 + {}^3A_1 + {}^3A_2 + 2{}^3E + 2{}^3T_1 + 2{}^3T_2 + {}^5E$
$(t_2)^2(e)^2: 2{}^1A_1 + {}^1A_2 + 3{}^1E + {}^1T_1 + 3{}^1T_2 + {}^3A_2 + {}^3E + 3{}^3T_1 + 2{}^3T_2 + {}^5T_2$
$(t_2)^1(e)^3: {}^1T_1 + {}^1T_2 + {}^3T_1 + {}^3T_2$
$(e)^4: {}^1A_1$
$(t_2)^5: {}^2T_2$
$(t_2)^4(e)^1: {}^2A_1 + {}^2A_2 + 2{}^2E + 2{}^2T_1 + 2{}^2T_2 + {}^4T_1 + {}^4T_2$
$(t_2)^3(e)^2: 2{}^2A_1 + {}^2A_2 + 3{}^2E + 4{}^2T_1 + 4{}^2T_2 + {}^4A_1 + {}^4A_2 + 2{}^4E + {}^4T_1 + {}^4T_2 + {}^6A_1$
$(t_2)^2(e)^3: {}^2A_1 + {}^2A_2 + 2{}^2E + 2{}^2T_1 + 2{}^2T_2 + {}^4T_1 + {}^4T_2$
$(t_2)^1(e)^4: {}^2T_2$

With Dq large the ground state is $(t_{2g})^6\ {}^1A_{1g}$; but for Dq small it is $(t_{2g})^4(e_g)^2\ {}^5T_{2g}$. It is therefore the strength of the bonding of the ligands to the Co^{3+} ion which determines the ground state.

The crossing of the states having different spin-multiplicities may be studied using either an INDO molecular-orbital method or crystal-field theory. An INDO analysis[3] of CoF_6^{3-} reveals that at the equilibrium bond distance the high-spin state ${}^5T_{2g}$ is the ground state. Conversely for $Co(H_2O)_6^{3+}$ the low-spin state ${}^1A_{1g}$ is stable at the equilibrium bond distance by 2.1 eV over ${}^5T_{2g}$. The calculated potential curves show that larger equilibrium bond distances correspond to an increase in the occupancy number of electrons in the e_g shell, thereby confirming an old proposal of van Santen and van Wieringen.[4] States with different spin-multiplicities but arising from the same electronic configuration $(t_{2g})^m(e_g)^n$ were further found[3] to have the same shape and equilibrium bond distance.

Using the phenomenological parameters of crystal-field theory the evaluations of the state energies are easy. Consider again the octahedral system $[Co^{3+}L_6]$. With

$$|\overset{+}{\psi}_{xy}\overset{-}{\psi}_{xy}\overset{+}{\psi}_{xz}\overset{-}{\psi}_{xz}\overset{+}{\psi}_{yz}\overset{-}{\psi}_{yz}|$$

the wave function of ${}^1A_{1g}$ and

$$|\overset{+}{\psi}_{xy}\overset{-}{\psi}_{xy}\overset{+}{\psi}_{xz}\overset{+}{\psi}_{yz}\overset{+}{\psi}_{x^2-y^2}\overset{+}{\psi}_{z^2}|$$

Table 5-2 Table of promotional energies in octahedral complexes

	High spin	Low spin	Promotional energy
d^4	${}^5E_g(-6Dq)$	${}^3T_{1g}(t_{2g})^4$	$6F_2 + 145F_4 - 10Dq$
d^5	${}^6A_{1g}(0Dq)$	${}^2T_{2g}(t_{2g})^5$	$15F_2 + 275F_4 - 20Dq$
d^6	${}^5T_{2g}(-4Dq)$	${}^1A_{1g}(t_{2g})^6$	$5F_2 + 255F_4 - 20Dq$
d^7	${}^4T_{1g}(-8Dq)$	${}^2E_g(t_{2g})^6(e_g)$	$7F_2 + 105F_4 - 10Dq$

a component of ${}^5T_{1g}$, we get easily using Table 2-3

$$W({}^5T_{2g}) = -4Dq + 15F_0 - 35F_2 - 315F_4 \tag{5-1}$$

$$W({}^1A_{1g}) = -24Dq + 15F_0 - 30F_4 - 60F_4 \tag{5-2}$$

Hence the ${}^1A_{1g}$ state will be the ground state provided $20Dq > 5(F_2 + 51F_4)$. The promotional energies of the four cases of interest for octahedral complexes are given in Table 5-2.

It is obvious that with a lower effective symmetry than O_h further state splittings occur and more unknown parameters will have to be introduced in attempts to prognosticate the ground state. Crystal-field considerations are anyhow only good for estimating orders of magnitude, and an overelaboration of parameter values serves no purpose.

5-2 STATIC JAHN–TELLER EFFECTS

In cases where the symmetry classification of the ground state predicts orbital degeneracy a Jahn–Teller effect is operative. The best-documented static Jahn–Teller cases are found for certain octahedral d^9 systems and we shall consider such a system.

Classifying the orbitals according to their transformation properties in the octahedral point group O_h, the ground state for a d^9 system transforms like ${}^2E$ with the orbital components

$$\Psi^0_{z^2} = \left|(\overset{+}{x}z)(\overset{-}{x}z)(\overset{+}{x}y)(\overset{-}{x}y)(\overset{+}{y}z)(\overset{-}{y}z)(x^2 \overset{+}{-} y^2)(x^2 \overset{-}{-} y^2)(\overset{+}{z^2})\right| \tag{5-3}$$

$$\Psi^0_{x^2-y^2} = \left|(\overset{+}{x}z)(\overset{-}{x}z)(\overset{+}{x}y)(\overset{-}{x}y)(\overset{+}{y}z)(\overset{-}{y}z)(x^2 \overset{+}{-} y^2)(\overset{+}{z^2})(\overset{-}{z^2})\right| \tag{5-4}$$

The interaction of an E state with an ε vibration has been dealt with in section 1-6. Considering only a linear Jahn–Teller coupling, the wave function for the lower potential sheet can be written (compare Eqs. (1-89), (1-73), and (1-74))

$$|0\rangle = \sin\phi/2\,\Psi^0_{z^2} + \cos\phi/2\,\Psi^0_{x^2-y^2} \tag{5-5}$$

ited states are given by

$$\Psi^0_{xz} = \left|(\overset{+}{x}z)(\overset{+}{x}y)(\overset{-}{x}y)(\overset{+}{y}z)(\overset{-}{y}z)(x^2 \overset{+}{-} y^2)(x^2 \overset{-}{-} y^2)(\overset{+}{z^2})(\overset{-}{z^2})\right| \tag{5-6}$$

$$\Psi^0_{yz} = |(\overset{+}{xz})(\overset{-}{xz})(\overset{+}{xy})(\overset{-}{xy})(\overset{+}{yz})(x^2 \overset{+}{-} y^2)(x^2 \overset{-}{-} y^2)(\overset{+}{z^2})(\overset{-}{z^2})| \tag{5-7}$$

$$\Psi^0_{xy} = |(\overset{+}{xz})(\overset{-}{xz})(\overset{+}{xy})(\overset{+}{yz})(\overset{-}{yz})(x^2 \overset{+}{-} y^2)(x^2 \overset{-}{-} y^2)(\overset{+}{z^2})(\overset{-}{z^2})| \tag{5-8}$$

spanning a T_2 state in O_h symmetry. The properties of the lower potential sheet can now be illuminated by considering the behavior of the g factors. From the definition of Eq. (3-59) we have $g_{ii} = 2(1 - \lambda k \Lambda_{ii})$, with Λ_{ii} given by Eq. (3-53).

Using Table 3-1 we get

$$\hat{L}_x |0\rangle = -i(\sqrt{3} \sin \phi/2 + \cos \phi/2) \Psi^0_{yz} \tag{5-9}$$

$$\hat{L}_y |0\rangle = i(\sqrt{3} \sin \phi/2 - \cos \phi/2) \Psi^0_{xz} \tag{5-10}$$

$$\hat{L}_z |0\rangle = 2i \cos \phi/2 \, \Psi^0_{xy} \tag{5-11}$$

The d-shell is more than half filled and $S = \frac{1}{2}$ so λ is equal to $-\zeta_M$, the molecular spin-orbit coupling constant. Hence[5]

$$g_{xx} = 2 + \frac{2k_{\sigma,\pi}\zeta_M \left[\sqrt{3} \sin \dfrac{\phi}{2} + \cos \dfrac{\phi}{2}\right]^2}{W_{yz} - W_0} \tag{5-12}$$

$$g_{yy} = 2 + \frac{2k_{\sigma,\pi}\zeta_M \left[\sqrt{3} \sin \dfrac{\phi}{2} - \cos \dfrac{\phi}{2}\right]^2}{W_{xz} - W_0} \tag{5-13}$$

$$g_{zz} = 2 + \frac{2k_{\sigma,\pi}\zeta_M \left[2 \cos \dfrac{\phi}{2}\right]^2}{W_{xy} - W_0} \tag{5-14}$$

where $k_{\sigma,\pi}$ is the orbital reduction factor.

In the theory for the linear Jahn–Teller effect ϕ is a cyclic coordinate. To second order this is no longer the case and the minima on the lower potential surface are determined by $\cos 3\phi = 1$ or -1 (see Eq. (1-79)), and the saddle points at $\cos 3\phi = -1$ or 1. At high temperatures where the system can override the barriers ϕ can, however, be considered effectively cyclic. Averaging the g factors over the vibrational wave functions we get in the high-temperature limit an isotropic g factor

$$\bar{g}_{xx} = \bar{g}_{yy} = \bar{g}_{zz} = 2 + \frac{4k_{\sigma,\pi}\zeta_M}{W_{T_2} - W_0} \tag{5-15}$$

In the low-temperature limit it is, however, possible to trap the system in one of the three valleys on the lowest potential surface. Let the positions of the valleys be given by $\cos 3\phi = -1$. Using $\phi = \pi$ we get

$$g_{xx} = g_{yy} = g_{\perp} = 2 + \frac{6k_{\sigma,\pi}\zeta_M}{W_{T_2} - W_0} \tag{5-16}$$

$$g_{zz} = g_{\parallel} = 2 \tag{5-17}$$

With $\cos 3\phi = 1$ the result is, taking $\phi = 0$,

$$g_{xx} = g_{yy} = g_{\perp} = 2 + \frac{2k_{\sigma,\pi}\zeta_M}{W_{T_2} - W_0} \tag{5-18}$$

$$g_{zz} = g_{\parallel} = 2 + \frac{8k_{\sigma,\pi}\zeta_M}{W_{T_2} - W_0} \tag{5-19}$$

Low-temperature resonance spectra of copper fluosilicate hexahydrate $CuSiF_6 \cdot 6H_2O$ showed[6] the 90°K spectrum to be nearly isotropic with $g_{\parallel} = g_{\perp} = 2.24$. At 20°K, however, $g_{\parallel} = 2.46$ and $g_{\perp} = 2.11$ in beautiful accord with theory provided $\cos 3\phi = 1$. For $\phi = 0$ we have from Eq. (5-5) that $|0\rangle = \Psi^0_{x^2-y^2}$ is the ground state. In this state function the antibonding orbital $\psi^*(x^2 - y^2)$ is only singly occupied, whereas $\psi^*(z^2)$ is doubly occupied. The indications are therefore that antibonding orbital $\psi^*(x^2 - y^2)$ is higher in energy than is $\psi^*(z^2)$, pointing towards a stronger bonding in the molecular xy plane than in the z direction. Hence the Jahn–Teller-stabilized molecular configuration should correspond to an elongated octahedron.

With a Kramers doublet, $E_{1/2}$ or $E_{5/2}$, as ground state an octahedral system is of course Jahn–Teller-resistant. A strong spin-orbit coupling may therefore stabilize an otherwise Jahn–Teller-unstable conformation. The fourfold degenerate octahedral double group representation G is, however, subject to a Jahn–Teller effect.[7]

The antisymmetric product $G \times G$ equals $A_1 + E + T_2$. The fourfold dimensional representations are therefore unstable with respect to displacements of ε_g and τ_{2g} symmetries. From symmetry considerations we may write for $\mathscr{H}^{(1)}$

$$\mathscr{H}^{(1)} = c_\varepsilon(\xi_{\varepsilon 1}\rho_1 + \xi_{\varepsilon 2}\rho_2) + c_\tau(\xi_{\tau 1}\sigma_1 + \xi_{\tau 2}\sigma_2 + \xi_{\tau 3}\sigma_3)$$

where the vector matrices $\boldsymbol{\rho}$ and $\boldsymbol{\sigma}$ are the Dirac 4×4 matrices. The potential sheets are doubly degenerate. They are given by

$$V = \tfrac{1}{2}k_\varepsilon r^2 + \tfrac{1}{2}k_\tau R^2 \pm \sqrt{c_\varepsilon^2 r^2 + c_\tau^2 R^2}$$

$$r^2 = \xi_{\varepsilon 1}^2 + \xi_{\varepsilon 2}^2, \qquad R^2 = \xi_{\tau 1}^2 + \xi_{\tau 2}^2 + \xi_{\tau 3}^2$$

Their minima lie on the spherical surface

$$r = 0, \qquad R = \frac{|c_\tau|}{k_\tau}$$

and on the circle

$$R = 0, \qquad r = \frac{|c_\varepsilon|}{k_\varepsilon}$$

respectively. The stabilization energies are

$$\Delta V^{(\varepsilon)}_{J-T} = \frac{c_\varepsilon^2}{2k_\varepsilon} \quad \text{or} \quad \Delta V^{(\tau)}_{J-T} = \frac{c_\tau^2}{2k_\tau}$$

The former are stable when $c_\tau^2/k_\tau > c_\varepsilon^2/k_\varepsilon$ and the latter are stable when the opposite holds true. The distortions are therefore either of the trigonal type, symmetry D_3, or of the tetragonal, D_4, type. Provided the stabilization energy is large compared to a vibrational quantum, a lower than octahedral symmetry may therefore be met with.

The infrared band system $G \rightarrow E_{5/2}$ in ReF_6 has a magnetic dipole–allowed origin and two electric dipole vibronic origins associated with single quanta of $\nu_4(\tau_{1u})$ and $\nu_6(\tau_{2u})$. On each of these three origins are found two or three members of a τ_{2g} progression. These are harmonic and show the operation of a trigonal Jahn–Teller effect in the G state. The data can be accounted for[8] by $\nu_5 = 176\ \text{cm}^{-1}$ and $\Delta V_{J-T}^{(\tau)} = 74\ \text{cm}^{-1}$.

5-3 g-FACTORS IN AN ORBITALLY DEGENERATE STATE

When the ground state of a complex is orbitally degenerate the g-factor may differ considerably from the spin value of 2. We have already seen an example of this for a $(t_{2g})^1$, ${}^2T_{2g}$ state (section 3-4). As a further example we shall consider the g-factor in the ${}^4T_{1g}$ state, found as ground state in high-spin octahedral Co^{2+} complexes.[9,10]

A ${}^4T_{1g}$ state will exhibit a first-order spin-orbit coupling. The twelvefold degeneracy of ${}^4T_{1g}$ will be lifted into three sets; one set being twofold degenerate, one set fourfold degenerate, and one set sixfold degenerate. Introducing a fictitious angular momentum $\mathbf{L} = 1$ (compare section 1-7) we use a representation in which $\hat{L}_z$ and $\hat{S}_z$ are diagonal. Writing $|M_L, M_S\rangle$ to characterize the components of a state we have, for example, for the $(t_{2g})^5(e_g)^2\ {}^4T_{1g}$, $M_J = \frac{5}{2}$ component.

$$|1, \tfrac{3}{2}\rangle = |\overset{+}{\omega}_1 \overset{-}{\omega}_1 \overset{+}{\omega}_0 \overset{-}{\omega}_0 \overset{+}{\omega}_{-1} \overset{+}{e}_a \overset{+}{e}_b|$$

Here ω_1, ω_0, and ω_{-1} are the t_{2g} orbitals of Eq. (1-128) and e_a and e_b the two orbitals spanning e_g.

The spin-orbit coupling operator for the ${}^4T_{1g}$ state is taken as $\mathscr{H}^{(1)} = \alpha\lambda\, \mathbf{L}\cdot\mathbf{S}$. Looking at the component $|1, \frac{3}{2}\rangle$ we get $\alpha\lambda\hat{L}_z\hat{S}_z|1, \frac{3}{2}\rangle = \frac{3}{2}\lambda\alpha|1, \frac{3}{2}\rangle$. Further using Eq. (1-130) we have

$$\sum_{n=1}^{7} \zeta_M \hat{l}_{zn}\hat{s}_{zn} |\overset{+}{\omega}_1 \overset{-}{\omega}_1 \overset{+}{\omega}_0 \overset{-}{\omega}_0 \overset{+}{\omega}_{-1} \overset{+}{e}_a \overset{+}{e}_b| = \tfrac{1}{2}\zeta_M |\overset{+}{\omega}_1 \overset{-}{\omega}_1 \overset{+}{\omega}_0 \overset{-}{\omega}_0 \overset{+}{\omega}_{-1} \overset{+}{e}_a \overset{+}{e}_b|$$

Comparing the two expressions we get $\lambda\alpha = \frac{1}{3}\zeta_M$. The spin-orbit coupling hamiltonian is therefore

$$\mathscr{H}^{(1)} = \tfrac{1}{3}\zeta_M(\hat{L}_z\hat{S}_z + \tfrac{1}{2}\hat{L}_+\hat{L}_- + \tfrac{1}{2}\hat{L}_-\hat{S}_+) \tag{5-20}$$

The spin-orbit coupling energies, measured in units of ζ_M for the ${}^4T_{1g}(t_{2g})^5(e_g)^2$ state are given by

$M_J = \frac{1}{2}$.

$$\begin{array}{ccc} |-1, \frac{3}{2}\rangle & |0, \frac{1}{2}\rangle & |1, -\frac{1}{2}\rangle \end{array}$$

$$\begin{vmatrix} -\dfrac{1}{2} - W^{(1)} & \sqrt{\dfrac{1}{6}} & 0 \\ \sqrt{\dfrac{1}{6}} & -W^{(1)} & \dfrac{\sqrt{2}}{3} \\ 0 & \dfrac{\sqrt{2}}{3} & -\dfrac{1}{6} - W^{(1)} \end{vmatrix} = 0 \qquad (5\text{-}21)$$

$M_J = \frac{3}{2}$.

$$\begin{array}{cc} |0, \frac{3}{2}\rangle & |1, \frac{1}{2}\rangle \end{array}$$

$$\begin{vmatrix} -W^{(1)} & \sqrt{\dfrac{1}{6}} \\ \sqrt{\dfrac{1}{6}} & \dfrac{1}{6} - W^{(1)} \end{vmatrix} = 0 \qquad (5\text{-}22)$$

$M_J = \frac{5}{2}$.

$$|1, \tfrac{3}{2}\rangle$$

$$|\tfrac{1}{2} - W^{(1)}| = 0 \qquad (5\text{-}23)$$

The solutions to Eqs. (5-21), (5-22), and (5-23) are

$$W^{(1)} = \tfrac{1}{2} \qquad M_J = \tfrac{1}{2}, \tfrac{3}{2}, \tfrac{5}{2} \qquad (5\text{-}24)$$

$$W^{(1)} = -\tfrac{1}{3} \qquad M_J = \tfrac{1}{2}, \tfrac{3}{2} \qquad (5\text{-}25)$$

$$W^{(1)} = -\tfrac{5}{6} \qquad M_J = \tfrac{1}{2} \qquad (5\text{-}26)$$

The lowest spin-orbit component is therefore the twofold degenerate level having $W^{(1)} = -\frac{5}{6}\zeta_M$. Solving Eq. (5-21) with this value of $W^{(1)}$ leads to the Kramers doublet, transforming like Γ_6.

$$\Gamma_6^{a,b} = \begin{cases} \sqrt{\frac{1}{2}}\,|-1, \frac{3}{2}\rangle - \sqrt{\frac{1}{3}}\,|0, \frac{1}{2}\rangle + \sqrt{\frac{1}{6}}\,|1, -\frac{1}{2}\rangle \\[2ex] \sqrt{\frac{1}{2}}\,|1, -\frac{3}{2}\rangle - \sqrt{\frac{1}{3}}\,|0, -\frac{1}{2}\rangle + \sqrt{\frac{1}{6}}\,|-1, \frac{1}{2}\rangle \end{cases} \qquad (5\text{-}27)$$

The isotropic g factor for the octahedral complex is calculated from

$$g = 2\langle \Gamma_6^a |\, \alpha \hat{L}_z + 2\hat{S}_z \,| \Gamma_6^a \rangle = -\tfrac{2}{3}\alpha + \tfrac{10}{3} \qquad (5\text{-}28)$$

where α is the effective Landé factor for the $(t_{2g})^5(e_g)^2\ {}^4T_{1g}$ state. Using Eqs. (1-130) and (3-24*a*) we get

$$\alpha \hat{L}_z\,|1, \tfrac{3}{2}\rangle = \alpha\,|1, \tfrac{3}{2}\rangle \equiv \sum_{n=1}^{7} \hat{l}_{zn}\,|\overset{+}{\omega}_1 \overset{-}{\omega}_1 \overset{+}{\omega}_0 \overset{-}{\omega}_0 \overset{+}{\omega}_{-1} \overset{+}{e}_a \overset{+}{e}_b| = -k_{\pi,\pi}\,|\overset{+}{\omega}_1 \overset{-}{\omega}_1 \overset{+}{\omega}_0 \overset{-}{\omega}_0 \overset{+}{\omega}_{-1} \overset{+}{e}_a \overset{+}{e}_b|$$

leading to $\alpha = -k_{\pi,\pi}$, where $k_{\pi,\pi}$ is an orbital reduction factor. The value of α is, however, dependent upon the amount of configuration interaction under the $1/r_{12}$ terms between the two ${}^4T_{1g}$ ligand-field states found in the system.

Writing

$$|{}^4T_{1g}\rangle = c_1\,|{}^4T_{1g}(t_{2g})^5(e_g)^2\rangle + c_2\,|{}^4T_{1g}(t_{2g})^4(e_g)^3\rangle \tag{5-29}$$

we find (compare Eq. (3-31))

$$\alpha = -(c_1^2 - \tfrac{1}{2}c_2^2)k_{\pi,\pi} - 2c_1c_2k_{\sigma,\pi} \tag{5-30}$$

where $k_{\pi,\pi}$ and $k_{\sigma,\pi}$ are the orbital reduction factors. Putting $k_{\pi,\pi} = k_{\sigma,\pi} = 1$, $c_1 = \sqrt{\tfrac{4}{5}}$, and $c_2 = \sqrt{\tfrac{1}{5}}$ valid for the electronic ground state in the limit $Dq = 0$, we find $\alpha = -\tfrac{3}{2}$. With this value of α Eq. (5-28) gives $g = 4.333$. Experiments[10] with Co^{2+} dissolved in MgO gave an experimental value of $g = 4.278$.

A refinement in the calculation of g would be to consider the mixing under the spin-orbit coupling operator of the ground state Γ_6 with the excited states of Γ_6 symmetry. The nearest such species is found in the manifold of spin-orbit states contained in the ${}^4T_{2g}$ state, located some $10Dq$ above the ground Γ_6. Mixing of the two Γ_6 states is estimated to change the calculated g value by some 4 percent. However, to introduce an additional small correction into the expression for g, which already through α depends upon the three parameters c_1, $k_{\pi,\pi}$, and $k_{\sigma,\pi}$ serves little purpose.

A lowering of the symmetry of a complex from octahedral to, say, tetragonal or trigonal can bring about quite drastic changes in the calculated g-factors. Consider the case of the octahedral state ${}^2T_{2g}(t_{2g})^1$. In the lower symmetry, two of the orbital components ω_1 and ω_{-1} will still be degenerate, and we take them to be separated by Δ from the nondegenerate component ω_0. Using a fictitious angular momentum $\mathbf{L} = 1$ the spin-orbit coupling operator is taken to be $\mathscr{H}^{(1)} = \alpha\lambda\mathbf{L}\cdot\mathbf{S}$. We have

$$\alpha\lambda\hat{L}_z\hat{S}_z\,|1,\tfrac{1}{2}\rangle = \tfrac{1}{2}\lambda\alpha\,|1,\tfrac{1}{2}\rangle \equiv \zeta_M\hat{l}_z\hat{s}_z\,|\overset{+}{\omega}_1| = -\tfrac{1}{2}\zeta_M\,|\overset{+}{\omega}_1|$$

Therefore

$$\mathscr{H}^{(1)} = -\zeta_M(\hat{L}_z\hat{S}_z + \tfrac{1}{2}\hat{L}_+\hat{S}_- + \tfrac{1}{2}\hat{L}_-\hat{S}_+) \tag{5-31}$$

with the energies being given by

$M_J = \tfrac{1}{2}$. $\qquad |0,\tfrac{1}{2}\rangle \qquad |1,-\tfrac{1}{2}\rangle$

$$\begin{vmatrix} -W^{(1)} & -\dfrac{\sqrt{2}}{2}\zeta_M \\ -\dfrac{\sqrt{2}}{2}\zeta_M & \Delta + \tfrac{1}{2}\zeta_M - W^{(1)} \end{vmatrix} = 0 \tag{5-32}$$

$M_J = \tfrac{3}{2}$. $\qquad |1,\tfrac{1}{2}\rangle$

$$|\Delta - \tfrac{1}{2}\zeta_M - W^{(1)}| = 0 \tag{5-33}$$

leading to

$$W^{(1)} = \begin{cases} \tfrac{1}{2}\Delta + \tfrac{1}{4}\zeta_M \pm \tfrac{1}{2}\sqrt{(\tfrac{3}{2}\zeta_M + \tfrac{1}{3}\Delta)^2 + \tfrac{8}{9}\Delta^2} \\ \Delta - \tfrac{1}{2}\zeta_M \end{cases} \tag{5-34}$$

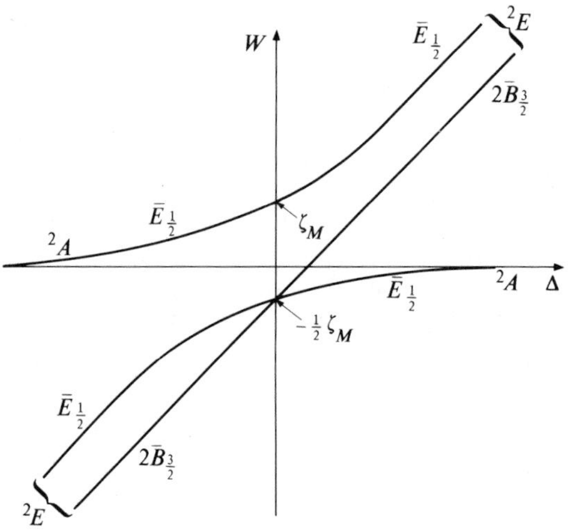

Figure 5-1 The splitting of $^2T_{2g}$ in trigonal symmetry, as characterized in C_3 symmetry.

The energies of the spin-orbit coupling components of $^2T_{2g}$ as a function of Δ are pictured in Fig. 5-1.

In the case of Δ negative, we get by evaluating the Zeeman energies for the ground state doublet $|1, \frac{1}{2}\rangle$, $|-1, -\frac{1}{2}\rangle$ with H parallel to the three- or fourfold symmetry axis

$$\Delta W = 2\beta H_{\parallel} \langle 1, \tfrac{1}{2}| \alpha\hat{L}_z + 2\hat{S}_z | 1, \tfrac{1}{2}\rangle = 2\beta H_{\parallel}(-k_{\pi,\pi} + 1) \tag{5-35}$$

However, the selection rules forbid a transition between the two split levels in the parallel case. With H perpendicular to the three- or fourfold symmetry axis $\Delta W = 0$.

For Δ positive the eigenfunctions for the ground state doublet take the form

$$|\tfrac{1}{2}\rangle = c_1 |0, \tfrac{1}{2}\rangle + c_2 |1, -\tfrac{1}{2}\rangle$$
$$|-\tfrac{1}{2}\rangle = c_1 |0, -\tfrac{1}{2}\rangle + c_2 |-1, \tfrac{1}{2}\rangle$$

where c_1 and c_2 are simple functions of Δ and ζ_M. Finding expectation values for $\alpha\hat{L}_z + 2\hat{S}_z$ and for $\alpha\hat{L}_x + 2\hat{S}_x$, the g values are evaluated as

$$g_{\parallel} = 2\langle \tfrac{1}{2}| \alpha\hat{L}_z + 2\hat{S}_z |\tfrac{1}{2}\rangle = 2[c_1^2 - c_2^2(k_{\pi,\pi} + 1)] \tag{5-36}$$

and

$$g_{\perp} = 2\langle \tfrac{1}{2}|\alpha\hat{L}_x + 2\hat{S}_x| -\tfrac{1}{2}\rangle = 2(c_1^2 - c_1 c_2 k_{\pi,\pi}\sqrt{2}) \tag{5-37}$$

Insertion of the functional dependence of c_1 and c_2 in Eqs. (5-36) and (5-37) lead to[11,12]

$$g_{\parallel} = -k_{\pi,\pi} + \frac{(\Delta + \zeta_M/2)(2 + k_{\pi,\pi})}{\sqrt{(\Delta + \zeta_M/2)^2 + 2\zeta_M^2}} \tag{5-38}$$

$$g_\perp = 1 + \frac{\Delta + \zeta_M/2 - 2\zeta_M k_{\pi,\pi}}{\sqrt{(\Delta + \zeta_M/2)^2 + 2\zeta_M^2}} \tag{5-39}$$

The expression for $g_\parallel$ reduces to $g = \frac{2}{3}(1 - k_{\pi,\pi})$ when $\zeta_M \gg \Delta$. The tetragonal perturbation is then suppressed and the effective symmetry is octahedral. The expression for $g_\perp$ should of course reduce to the same value. However, for $\zeta_M \gg \Delta$ Eq. (5-39) does not converge into g (octahedral) due to the neglect of the matrix element $\langle 1, \frac{1}{2} | \alpha\hat{L}_x + 2\hat{S}_x | 0, \frac{1}{2} \rangle$ in our calculation of $g_\perp$. (Compare Eqs. (3-102), (3-105), (3-106), and (3-108)). At the other extreme $\Delta \gg \zeta_M$ we get $g_\parallel = g_\perp = 2$.

The g factors for caesium titanium alum, which contain the units $Ti(H_2O)_6^{3+}$ are[13] $g_\parallel = 1.25$ and $g_\perp = 1.14$. The Ti^{3+} ion has a ζ_M value of 154 cm^{-1}. The Eqs. (5-38) and (5-39) can then be fitted by taking $\Delta = 140$ cm^{-1} and $k = 0.57$. In deriving Eqs. (5-38) and (5-39) we have, however, neglected the mixing under the spin-orbit coupling of the ground state doublet with the excited doublets contained in ^2E placed at some $10Dq$. The above parameter values are therefore only orders of magnitude. There is, however, no doubt that the electronic ground state has been properly identified as the Kramers doublet $|\pm\frac{1}{2}\rangle$.

5-4 THE ZERO-FIELD SPLITTINGS

The constants D and E in the spin hamiltonian, Eq. (3-74), reflect the zero-field splittings of an orbitally nondegenerate ^{2S+1}X, $S \geq 1$, ground state. They are measured experimentally using paramagnetic resonance methods, far-infrared spectroscopy, heat-capacity estimates, and magnetic-anisotropy measurements.

The experimental values of D and E may be sensitive to temperature. In the trigonal complex $Ni(H_2O)_6SiF_6$ the value of D for the $^3A_{2g}$ ground state is[14] -0.32 cm^{-1} at 195°K, but this value falls to -0.12 cm^{-1} at 14°K. Hence the spin doublet $M_S = \pm 1$ is the lowest, contrary to what is shown on Fig. 3-1. For the g-values $g_\parallel = g_\perp = 2.29$ at 195°K.

The ground state of the Ni^{2+} complex, $^3A_{2g}$, is coupled to the excited state $^3T_{2g}$ placed at some $10Dq$ via Λ_{ab}. In trigonal symmetry $^3T_{2g}$ splits up into a ^3E_g and a $^3A_{1g}$. Denoting their energies Δ' and Δ'' respectively, we get easily from Eqs. (3-53) and (3-61)

$$D = \lambda^2\left(\frac{4}{\Delta'} - \frac{4}{\Delta''}\right) = \frac{4\lambda^2(\Delta'' - \Delta')}{\Delta'\Delta''} \tag{5-40}$$

The corresponding formulae for the g values are obtained using Eq. (3-59)

$$g_\parallel = 2 - \frac{8k_{\sigma,\pi}\lambda}{\Delta''} \tag{5-41}$$

$$g_\perp = 2 - \frac{8k_{\sigma,\pi}\lambda}{\Delta'} \tag{5-42}$$

Putting $k_{\sigma,\pi}$ equal to 1 and[15] $\Delta' \approx \Delta'' = 8{,}000\ \text{cm}^{-1}$, we get from Eqs. (5-41) and (5-42) that $\lambda \approx -290\ \text{cm}^{-1}$. Equation (5-40) then gives $(\Delta'' - \Delta') \approx -60\ \text{cm}^{-1}$ assuming $D = -0.32\ \text{cm}^{-1}$. The values for the derived parameters are, however, only orders of magnitude. The formula for D has only considered interactions of the ground state with the excited spin triplet states. The influence of the spin-singlet states has been left out. By putting $k_{\sigma,\pi} = 1$ considerable uncertainty as to the λ value to use in Eq. (5-40) has further been introduced.

Measurements of the specific heat tail between 1 and 20°K of α-$NiSO_4 \cdot 6H_2O$ containing the complex $Ni(H_2O)_6^{2+}$ has been performed by Stout and Hadley.[16] Their heat capacity measurements could be reproduced using a ground manifold of three electronic levels placed at $0\ \text{cm}^{-1}$, $4.48 \pm 0.07\ \text{cm}^{-1}$ and $5.05 \pm 0.07\ \text{cm}^{-1}$. These values give (compare Eq. (3-119)) $D = 4.77\ \text{cm}^{-1}$ and $E = 0.29\ \text{cm}^{-1}$.

Cubic octahedral high-spin Fe^{3+} and Mn^{2+} complexes have the ground state $^6A_{1g}$. The double group representations contained in $^6A_{1g}$ are $E_{5/2}$ and G. In cubic symmetry the zero-field splitting operator of Eq. (3-70) is $-\lambda^2\Lambda_{ij}\hat{S}_i\hat{S}_j \equiv -KS(S+1)$ which cannot separate $E_{5/2}$ and G. The simplest operator which can do this is

$$\mathscr{H}^{(1)} = \frac{a}{6}\left[\hat{S}_x^4 + \hat{S}_y^4 + \hat{S}_z^4 - \frac{1}{5}S(S+1)(3S^2 + 3S - 1)\right] \tag{5-43}$$

This operator has been explained by Van Vleck and Penney[17] as reflecting a fifth-order perturbation. With ΔW as the energy difference between the ground state $^6A_{1g}$ and the first excited spin quartet states, and ζ_M the spin-orbit coupling constant, we get for an order of magnitude

$$a \approx \frac{\zeta_M^4 \times 10Dq}{(\Delta W)^4} \tag{5-44}$$

With $\zeta_M = 300\ \text{cm}^{-1}$, $10Dq = 10^4\ \text{cm}^{-1}$, and $\Delta W = 20{,}000\ \text{cm}^{-1}$ valid for Mn^{2+} complexes, we find $a = 5 \times 10^{-4}\ \text{cm}^{-1}$.

For an axial complex with $S = \frac{5}{2}$ we can use[18] the spin hamiltonian

$$\mathscr{H} = g_{\parallel}\beta H_z\hat{S}_z + g_{\perp}\beta H_x\hat{S}_x + \frac{a}{6}\left(\hat{S}_x^4 + \hat{S}_y^4 + \hat{S}_z^4 - \frac{707}{16}\right) + D\left(\hat{S}_z^2 - \frac{35}{12}\right) + b\left(\hat{S}_z^4 - \frac{95}{14}\hat{S}_z^2 + \frac{81}{16}\right) \tag{5-45}$$

The D and b terms reflect the strength of the axial perturbation measured by Ds. The leading term appears to be a fourth-order perturbation

$$D \approx \frac{Ds \times \zeta_M^2 \times 10Dq}{(\Delta W)^3} \tag{5-46}$$

With $Ds = 300\ \text{cm}^{-1}$ we get $D = 3 \times 10^{-2}\ \text{cm}^{-1}$.

For octahedral Mn^{2+} complexes $g_{\parallel} \approx g_{\perp} \approx 2$. Both the terms in $\hat{S}_i^4$ in Eq. (5-45) may safely be neglected in practice. For octahedral Fe^{3+} complexes both the a and D terms are larger than those for Mn^{2+} complexes. Treating the ground state $^6A_{1g}$ of $K_3Fe(oxalate)_3 \cdot 3H_2O$ as a spin sextet, measurements of the paramagnetic anisotropy gave[19] a D value of -0.55 cm^{-1}. The ground state is therefore the Kramers doublet $E_{5/2}$.

5-5 MAGNETIC SUSCEPTIBILITIES

Historically the oldest method used to characterize the ground state of a complex is the measurement of the magnetic susceptibility of the compound. Using refined experimental techniques this is still a very effective procedure. In treating some examples of its use we shall first consider a diamagnetic complex.

For an axial complex with $S = 0$ the formulae Eqs. (3-84) and (3-91) give for the susceptibility of a powdered sample

$$\bar{\chi} = 2L_A k^2 \beta^2 \frac{\Lambda_\parallel + 2\Lambda_\perp}{3} - \frac{e^2 L_A}{6mc^2} \sum_N \langle r_N^2 \rangle \tag{5-47}$$

In a diamagnetic Co^{3+} complex the ground state $^1A_{1g}(t_{2g})^6$ can couple via Λ to $^1T_{1g}(t_{2g})^5(e_g)^1$. Taking $\Lambda_\parallel = \Lambda_\perp$ we get easily

$$\bar{\chi} = \frac{16L_A \beta^2 k_{\pi,\sigma}^2}{W(^1T_{1g}) - W(^1A_{1g})} - \frac{e^2 L_A}{6mc^2} \sum_N \langle r_N^2 \rangle \tag{5-48}$$

The value of the second term in Eq. (5-48) can be estimated by measuring the susceptibility of the corresponding Zn^{2+} complex. Here the first term can be neglected. For $Zn(en)_3^{2+}$ and $Co(en)_3^{3+}$ one finds[20] respectively $\bar{\chi}[Zn(en_3^{2+}] = -160 \times 10^{-6}$, and $\bar{\chi}[Co(en)_3^{3+}] = -91 \times 10^{-6}$. The first term in Eq. (5-48) equals therefore some 70×10^{-6}. With $W(^1T_{1g}) - W(^1A_{1g}) = 21{,}000$ cm^{-1} we can estimate $k_{\pi,\sigma} \approx 0.6$.

Next we shall look at the cubic octahedral system $(t_{2g})^1$, $^2T_{2g}$. The susceptibility of this state has already been calculated in Eq. (3-110) under the assumption that $\zeta_M \ll 10Dq$. However, in a more realistic calculation we have to consider the mixing of the ground state $^2T_{2g}(G)$ with the excited state $^2E_g(G)$ under the spin-orbit coupling,[21] before evaluating the paramagnetic susceptibility.

With $\Delta = 0$ we get from Eqs. (5-32) and (5-33) in a M_L, M_S scheme using Bethe's double group nomenclature $E_{1/2} \equiv \Gamma_6$, $E_{5/2} \equiv \Gamma_7$ and $G \equiv \Gamma_8$

$$\Gamma_8^a(^2T_{2g}) = \left| -\frac{1}{\sqrt{2}}(yz + ixz), \frac{1}{2} \right\rangle \tag{5-49}$$

$$\Gamma_8^b(^2T_{2g}) = \sqrt{\frac{2}{3}}\left| xy, \frac{1}{2} \right\rangle + \sqrt{\frac{1}{3}}\left| -\frac{1}{\sqrt{2}}(yz + ixz), -\frac{1}{2} \right\rangle \tag{5-50}$$

$$\Gamma_8^c(^2T_{2g}) = \sqrt{\frac{2}{3}}\left| xy, -\frac{1}{2} \right\rangle + \sqrt{\frac{1}{3}}\left| \frac{1}{\sqrt{2}}(yz - ixz), \frac{1}{2} \right\rangle \tag{5-51}$$

$$\Gamma_8^d(^2T_{2g}) = \left| \frac{1}{\sqrt{2}}(yz - ixz), -\frac{1}{2} \right\rangle \tag{5-52}$$

For $\Gamma_8(^2E_g)$ the wave functions can be taken as

$$\Gamma_8^a(^2E_g) = \mathrm{i}\,|z^2, -\tfrac{1}{2}\rangle \tag{5-53}$$

$$\Gamma_8^b(^2E_g) = \mathrm{i}\,|x^2 - y^2, \tfrac{1}{2}\rangle \tag{5-54}$$

$$\Gamma_8^c(^2E_g) = -\mathrm{i}\,|x^2 - y^2, -\tfrac{1}{2}\rangle \tag{5-55}$$

$$\Gamma_8^d(^2E_g) = -\mathrm{i}\,|z^2, \tfrac{1}{2}\rangle \tag{5-56}$$

With

$$\mathscr{H} = \mathscr{H}_{\mathrm{el}} + \sum \zeta_M \mathbf{l}\cdot\mathbf{s} \tag{5-57}$$

the energies of the two Γ_8 states are given by

$$\begin{vmatrix} -\zeta_M/2 - 4Dq - W & -\dfrac{\sqrt{6}}{2}\zeta_M \\ -\dfrac{\sqrt{6}}{2}\zeta_M & 6Dq - W \end{vmatrix} = 0 \tag{5-58}$$

Defining[21] an angle θ such that

$$\tan\theta = \frac{-\sqrt{6}\zeta_M}{10Dq + \frac{1}{2}\zeta_M} \quad \text{where } \frac{\pi}{2} < \theta < \pi \tag{5-59}$$

we get for the ground-state wave functions

$$|\alpha\Gamma_8^j\rangle = \sin\frac{\theta}{2}\,\Gamma_8(^2\Gamma_{2g}^j) + \cos\frac{\theta}{2}\,\Gamma_8^j(^2E_g) \qquad j = a, b, c, d \tag{5-60}$$

Taking $\hat{\mu}_z = \beta H(k\hat{l}_z + 2\hat{s}_z)$ we have

$$\langle\Gamma_8^a|\hat{\mu}_z|\Gamma_8^a\rangle = \beta H\left[(1 - k_{\pi,\pi})\sin^2\frac{\theta}{2} - \cos^2\frac{\theta}{2}\right] \tag{5-61}$$

$$\langle\Gamma_8^b|\hat{\mu}_z|\Gamma_8^b\rangle = \beta H\left[\frac{1}{3}(1 - k_{\pi,\pi})\sin^2\frac{\theta}{2} - 2\sqrt{\frac{2}{3}}k_{\pi,\sigma}\sin\theta + \cos^2\frac{\theta}{2}\right] \tag{5-62}$$

$$\langle\Gamma_8^c|\hat{\mu}_z|\Gamma_8^c\rangle = \beta H\left[-\frac{1}{3}(1 - k_{\pi,\pi})\sin^2\frac{\theta}{2} + 2\sqrt{\frac{2}{3}}k_{\pi,\sigma}\sin\theta - \cos^2\frac{\theta}{2}\right] \tag{5-63}$$

$$\langle\Gamma_8^d|\hat{\mu}_z|\Gamma_8^d\rangle = \beta H\left[-(1 - k_{\pi,\pi})\sin^2\frac{\theta}{2} + \cos^2\frac{\theta}{2}\right] \tag{5-64}$$

With only $|\alpha\Gamma_8^j\rangle$ being populated we get for the magnetic susceptibility

$$\begin{aligned}\bar{\chi} = \frac{L_A\beta^2}{3k_BT}\bigg\{&\frac{1}{4}(1 + \cos\theta)[(3 + 16k_{\pi,\sigma}^2) + (3 - 16k_{\pi,\sigma}^2)\cos\theta - 4\sqrt{6}k_{\pi,\sigma}\sin\theta] \\ &- \frac{1}{12}(1 - k_{\pi,\pi})(1 - \cos\theta)[(1 + 5k_{\pi,\pi}) + (11 - 5k_{\pi,\pi})\cos\theta + 4\sqrt{6}k_{\pi,\sigma}\sin\theta]\bigg\} \\ &+ \frac{4}{27}\frac{L_A\beta^2(k_{\pi,\pi} + 2)^2}{\zeta_M} - \frac{e^2L_A}{6mc^2}\sum_N\langle r_N^2\rangle\end{aligned} \tag{5-65}$$

where the "high frequency" term has been calculated using Eqs. (3-102), (3-105), (3-106), and (3-108), assuming no mixing between $\Gamma_8(^2T_{2g})$ and $\Gamma_8(^2E_g)$.

The magnetic susceptibility of ReF_6 has been measured[22] in the range 4–296°K. The susceptibility can be expressed as

$$\bar{\chi} = \frac{78 \times 10^{-4}}{T} + 0.87 \times 10^{-4} \tag{5-66}$$

In order to correct for diamagnetism, WF_6 was measured, and a value of -0.53×10^{-4} obtained. With $k_{\pi,\pi} = 1$ we get, from the temperature-independent susceptibility, $\zeta_M \approx 2{,}700\ \text{cm}^{-1}$ which is of the right order of magnitude. The optical spectra yield[21] $\zeta_M \approx 3{,}400\ \text{cm}^{-1}$ and $10Dq = 30{,}000\ \text{cm}^{-1}$. Hence $\theta = 165.28°$ and with $k_{\pi,\pi} = 1$ and $k_{\pi,\sigma} = 0.6$ we reproduce the experimental result. Notice that by putting $k_{\pi,\pi} = 1$ all the temperature-dependent susceptibility arises from the mixing of the excited $\Gamma_8(^2E_g)$ into $\Gamma_8(^2T_{2g})$.

Next we shall calculate the magnetic susceptibility of an octahedral V^{3+} complex. The ground state is in O_h symmetry taken to be $^3T_{1g}(t_{2g})^2$. With a fictitious angular momentum $L = 1$ the nine states which span $^3T_{1g}$ are taken as $|M_L, M_S\rangle$ with $M_L = 1, 0, -1$ and $M_S = 1, 0, -1$. We have for instance $|1, 1\rangle = |\overset{+}{\omega}_1\overset{+}{\omega}_0|$ where $\omega_1 = 1/\sqrt{2}(yx + ixz)$ and $\omega_0 = xy$. Hence

$$\alpha\lambda\hat{L}_z\hat{S}_z|1, 1\rangle = \alpha\lambda|1, 1\rangle = \sum_n \zeta_M \hat{l}_{zn}\hat{s}_{zn}|\overset{+}{\omega}_1\overset{+}{\omega}_0| = -\tfrac{1}{2}\zeta_M|\overset{+}{\omega}_1\overset{+}{\omega}_0|$$

and

$$\alpha\hat{L}_z|1, 1\rangle = \sum_n \hat{l}_{zn}|\overset{+}{\omega}_1\overset{+}{\omega}_0| = -k_{\pi,\pi}|\overset{+}{\omega}_1\overset{+}{\omega}_0|$$

The hamiltonian to use inside the state $^3T_{1g}(t_{2g})^2$ is therefore

$$\mathscr{H} = -\tfrac{1}{2}\zeta_M \mathbf{L}\cdot\mathbf{S} + \beta H(-k_{\pi,\pi}\hat{L}_z + 2\hat{S}_z)$$

Putting $k_{\pi,\pi} = 1$ this leads to a Bohr magneton number μ_{eff} for the state $^3T_{1g}(t_{2g})^2$ equal to[23]

$$\mu_{\text{eff}}^2 = \frac{5(3x + 18)\exp(3x/2) + 3(x + 18)\exp(x/2) - 144}{2x(1 + 3\exp(x/2) + 5\exp(3x/2)} \tag{5-67}$$

with $x = \zeta_M/k_BT$. For $T = 0$, $\mu_{\text{eff}} = 1.22$ while for $T \to \infty$ we find $\mu_{\text{eff}} = 3.16$.

The measured Bohr magneton number of, for instance, $V(\text{urea})_6(ClO_4)_3$ is about 2.71 at room temperature.[24] It falls to 2.67 when the temperature is lowered to 80°K. Taking[25] $\zeta_M \approx 200\ \text{cm}^{-1}$ Eq. (5-67) predicts $\mu_{\text{eff}} = 2.72$ at room temperature and $\mu_{\text{eff}} = 2.00$ at 80°K. The experimental result is therefore incompatible with the assumption of a $^3T_{1g}$ ground state; the orbital degeneracy must be lowered. The site group of V^{3+} in $V(\text{urea})_6(ClO_4)_3$ is indeed also found[26] to be D_3. Using the D_3 symmetry $^3T_{1g}$ is split into $^3A_2 + {}^3E$. We take $^3A_2$ to be the ground state with ^3E placed so high in energy as not to be thermally populated.

Changing to a trigonal representation, we have for the trigonal t_2 orbitals (compare Table 1-2)

$$\begin{pmatrix} t_x \\ t_y \\ t_z \end{pmatrix} = \begin{pmatrix} \frac{2}{\sqrt{6}} & -\frac{1}{\sqrt{6}} & -\frac{1}{\sqrt{6}} \\ 0 & -\frac{1}{\sqrt{2}} & \frac{1}{\sqrt{2}} \\ \frac{1}{\sqrt{3}} & \frac{1}{\sqrt{3}} & \frac{1}{\sqrt{3}} \end{pmatrix} \begin{pmatrix} t_{xy} \\ t_{xz} \\ t_{yz} \end{pmatrix} \tag{5-68}$$

The orbital functions spanning ${}^3T_{1g}(t_{2g})^2$ are in D_3 symmetry

$$|A_2\rangle = |(\overset{+}{t_x})(\overset{+}{t_y})| \qquad |E_a\rangle = |(\overset{+}{t_z})(\overset{+}{t_x})| \qquad |E_b\rangle = |(\overset{+}{t_y})(\overset{+}{t_z})|$$

Using Eq. (3-59) we have

$$g_{\parallel} = 2(1 - \tfrac{1}{2}\zeta_M k_{\pi,\pi}\Lambda_{\parallel}) \tag{5-69}$$

$$g_{\perp} = 2(1 - \tfrac{1}{2}\zeta_M k_{\pi,\pi}\Lambda_{\perp}) \tag{5-70}$$

With the above wave functions we get $\Lambda_{\parallel} = 0$ and $\Lambda_{\perp} = 1/W(E) - W(A_2)$. Hence to first order[27]

$$g_{\parallel} = 2 \tag{5-71}$$

$$g_{\perp} = 2 - \frac{\zeta_M k_{\pi,\pi}}{W(E) - W(A_2)} \tag{5-72}$$

and

$$D = \frac{\zeta_M^2}{4[W(E) - W(A_2)]} \tag{5-73}$$

With $\zeta_M \approx 200\ \text{cm}^{-1}$ and[28] $W(E) - W(A_2) \approx 1{,}000\ \text{cm}^{-1}$ we find $D \approx 10\ \text{cm}^{-1}$. A zero-field splitting of this order of magnitude explains why no signal has been obtained in an electron spin-resonance experiment. Measurements of the magnetic susceptibility at very low temperatures[26] have further revealed that the ground state is a nonmagnetic state. In the D_3 double group the state ${}^3A_2$ span A_1 and E. Evidently A_1 is the absolute ground state. (See Fig. 5-2.)

Belonging to the same $(t_{2g})^2$ configuration as the ground state are the excited states ${}^1E_g$, ${}^1T_{2g}$, both expected at an energy $6F_2 + 40F_4 \approx 10{,}000\ \text{cm}^{-1}$, and ${}^1A_{1g}$ placed at $15F_2 + 100\ F_4 \approx 25{,}000\ \text{cm}^{-1}$. Pairs of lines separated by $6.0 \pm 0.5\ \text{cm}^{-1}$ appear in σ polarization located at some $9{,}900\ \text{cm}^{-1}$. The marked temperature dependence makes the "red" line disappear at 1.5°K. At this low temperature the π spectrum is likewise blank. Thus the zero-field splitting is $D = 6 \pm 0.5\ \text{cm}^{-1}$. The $g_{\parallel}$ factor has been measured in an optical Zeeman experiment, using the transition $E \rightarrow A_1$ found at $21{,}600\ \text{cm}^{-1}$ (see Fig. 5-2). The measurements yield[29] $g_{\parallel} = 1.9 \pm 0.1$.

The electronic structures and magnetic properties of the linear complexes UO_2^{2+} (5f)°, NpO_2^{2+} (5f)^1 and PuO_2^{2+} (5f)^2 have been looked into by Eisenstein and Pryce.[30,31] Using a pure crystal-field argument they considered the strong linear

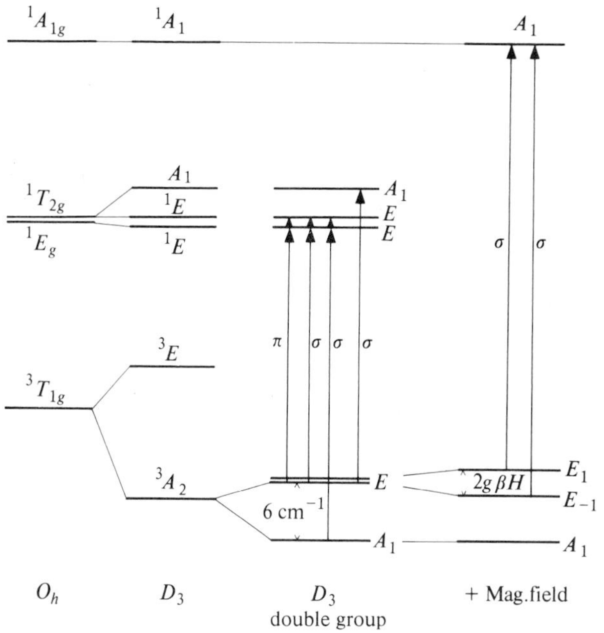

Figure 5-2 Level diagram for some low-lying states of $V(urea)_6(ClO_4)_3$.

field to repel the 5f electrons. The lowest electronic state of NpO_2^{2+} should therefore be one in which the "ligand field" electron is in that orbital which keeps it furthest away from the O–Np–O axis. This is a $\phi_u|\pm\frac{5}{2}\rangle$ level. The observed susceptibility of NpO_2^{2+} corrected for diamagnetism corresponds to $g_\parallel^2 + 2g_\perp^2 = 13.60$. Using an orbital reduction factor equal to 1 it is deduced that $g_\parallel = 3.55$ and $g_\perp = 0.70$. Taking $\zeta_M = 1{,}300$ cm^{-1} these numbers give us a spacing $\Delta = 4{,}225$ cm^{-1} to the next $|\pm\frac{7}{2}\rangle$ level.

The paramagnetism of complexes containing more than one metal center has received a great deal of attention. The interaction of two magnetic ions has been considered in section 3-6. A system that contains two weakly interacting chromophores each having $S = \frac{1}{2}$ is found in copper acetate monohydrate. In this dimer two Cu^{2+} ions are held together by acetate bridges.[32]

The g values for the coupled system[33] are related to the g_1 and g_2 values for the interacting chromophores. Equation (3-152) gives $g = g_1 = g_2$. Comparisons with g values for single Cu^{2+} chromophores confirm this. The experimental results can be fitted to a spin hamiltonian

$$\mathscr{H} = D\hat{S}_z^2 + E(\hat{S}_x^2 - \hat{S}_y^2) + \beta(g_zH_z\hat{S}_z + g_xH_x\hat{S}_x + g_yH_y\hat{S}_y) \qquad (5\text{-}74)$$

with $S = 1$, $D = 0.34 \pm 0.03$ cm^{-1}, $E = 0.01 \pm 0.005$ cm^{-1}, $g_z = g_\parallel = 2.42 \pm 0.03$ and $g_x = g_y = g_\perp = 2.08 \pm 0.03$. Adding a spin–spin coupling term of the form $J\mathbf{S}_1 \cdot \mathbf{S}_2$, $S_1 = S_2 = \frac{1}{2}$, to the hamiltonian, J was evaluated[33] to be 260 cm^{-1}. On the other hand, measuring the magnetic susceptibility of copper acetate as a function of temperature,[34] a value of $J = 286\ \mathrm{cm}^{-1}$ was found.

It appears that the electronic structures of the copper(II) dimer are best described using a coupled chromophore model.[35] The advantages of such an approach over models based on a direct copper–copper bond of σ, π, or δ type is that it offers a straightforward mechanism for a spin-singlet state to become the ground state and gives a physically appealing description of the excited states.

Complexes containing three Cu^{2+} ions are also known. In

$$Cu_3(C_6H_5N_2O)_3(OH)SO_4 \cdot 10\tfrac{1}{2}H_2O$$

the three copper ions form an equilateral triangle. We can here use the effective hamiltonian

$$\mathscr{H} = J(\mathbf{S}_A \cdot \mathbf{S}_B + \mathbf{S}_A \cdot \mathbf{S}_C + \mathbf{S}_B \cdot \mathbf{S}_C) \tag{5-75}$$

to describe the interaction between the three copper centers A, B, and C. Defining the total spin

$$\mathbf{S} = \mathbf{S}_A + \mathbf{S}_B + \mathbf{S}_C \tag{5-76}$$

the low-lying energies of the system are with $S_A = S_B = S_C = \frac{1}{2}$ given by

$$W(S) = \tfrac{1}{2}J[S(S+1) - 3 \cdot \tfrac{1}{2}(\tfrac{1}{2}+1)] \tag{5-77}$$

The quantum number S can be either $S = \frac{3}{2}$ or $S = \frac{1}{2}$. In a coupled chromophore model the wave functions are

$$\left.\begin{array}{l} |\overset{+}{A}\overset{+}{B}\overset{+}{C}| \\[2mm] \frac{1}{\sqrt{3}}[|\overset{-}{A}\overset{+}{B}\overset{+}{C}| + |\overset{+}{A}\overset{-}{B}\overset{+}{C}| + |\overset{+}{A}\overset{+}{B}\overset{-}{C}|] \\[2mm] \frac{1}{\sqrt{3}}[|\overset{-}{A}\overset{-}{B}\overset{+}{C}| + |\overset{-}{A}\overset{+}{B}\overset{-}{C}| + |\overset{+}{A}\overset{-}{B}\overset{-}{C}|] \\[2mm] |\overset{-}{A}\overset{-}{B}\overset{-}{C}| \end{array}\right\}\ {}^4A_2 \tag{5-78}$$

$$\left.\begin{array}{l} \frac{1}{\sqrt{2}}[|\overset{-}{A}\overset{+}{B}\overset{+}{C}| - |\overset{+}{A}\overset{-}{B}\overset{+}{C}|] \\[2mm] \frac{1}{\sqrt{6}}[|\overset{-}{A}\overset{+}{B}\overset{+}{C}| + |\overset{+}{A}\overset{-}{B}\overset{+}{C}| - 2|\overset{+}{A}\overset{+}{B}\overset{-}{C}|] \\[2mm] \frac{1}{\sqrt{2}}[|\overset{-}{A}\overset{-}{B}\overset{+}{C}| - |\overset{-}{A}\overset{+}{B}\overset{-}{C}|] \\[2mm] \frac{1}{\sqrt{6}}[|\overset{-}{A}\overset{-}{B}\overset{+}{C}| + |\overset{-}{A}\overset{+}{B}\overset{-}{C}| - 2|\overset{+}{A}\overset{-}{B}\overset{-}{C}|] \end{array}\right\}\ {}^2E \tag{5-79}$$

where for instance

$$\overset{+}{A} = |(\overset{+}{xz})_A(\overset{+}{xz})_A(\overset{-}{yz})_A(\overset{+}{yz})_A(\overset{-}{xy})_A(\overset{+}{xy})_A(\overset{-}{z^2})_A(\overset{+}{z^2})_A(\overset{-}{x^2 - y^2})_A| \tag{5-80}$$

Classifying the states in D_3 symmetry we find using Eq. (5-77) that $W(^4A_2) = \frac{3}{4}J$ and $W(^2E) = -\frac{3}{4}J$.

In D_3 symmetry the antisymmetric product $E \times E$ equals A_2. This is also how $\hat{L}_z$ transforms. The ^2E state will therefore exhibit first-order spin-orbit coupling. The operator to use to calculate the spin-orbit coupling splitting inside the ^2E state is evidently $\mathscr{H}^{(1)} = \lambda \hat{L}_z \hat{S}_z$. It will split ^2E into two Kramers doublets. An order of magnitude for the expectation value of $\langle ^2E|\hat{L}_z|^2E\rangle$ can be found from a measured value of $g_{\parallel} = 2\langle ^2E|L_z + 2\hat{S}_z|^2E\rangle$. We find

$$\langle \hat{L}_z \rangle = \frac{g_{\parallel} - 2}{2} \tag{5-81}$$

and hence the spin-orbit splitting $\Delta(^2E)$

$$\Delta(^2E) = \frac{g_{\parallel} - 2}{2}\lambda \tag{5-82}$$

The $^4A_2$ state, on the other hand, can only show second-order spin-orbit splitting.

Classifying the two Kramers doublets to which ^2E gives rise as Γ_1 and Γ_2 with a separation of Δ we get for the magnetic susceptibility per copper ion:

$$\bar{\chi} = \frac{L_A}{3}\frac{\beta^2}{4k_BT}\frac{g_1^2 + g_2^2 \exp(-\Delta/k_BT)}{1 + \exp(-\Delta/k_BT)} \tag{5-83}$$

With $g_1 \approx g_2 \approx 2$ we find

$$\bar{\chi} \approx \frac{L_A\beta^2}{3k_BT} \tag{5-84}$$

corresponding to a $\mu_{\text{eff}} = 1$ (compare Eq. (3-111)) per copper ion. If, on the other hand, the ground state had been $^4A_2$, we would have

$$\bar{\chi} = \frac{L_A}{3}\frac{\beta^2 g^2}{4k_BT} \times 5 \tag{5-85}$$

With $g \approx 2$ we find $\mu_{\text{eff}} \approx \sqrt{5}$ per copper ion.

Measurements[36] of the magnetic susceptibility of

$$Cu_3(C_6H_5N_2O)_3(OH)SO_4 \cdot 10\tfrac{1}{2}H_2O$$

between 105°K and 400°K yielded $\mu_{\text{eff}} = 1.0$, corresponding to ^2E being the ground state. A minimum value of $J \approx 600\ \text{cm}^{-1}$ was further found since μ_{eff} was equal to 1 over the whole measured' temperature interval. Measuring the magnetic susceptibility at lower temperatures, Tsukerblat et al.[37] found the spin-orbit splitting $\Delta = 12\ \text{cm}^{-1}$ and $g = 2.1$. With these values we notice that Eq. (5-82) gives us $\lambda \approx 240\ \text{cm}^{-1}$.

The magnetic susceptibility of a tetranuclear compound, the ion $Cr_4(OH)_6(en)_6^{6+}$, has been investigated by Gray and coworkers.[38] Using a rhomboid hamiltonian

$$\mathscr{H} = J(\mathbf{S}_1 \cdot \mathbf{S}_3 + \mathbf{S}_1 \cdot \mathbf{S}_4 + \mathbf{S}_2 \cdot \mathbf{S}_3 + \mathbf{S}_2 \cdot \mathbf{S}_4) + J_{12}\mathbf{S}_1\mathbf{S}_2 \tag{5-86}$$

with $S_i = \frac{3}{2}$, $i = 1, 2, 3, 4$, $J = 15\ \text{cm}^{-1}$, and $J_{12} = 28\ \text{cm}^{-1}$ the observed antiferromagnetism is satisfactorily explained assuming $g = 2.00$.

5-6 ELECTRONIC RAMAN AND EMISSION SPECTROSCOPY

The electronic Raman effect is of particular value when investigating the position of low-lying electronic states. As examples we first take Co^{2+} in six-coordinated surroundings. In O_h symmetry the high-spin ground state is a ${}^4T_{1g}$ state. The energies of the spin-orbit components are (see Eqs. (5-24), (5-25), (5-26), and (5-30)), with $-\frac{3}{2} \leq \alpha \leq -1$:

$$W(E_{1/2}) = \tfrac{5}{6}\alpha\zeta_M,\ W(G) = \tfrac{1}{3}\alpha\zeta_M, \text{ and } W(G, E_{5/2}) = -\tfrac{1}{2}\alpha\zeta_M$$

A symmetry lower than O_h will split the remaining degeneracies, so that we get six Kramers doublets. A distortion from cubic symmetry can be accounted for by adding a term $-\Delta(\hat{L}_z^2 - \frac{2}{3})$ to the hamiltonian Eq. (5-20). Δ is seen to be a measure of the splitting of ${}^4T_{1g}$ into a ${}^4E$ and ${}^4A_2$ state.

The ion Co^{2+} in $CoCl_2$ is octahedrally coordinated with a slight trigonal distortion. Electronic Raman spectroscopy shows[39] levels at 233 cm^{-1}, 551 cm^{-1}, 962 cm^{-1}, 984 cm^{-1}, and 1,014 cm^{-1}. These states can be reasonably well accounted for by taking $\alpha\zeta_M = -666\ \text{cm}^{-1}$ and $\Delta = -400\ \text{cm}^{-1}$.

The ground state for the ferrocenium cation, $Fe(cp)_2^+$, is expected to be either[40,41] $(e_{2g})^3(a_{1g})^2$, ${}^2E_{2g}$, or $(e_{2g})^4(a_{1g})^1$, ${}^2A_{1g}$. The energy difference between these two states was calculated to be[42]:

$$W({}^2A_{1g}) - W({}^2E_{2g}) = \bar{h}_{xy,xy} - \bar{h}_{z^2,z^2} + 20(F_2 - 5F_4) \tag{5-87}$$

It was taken to be 16,300 cm^{-1}. However, measurements of the g factors show the two states to be nearly degenerate.

The spin-orbit coupling will split[43] the ${}^2E_{2g}$ state into two Kramers doublets, E'' and $(A_1' A_2')$ separated by $2\zeta_M \approx 800\ \text{cm}^{-1}$. The double group representation of ${}^2A_{1g}$ is E'. A lower symmetry than D_{5d}, may, however, mix all three Kramers doublets. Taking the wave function for the lowest of the three Kramers doublets to be of the form

$$\Psi = c_1|E''\rangle + c_2|E'\rangle + c_3|A_1' A_2'\rangle \tag{5-88}$$

the g factors for this state can be evaluated. We get, with k being the orbital reduction factor defined as

$$k = \frac{1}{2i}\langle xy|\,\hat{l}_z\,|x^2 - y^2\rangle \tag{5-89}$$

that

$$g_{\parallel} = 2 + 4k(c_1^2 - c_3^2) \tag{5-90}$$

$$g_{\perp} = 2 - (c_1 - c_3)^2 \tag{5-91}$$

The g factors are measured[44] to be $g_{\parallel} = 4.35$ and $g_{\perp} = 1.26$, indicating a strong mixing of $^2E_{2g}(E'')$ and $^2A_{1g}(E')$.

Magnetic susceptibility measurements[45] show that the first excited Kramers doublet can be thermally populated. A transition between this state and the ground state is allowed in Raman scattering, and an observed shift of 213 cm^{-1} is assigned[46] to this transition. The three low-lying Kramers doublets in $Fe(cp)_2^+$ are therefore placed at 0 cm^{-1}, 213 cm^{-1}, and $\sim$800 cm^{-1}.

For transition-metal ions doped into host lattices it is sometimes possible to observe low-lying electronic states in emission. The fluorescence spectrum of Co^{2+} in MgF_2 has for instance been measured by pumping into the higher-lying electronic states. Six electronic states placed at 0 cm^{-1}, 152 cm^{-1}, 798 cm^{-1}, 1,087 cm^{-1}, 1,256 cm^{-1}, and 1,398 cm^{-1} corresponding to the six Kramers doublets originating from the $^4T_{1g}$ octahedral state were observed.[47]

The fluorescence of ruby ($Al_2O_3 + Cr^{3+}$) is well known, and high-resolution spectroscopy shows[48] the zero field splitting of the ground state $^4A_{2g}$ to be 0.40 cm^{-1}. A rare case of a pure compound showing a detectable electronic splitting in emission is found in $V(urea)_6(ClO_4)_3$. The 6 cm^{-1} splitting of $^3A_2$ (see Fig. 5-2) is also observed by Flint and Greenough[49] in the luminescence spectrum of the $V(urea)_6^{3+}$ ion.

As an example of a vibronic intensity giving mechanism in emission we take[50] the complex $ReBr_6^{2-}$ coped into single crystals of Cs_2ZrBr_6. The ground state of $ReBr_6^{2-}$ is $\Gamma_8[^4A_{2g}(t_{2g})^3]$ and the emitting state is $\Gamma_7[^2T_{2g}(t_{2g})^3]$. The transition is forbidden by parity to appear as an electric dipole transition, but it is allowed as magnetic dipole. The 0–0 line of $\Gamma_7 \rightarrow \Gamma_8$ is observed at 13,144 cm^{-1}. Built upon that line, the emission spectrum shows three lines at 85 cm^{-1}, 116 cm^{-1}, and 217 cm^{-1}. The odd vibrations in an octahedral MX_6 system are in Herzberg's[51] notation $\nu_3(\tau_{1u})$, $\nu_4(\tau_{1u})$, and $\nu_6(\tau_{2u})$. For $ReBr_6^{2-}$ infrared spectroscopy[52] has identified $\nu_3 = 217$ cm^{-1} and $\nu_4 = 118$ cm^{-1}. Clearly we may identify two of the observed lines with one quantum of ν_3 and ν_4 while the third line at 85 cm^{-1} can be taken as ν_6.

REFERENCES

1. L. E. Orgel, *J. Chem. Phys.*, **23**: 1819, 1955.
2. J. S. Griffith, *J. Inorg. Nucl. Chem.*, **2**: 1, 229, 1956.
3. D. W. Clack and W. Smith, *J. Chem. Soc. Dalton*, 2015, 1974.
4. J. H. van Santen and J. S. van Wieringen, *Rec. Trav. Chim. Pays Bas*, **71**: 420, 1952.
5. A. Abragam and M. H. L. Pryce, *Proc. Phys. Soc.*, **A63**: 409, 1950.
6. B. Bleaney and K. D. Bowers, *Proc. Phys. Soc.*, **A65**: 667, 1952.
7. W. Moffitt and W. Thorsen, *Phys. Rev.*, **108**: 1251, 1957.
8. J. C. D. Brand, G. L. Goodman and B. Weinstock, *J. Mol. Spectroscopy*, **38**: 449, 1971.

9. A. Abragam and M. H. L. Pryce, *Proc. Roy. Soc.*, **A206**: 173, 1951.
10. W. Low, *Phys. Rev.*, **109**: 256, 1958.
11. B. Bleaney, *Proc. Phys. Soc.*, **A63**: 407, 1950.
12. K. W. H. Stevens, *Proc. Roy. Soc.*, **A219**: 542, 1953.
13. B. Bleaney, G. S. Bogle, A. H. Cooke, R. J. Duffus, M. C. M. O'Brien, and K. W. H. Stevens, *Proc. Phys. Soc.*, **A68**: 57, 1955.
14. R. P. Penrose and K. W. H. Stevens, *Proc. Phys. Soc.*, **A63**: 29, 1950.
15. E. I. Solomon and C. J. Ballhausen, *Mol. Phys.*, **29**: 279, 1975.
16. J. W. Stout and W. B. Hadley, *J. Chem. Phys.*, **40**: 55, 1964.
17. J. H. Van Vleck and W. G. Penney, *Phil. Mag.*, **17**: 55, 1934.
18. B. Bleaney and K. W. H. Stevens, *Rep. Prog. Phys.*, **16**: 108, 1953.
19. M. Gerloch, J. Lewis and R. C. Slade, *J. Chem. Soc.*, (A) **1969**, 1422.
20. C. J. Ballhausen and R. W. Asmussen, *Acta Chem. Scand.*, **11**: 479, 1957.
21. W. Moffitt, G. L. Goodman, M. Fred, and B. Weinstock, *Mol. Phys.*, **2**: 109, 1959.
22. H. Selig, F. A. Cafasso, D. M. Gruen, and J. G. Malm, *J. Chem. Phys.*, **36**: 3440, 1962.
23. M. Kotani, *J. Phys. Soc. Japan*, **4**: 293, 1949.
24. D. J. Machin and K. S. Murray, *J. Chem. Soc.*, (A) **1967**, 1498.
25. C. E. Moore, *Atomic Energy Levels*, National Bureau of Standards Circular 467, vol. 2, 1952.
26. R. Dingle, P. J. McCarthy, and C. J. Ballhausen, *J. Chem. Phys.*, **50**: 1957, 1969.
27. A. Abragam and M. H. L. Pryce, *Proc. Roy. Soc.*, **A205**: 135, 1951.
28. M. H. L. Pryce and W. A. Runciman, *Diss. Far. Soc.*, **26**: 34, 1958.
29. C. J. Ballhausen and I. Trabjerg, *13. Nordiske Kemikermøde*, 1968, 182.
30. J. C. Eisenstein and M. H. L. Pryce, *Proc. Roy. Soc.*, **A229**: 20, 1955.
31. J. C. Eisenstein and M. H. L. Pryce, *Proc. Roy. Soc.*, **A238**: 31, 1956.
32. J. N. van Niekerk and F. R. L. Schoening, *Acta Cryst.*, **6**: 227, 1953.
33. B. Bleaney and K. D. Bowers, *Proc. Roy. Soc.*, **A214**: 451, 1952.
34. B. N. Figgis and R. L. Martin, *J. Chem. Soc.*, **1956**, 3837.
35. Aa. E. Hansen and C. J. Ballhausen, *Trans. Far. Soc.*, **61**: 631, 1965.
36. R. Beckett, R. Colton, B. F. Hoskins, R. L. Martin, and D. G. Vince, *Aust. J. Chem.*, **22**: 2527, 1969.
37. B. S. Tsukerblat, B. Ya. Kuyavskaya, M. I. Belinskii, A. V. Ablov, V. M. Novotortsev, and V. T. Kalinnikov, *Theoret. Chim. Acta.*, **38**: 131, 1975.
38. M. T. Flood, C. G. Barraclough, and H. B. Gray, *Inorg. Chem.*, **8**: 1855, 1969.
39. J. H. Christie and D. J. Lockwood, *Chem. Phys. Lett.*, **8**: 120, 1971.
40. W. Moffitt, *J. Am. Chem. Soc.*, **76**: 3386, 1954.
41. J. Dunitz and L. E. Orgel, *J. Chem. Phys.*, **23**: 954, 1955.
42. D. A. Levy and L. E. Orgel, *Mol. Phys.*, **4**: 93, 1961.
43. A. H. Maki and T. E. Berry, *J. Am. Chem. Soc.*, **87**: 4437, 1965.
44. R. Prins, *Mol. Phys.*, **19**: 603, 1970.
45. Y. S. Sohn, D. N. Hendrickson, and H. B. Gray, *J. Am. Chem. Soc.*, **92**: 3233, 1970; and *Inorg. Chem.*, **10**: 1559, 1971.
46. B. F. Gächter, J. A. Koningstein, and V. T. Aleksanjan, *J. Chem. Phys.*, **62**: 4628, 1975.
47. L. F. Johnson, R. E. Dietz, and H. J. Guggenheim, *Appl. Phys. Lett.*, **5**: 21, 1964.
48. W. Kaiser, S. Sugano, and D. L. Wood, *Phys. Rev. Lett.*, **6**: 605, 1961.
49. C. D. Flint and P. Greenough, *Chem. Phys. Lett.*, **16**: 369, 1972.
50. H. H. Patterson, J. L. Nims, and C. M. Valencia, *J. Mol. Spect.*, **42**: 567, 1972.
51. G. Herzberg, *Infrared and Raman Spectra of Polyatomic Molecules*, Van Nostrand, 1945.
52. L. A. Woodward and M. J. Ware, *Spectrochim. Acta*, **20**: 711, 1964.

CHAPTER

SIX

THE CHARACTERIZATION OF EXCITED STATES. SPECIFIC EXAMPLES

6-1 THE ENERGIES

The construction of correlation diagrams which show the quantitative behavior of excited ligand field states as a function of $10Dq$ has been dealt with in section 2-6. Among molecular theories these diagrams are unsurpassed in their power of dealing with the energies of excited states; indeed a great many state identifications hinge upon the trustworthiness of the diagrams.

The correlation diagrams are completely general in the sense that they can be constructed whenever we have a complex where the d and f shells are partly filled. However, moving from 3d to 4d and 5d shells causes the group overlap integrals G_{ML} to increase. Hence the orbital splittings get larger and the correlative effects comparatively smaller. At the same time spin-orbit coupling assumes more importance. We move therefore from a situation where the orbital splittings are comparable with the multiplet splittings, but large compared with the spin-orbit coupling, to a situation where spin-orbit coupling and core splittings are of the same order of magnitude and large compared to the correlative interactions between the electrons. At the same time the charge-transfer transitions move down in energy, thereby often blotting out the excited ligand field states. Due to these complications, the application of the simple correlation diagrams meets with increasing difficulties as we move from 3d complexes to 4d and 5d compounds.

The success of the diagrams depends upon the fact that the orders of magnitude are known for the two-electron integrals and that we have only one core parameter, $10Dq$. A lower than cubic symmetry causes trouble in that more core parameters are needed in order to specify the core energies of the states. Unfortunately it is not possible to find a unique expression for the orbital energies in terms of the observable band splitting parameters.[1] Furthermore, even though the full energy-level scheme in lower symmetries than cubic may be handled on a computer, we do not possess enough experimental details to determine the core parameters. An elaboration of theoretical parameters based upon fittings of experimental spectra to correlation diagrams is therefore not recommended. However, used with discretion a cubic energy-level diagram may also in cases of low symmetry provide us with suggestions for band assignments.

In Fig. 6-1 we have pictured the cubic octahedral energy level diagram for a Cr^{3+} complex as adapted from the calculations of Liehr.[2] The ground state is $^4A_{2g}$. For $Dq < 1{,}200$ cm^{-1} the first excited state is the spin-orbit split $^4T_{2g}$ state. For $Dq > 1{,}500$ cm^{-1} it is ^2E_g. The transition $^4A_{2g} \rightarrow {}^4T_{2g}$ is expected to be broad since in the transition a t_{2g} electron is transferred to an e_g shell. In contrast the

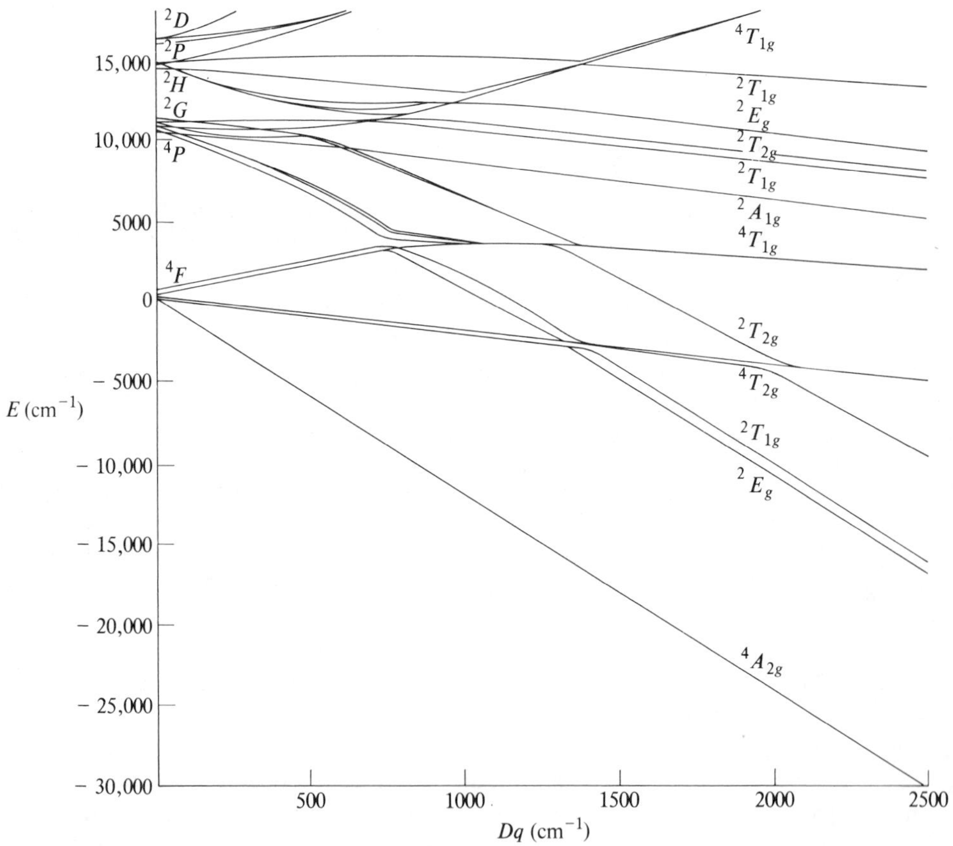

Figure 6-1 Correlation energy level diagram for cubic, octahedral Cr^{3+} complexes. The parameters used in the construction of the diagram are $F_2 = 1{,}069$ cm^{-1}, $F_4 = 78$ cm^{-1}, and $\zeta_M = 175$ cm^{-1}. Reproduced with permission from *Semiempirical Methods of Electronic Structure Calculation*, Part B, Plenum Press, 1977.

spin-forbidden transition $^4A_{2g} \rightarrow {}^2E_g$ is expected to be sharp since both of these states belong to the same electronic configuration, $(t_{2g})^3$. The intensity of the transition should be small and concentrated in one or two lines.

The absorption spectrum of Cr^{3+} doped into Al_2O_3 has been measured by McClure.[3] In corundum each Cr^{3+} ion is surrounded by six oxygen ions having a distorted octahedral array. Due to the lower than O_h symmetry, the broad bands are slightly structured and quite polarized. The first band system is constituted of some sharp lines located at 14,500 cm^{-1}. They are followed by a broad band having its maximum around 18,000 cm^{-1}. At 21,000 cm^{-1} some further sharp lines are seen, followed by a broad band topping at 25,000 cm^{-1}. Finally a third broad band is observed at 39,000 cm^{-1}. This is partly covered by the rising continuum.

We now use an effective cubic octahedral symmetry to characterize the electronic states of the Cr^{3+} chromophore. The first broad absorption band at 18,000 cm^{-1} is tentatively assigned to the first spin-allowed transition $^4A_{2g} \rightarrow {}^4T_{2g}$ (compare Fig. 6-1). This gives us $10Dq = 18{,}000$ cm^{-1}, in complete accord with Eqs. (2-83) and (2-84). At this value of Dq the correlation diagram Fig. 6-1 shows the transitions $^4A_{2g} \rightarrow {}^2E_g$, $^2T_{1g}$ to be placed at 13,000 cm^{-1}; $^4A_{2g} \rightarrow {}^2T_{2g}$ at 20,000 cm^{-1}; $^4A_{2g} \rightarrow {}^4T_{1g}(1)$ at 25,000 cm^{-1}; and finally $^4A_{2g} \rightarrow {}^4T_{1g}(2)$ at 40,000 cm^{-1}. The agreement of these predicted values with the observed energy intervals is so close as to leave no doubts as to the correctness of the identifications.

The potential surfaces of the excited states of an octahedral Cr^{3+} complex as a function of a totally symmetric ξ_1 vibrational coordinate probably look as pictured in Fig. 6-2. The low-lying spin doublet states will have their potential surface minima displaced vertically above the minimum of the ground state since these states arise from the same electronic configuration. The higher spin doublet and quartet states have, however, electronic configurations which differ from the ground-state configuration in that a transfer of electron(s) from the t_{2g} shell to the e_g shell has taken place. Their minima are consequently displaced, corresponding

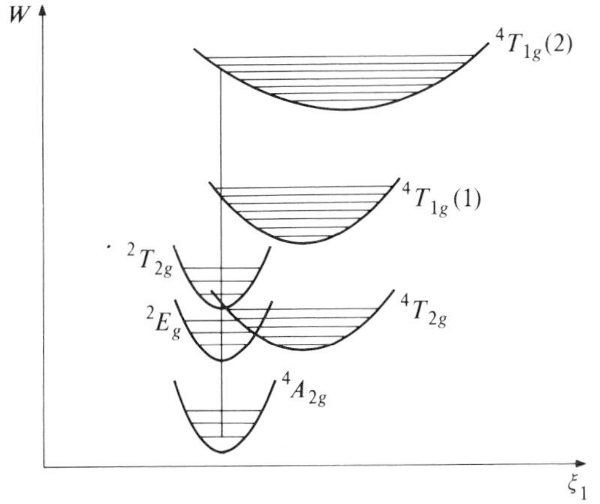

Figure 6-2 Potential surfaces for octahedrally coordinated Cr^{3+} ions pictured as a function of the totally symmetric breathing coordinate.

to a larger equilibrium bond distance. A vertical transition from the ground state will therefore hit them at a high vibrational quantum number. The consequences for the band shape have been dealt with in sections 4-7 to 4-12.

As a second example of an analysis of the transitions found in a Cr^{3+} complex we take the low-temperature spectrum of $2[Cr(en)_3]Cl_3 \cdot KCl \cdot 6H_2O$ as measured by McCarthy and Vala.[4] The chromophore is $Cr(en)_3^{3+}$, where (en) stands for ethylenediamine. The maxima of the two first spin-allowed transitions $^4A_{2g} \rightarrow {}^4T_{2g}$ and $^4T_{1g}(1)$ were found at some 22,300 cm^{-1} and 28,700 cm^{-1}, respectively. The position of the first spin-allowed band indicates $10Dq =$ 22,300 cm^{-1}. Using this Dq value the correlation diagram Fig. 6-1 places the transition $^4A_{2g} \rightarrow {}^4T_{1g}(1)$ at 28,000 cm^{-1}. The $^4A_{2g} \rightarrow {}^2E_g$, $^2T_{1g}$ system is expected at some 14,000 cm^{-1}.

Starting at 14,700 cm^{-1}, thick crystals of the compound show more than 80 separate peaks. From the vibrational analysis the prominent doublet at 14,883, 14,901 cm^{-1} is assigned as $(0, 0, \ldots \rightarrow 0, 0, \ldots)$ lines of $^4A_{2g}$ to the split components of ^2E_g. The 18-cm^{-1} splitting of ^2E_g is interpreted as the result of a combined spin-orbit and trigonal perturbation.

Diagonalizing the crystal-field matrices, the positions of the three identified states ^2E_g, $^4T_{2g}$, and $^4T_{1g}(1)$ were used to determine the three parameters $10Dq = 22{,}350$ cm^{-1}, $F_2 = 1344$ cm^{-1}, and $F_4 = 99$ cm^{-1}. The two states $^2T_{1g}(t_{2g})^3$ and $^2T_{2g}(t_{2g})^3$ were predicted with these parameters to occur at 15,300 cm^{-1} and 22,900 cm$^-$1, respectively. The very complex spectrum in the 14,700 to 18,900 cm^{-1} region shows a conspicuous group of levels around 15,500 cm^{-1}. These lines are identified as the $(0, 0, \ldots \rightarrow 0, 0, \ldots)$ levels of the transition from $^4A_{2g}$ to the split components of $^2T_{1g}$.

The calculated energy of $^2T_{2g}$ shows this state to be almost degenerate with the $^4T_{2g}$ state. It is suggested that an interference effect takes place[4] between $^4T_{2g}$ and $^2T_{2g}$ leading to an "antiresonance" as described in section 4-4. Indeed, the observed intensity envelope of the $^4T_{2g}$ band showing as it does a dip (or alternatively two peaks) clearly indicates the presence of $^2T_{2g}$. That we are dealing with an antiresonance seems reasonable in view of the corresponding features found in the isoelectronic VF_6^{4-} system.[5]

As a third example of the use of a correlation diagram, we consider the first spin-allowed transition in the $Ni(H_2O)_6^{2+}$ chromophore, $^3A_{2g} \rightarrow {}^3T_{2g}$, which is observed to have its maximum intensity at some 9,000 cm^{-1}. This energy difference is approximately equal to $10Dq$. Reference to Fig. 2-3, the crystal-field energy diagram for d^8 systems, tells us that additional spin-allowed transitions are expected at 15,000 cm^{-1} $(^3A_{2g} \rightarrow {}^3T_{1g}(1))$ and 26,000 cm^{-1} $(^3A_{2g} \rightarrow {}^3T_{1g}(2))$. This is in excellent accord with experiments.

It is interesting that the energy region around 15,000 cm^{-1} is distinguished by having a level of ^1E_g symmetry crossing over the states belonging to $^3T_{1g}(1)$. This feature is seen in the $Ni(H_2O)_6^{2+}$ spectrum already at room temperature in that the absorption band $^3A_{2g} \rightarrow {}^3T_{1g}(1)$ shows a double-peaked band envelope.

For a Dq value of 900 cm^{-1} the correlation diagram predicts the following ordering of the states[6] using the Bethe double-group nomenclature.

$$\Gamma_3({}^1E_g) < \Gamma_1({}^3T_{1g}) < \Gamma_4({}^3T_{1g}) < \Gamma_5({}^3T_{1g}) < \Gamma_3({}^3T_{1g})$$

Note that this level order refers to the excitation energies for a vertical transition from the ground ${}^3A_{2g}(\Gamma_5)$ state. Electronic transitions to these states are formally parity-forbidden, but the temperature dependence of the intensity demonstrates that the parity-forbiddenness is overcome by odd molecular vibrations.

Consider for instance $Ni(H_2O)_6 \cdot (BrO_3)_2$. The site symmetry of Ni^{2+} is approximately O_h, and the crystal spectrum shows[7] a 500-cm^{-1} progression originating from two sharp vibronic origins, centered at 15,900 cm^{-1} and separated by 137 cm^{-1}. Studying the splittings and behavior of these origins when $Ni(H_2O)_6^{2+}$ is put into a large variety of molecular crystals, one may conclude that the two lines are associated with the electronic origin[7,8] of ${}^3T_{1g}(\Gamma_3)$.

The intensity of the spin-forbidden transition ${}^3A_{2g} \rightarrow {}^1E_g$ depends critically upon the spin-orbit scrambling of ${}^1E_g(\Gamma_3)$ with ${}^3T_{1g}(\Gamma_3)$. This mixing is expected to be very affected by small variations in Dq. In order to identify the low energy ${}^1E_g(\Gamma_3)$ state the intensity variations of the broad band envelope were studied with $Ni(H_2O)_6^{2+}$ imbedded in a large number of different molecular crystals. It was concluded[7] that the second Γ_3 state was located at some 14,000 cm^{-1}.

Careful study of the spectrum of $Ni(H_2O)_6SiF_6$ led to the conclusion[7] that the line at 12,728 cm^{-1}, which marks the onset of the band envelope, could be identified as an odd vibration built upon the electronic level $\Gamma_1({}^3T_{1g})$. A line at 13,124 cm^{-1} was further found to be the same odd vibration built upon $\Gamma_4({}^3T_{1g})$. Using these results, together with the calculated energy spacings,[6] the potential surfaces as a function of a totally symmetric ξ_1 vibrational coordinate have been outlined in Fig. 6-3.

A nice example of a vibrational-electronic coupling can be seen in the absorption spectrum of $ReCl_6^{2-}$. Re^{4+} has a $(5d)^3$ configuration. Neglecting for the moment spin-orbit coupling we have (compare Fig. 6-1), taking only the $(t_{2g})^3$ configuration into account,

$$W({}^2E_g, {}^2T_{1g}) - W({}^4A_{2g}) = 3K_{xz,yz}^{\text{atomic}} = 3(3F_2 + 20F_4) \tag{6-1}$$

$$W({}^2T_{2g}) - W({}^4A_{2g}) = 5K_{xz,yz}^{\text{atomic}} = 5(3F_2 + 20F_4) \tag{6-2}$$

A complete calculation with spin-orbit coupling included gave[9] with $F_2 = 630$ cm^{-1}, $F_4 = 50$ cm^{-1}, and $\zeta_M = 2{,}300$ cm^{-1}: ${}^4A_{2g}(\Gamma_8)$, 0 cm^{-1}; ${}^2T_{1g}(\Gamma_8)$, 7,895 cm^{-1}; ${}^2E_g(\Gamma_8)$, 8,798 cm^{-1}; ${}^2T_{1g}(\Gamma_6)$, 9,167 cm^{-1}; ${}^2T_{2g}(\Gamma_7)$, 14,653 cm^{-1}; and ${}^2T_{2g}(\Gamma_8)$, 15,723 cm^{-1}. Using this calculation as a guide, the absorption lines observed[10] at 9,029 cm^{-1} and 9,471 cm^{-1} in single crystals of K_2PtCl_6 containing K_2ReCl_6 were assigned as transitions to ${}^2E_g(\Gamma_8)$ and ${}^2T_{1g}(\Gamma_6)$, respectively.

Zero phonon lines were further seen[10] at 13,840 cm^{-1} and 15,299 cm^{-1}. They are identified as transitions to the Γ_7 and Γ_8 states. In the magnetic dipole–allowed transition ${}^4A_{2g}(\Gamma_8) \rightarrow {}^2T_{2g}(\Gamma_7)$, four lines are observed. The energy differences between the lowest energy line and the other three are 124 cm^{-1}, 167 cm^{-1}, and 319 cm^{-1}. For the ground state of $ReCl_6^{2-}$ Woodward and Ware[11] assigned the odd vibrations $\nu_3(\tau_{1u}) = 313$ cm^{-1} and $\nu_4(\tau_{1u}) = 172$ cm^{-1}.

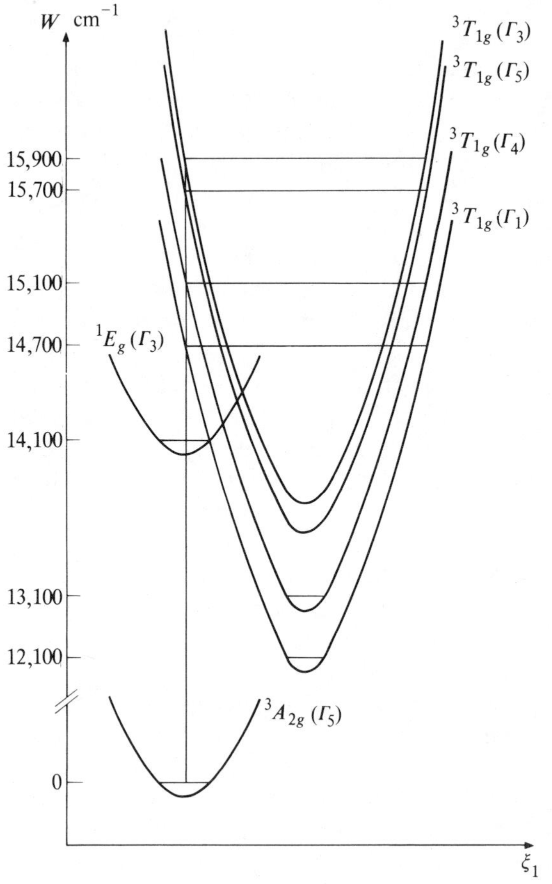

Figure 6-3 Some potential-energy surfaces for $Ni(H_2O)_6^{2+}$ as functions of a totally symmetric vibration coordinate ξ_1.

The line at 124 cm^{-1} built upon the 0–0 line is therefore assigned to the third possible odd mode $\nu_6(\tau_{2u})$. Hence the pure electronic transition $^4A_{2g}(\Gamma_8) \rightarrow {}^2T_{2g}(\Gamma_7)$ at 13,840 cm^{-1} has coupled to it the three fundamental odd vibrations of an MX_6 compound of O_h symmetry.

The low-lying electronic states of the complexes $ReF_6(5d)^1$, $OsF_6(5d)^2$, $IrF_6(5d)^3$, and $PtF_6(5d)^4$ have been considered by Moffitt et al.[12] Looking at the electronic configurations $(t_{2g})^n$, $n = 1$ to 4, using only two parameter values $\zeta_M = 3{,}400\ cm^{-1}$ and $K_{xz,yz} = 3F_2 + 20F_4 = 2{,}400\ cm^{-1}$, these authors succeeded in accounting for thirteen band systems found in the four hexafluorides. Similarly, the low-lying absorption bands of the series $TcF_6(4d)^1$, $RuF_6(4d)^2$, $RhF_6(4d)^3$, and $PdF_6(4d)^4$ can be assigned[13] using $\zeta_M = 1{,}280\ cm^{-1}$ and $K_{xz,yz} = 3{,}000\ cm^{-1}$.

Next we shall briefly consider the complexes of the rare earths and actinides. In the $(4f)^n$ rare-earth compounds the 4f electrons of the metal ion are shielded from the ligand electrons by the filled metal 5s and 5p shells. Consequently a very small value is found for the group overlap integral, $G_{4f,L}$. Hence the ligand field core splittings are orders of magnitude less than both the multiplet splittings of the metal ion and the spin-orbit coupling. Observations of the excited electronic states are therefore effectively giving us the atomic line spectra of the rare-earth ions.

A completely different situation is encountered in the complexed actinide ions where the 5f shell is being filled. At least two ligand field-splitting parameters in addition to the spin-orbit coupling parameter will have to be considered. These three quantities are all of comparable magnitude. The correlation diagrams for the coordinated $(5f)^n$ systems are therefore very complicated.

In O_h the seven f-orbitals span a_{2u}, t_{1u}, and t_{2u} representations. The energy matrices in this representation have been given by Eisenstein and Pryce.[14] We place the core energies at $\overline{h}(t_{2u}) = 0$, $\overline{h}(a_{2u}) = -\Delta$, and $\overline{h}(t_{1u}) = \Theta$. The level order for, say, UCl_6^- $(5f)^1$ is then $W(\Gamma_7) < W(\Gamma_8) < W(\Gamma_8') < W(\Gamma_6)$. The absorption spectrum can be fitted[15] with $\zeta_M = 1{,}940\ \text{cm}^{-1}$, $\Delta = 1{,}940\ \text{cm}^{-1}$, and $\Theta = 3{,}710\ \text{cm}^{-1}$. One should, however, notice that the values of Δ and Θ as evaluated from fitting the absorption spectra are extremely sensitive to small variations in the chosen value of ζ_M.

Turning now to the excited states of inorganic complexes which do not possess d- or f-electrons, the situation is rather unsatisfactory. Even for one of the best-investigated systems, the permanganate ion MnO_4^-, the calculated excited state energies are in very poor agreement with the experimentally observed so-called charge-transfer transitions. Unfortunately, the use of the virtual orbitals emerging from a ground state Hartree–Fock calculation in combination with Eqs. (2-47) and (2-48) produces quite unreliable results. *Ab initio* SCF LCAO MO calculations on MnO_4^- with near Hartree–Fock accuracy[16] yielded only a 4,000-cm^{-1} energy difference between the two lowest virtual orbitals of $7t_2$ and $2e$ symmetry. This indicates that only by performing detailed and accurate independent Hartree–Fock calculations of each excited state can we hope to place these at their right energy.

This conclusion is borne out by the calculations of Hillier and Saunders,[17] Wood,[18] and Hsu et al.[19] Using the virtual orbitals of a better-than-minimal basis set for MnO_4^-, and with extensive configuration interaction included, the first excited state of $^1T_1$ symmetry was placed[17] at 24,900 cm^{-1}, and the first excited $^1T_2$ state at 27,600 cm^{-1}. It is now generally agreed (see section 6-5) that $^1T_1$ is experimentally found at 14,500 cm^{-1} and $^1T_2$ at 18,600 cm^{-1}. The most extensive calculations[18,19] again using a configuration interaction approach are even worse than the above.

Using the SCF-Xα approach Johnson[20] has considered the levels involved in the lowest optical transitions in MnO_4^-. In calculating the transition state one-half of a unit of electronic charge is removed from the initial orbital and added to the charge of the final orbital. His calculation gives an energy excitation equal to 18,600 cm^{-1} for $(1t_1) \rightarrow (2e)$. Notice, however, that no splittings of the multiplets to which the $(t_1)^{-1}(e)^1$ configuration gives rise has been considered.

The excited states of $Ni(CN)_4^{2-}$ have been calculated by Demuynck and Veillard.[21] $Ni(CN)_4^{2-}$ is a square planar ion, possessing D_{4h} symmetry. The electronic ground state is $^1A_{1g}$. The calculated orbital configuration of the ground state shows all of the uppermost occupied orbitals to be ligand π orbitals. The same holds true for the first virtual orbitals. Separate self-consistent field calculations were performed for the lowest excited spin-singlet and spin-triplet states of $Ni(CN)_4^{2-}$. All the lowest excited states turned out to correspond to d–d transitions. This shows very clearly the importance of electronic relaxation upon excitation and the role played by the electronic repulsion terms. Putting $W(^1A_{1g}) = 0$ the lowest computed transition energies are given in Table 6-1. The agreement between the calculated and assigned "ligand field states" is poor. On the other hand, the "charge transfer states" seem reasonably well accounted for.[22]

All in all, we must conclude that molecular orbital calculations of complexes cannot be relied upon when it comes to the assessment of the position of the excited states. This state of affairs holds also for the less sophisticated, heavily scaled calculations. The use of the calculations lies in their ability to point towards a limited number of alternatives. One can then explore these and thus guided, perform experiments which hopefully can decide the issue.

Charge-transfer transitions in which an electron is transferred from an orbital of predominantly ligand character into a vacant antibonding orbital mainly located on the metal, are usually found at higher energies than the energetically lowest ligand-field transitions. It is remarkable that the presence of the charge-transfer states do not exert a greater influence on the lower-frequency behavior of the ligand-field bands. This reflects that the associated electronic motions are separable to high approximation.

The allowed ligand-to-metal charge transfer bands in $IrCl_6^{2-}$ are quite interesting. The ground-state configuration in $IrCl_6^{2-}$ is $^2T_{2g}(t_{2u})^6(t_{2g})^5$ where the nonbonding t_{2u} orbitals are entirely located on the six chlorine ligands. A charge transfer gives rise to the electronic state $^2T_{2u}(t_{2u})^5(t_{2g})^6$, which in turn is split by

Table 6-1 Calculated[18] and experimental[19] energies of the excited states of $Ni(CN)_4^{2-}$

Excited states		Calculated	Experimental
$(z^2) \rightarrow (x^2 - y^2)$	$^1B_{1g}$	20,600 cm^{-1}	
	$^3B_{1g}$	4,000 cm^{-1}	
$(xz), (yz) \rightarrow (x^2 - y^2)$	^1E_g	21,900 cm^{-1}	31,700 cm^{-1}
	^3E_g	7,700 cm^{-1}	
$(xy) \rightarrow (x^2 - y^2)$	$^1A_{2g}$	22,500 cm^{-1}	
	$^3A_{2g}$	14,700 cm^{-1}	
$(z^2) \rightarrow \pi^*$	$^1A_{2u}$	33,900 cm^{-1}	34,400 cm^{-1}
	$^3A_{2u}$	30,300 cm^{-1}	
$(xz), (yz) \rightarrow \pi^*$	^1E_u	37,900 cm^{-1}	36,700 cm^{-1}
	^3E_u	36,000 cm^{-1}	35,800 cm^{-1}

the spin-orbit coupling. The operator of Eq. (1-122)

$$\mathscr{H}^{(1)} = \sum_{\mu} \sum_{j} \zeta_{\mu}(r_{\mu j}) \mathbf{l}_{\mu j} \cdot \mathbf{s}_j \tag{6-3}$$

gives for the splitting of the $^2T_{2u}$ state

$$W(G) - W(E_{5/2}) = \tfrac{3}{4}\zeta_{3p}(\mathrm{Cl}) \tag{6-4}$$

The expected spin-orbit splitting of some 440 cm^{-1} is, however, quenched by the operation of the Ham effect[23] (see section 6-3). The $^2T_1$ state found in the $TiBr_4^+$ ion also exhibits a spin-orbit splitting of some $\tfrac{3}{4}\zeta_{4p}(\mathrm{Br})$. Measurements of the photo-electron spectrum[24] gave a splitting of 0.24 eV, in good agreement with the calculated value of 0.243 eV.

The charge-transfer states of $CoBr_4^{2-}$ have been investigated by Bird and Day.[25] The ground state is $^4A_2(t_1)^6(3t_2)^6(2e)^4(4t_2)^3$. The excited charge-transfer configurations $(t_1)^5\,(3t_2)^6\,(2e)^4\,(4t_2)^4$ and $(t_1)^6\,(3t_2)^5\,(2e)^4\,(4t_2)^4$ each contain one $^4T_1$ state, accessible by an electric dipole transition. The spin-orbit coupling will split the $^4T_1$ state into three levels, as described in Eq. (1-131). The energies are given by $W(E_{5/2}, G) = \tfrac{3}{2}\lambda\alpha$, $W(G) = -\lambda\alpha$, and $W(E_{1/2}) = -\tfrac{5}{2}\lambda\alpha$. Reducing the proportionality constants to one-electron matrix elements, $\lambda\alpha$ was found for the $^4T_1(t_1 \rightarrow 4t_2)$ charge-transfer state

$$\lambda\alpha = -\tfrac{1}{18}[\zeta(t_1) + \zeta(4t_2)] \tag{6-5a}$$

and for $^4T_1(3t_2 \rightarrow 4t_2)$

$$\lambda\alpha = \tfrac{1}{18}[\zeta(3t_2) + \zeta(4t_2)] \tag{6-5b}$$

Very careful MCD studies of Rivoal and Briat[26] indicate that the lowest-lying charge-transfer transition at 34,400 cm^{-1} is associated with an effective $J = \tfrac{5}{2}$. The proportionality constant $\lambda\alpha$ must therefore be negative and the evidence shows that the first band system is associated with the orbital excitation $(t_1 \rightarrow 4t_2)$.

We have seen that the spin-orbit splittings of the charge-transfer states may be expressed as a sum of one-electron contributions from the sundry open shells. If the spin-orbit coupling constant ζ_M associated with the "ligand field" orbitals is much smaller than the ζ_L constant associated with an open shell of ligand orbitals, the spin-orbit splittings of the charge transfer bands will therefore be dominated by ζ_L. However, it may also happen that the various contributions to $\lambda\alpha$ tend to cancel each other.

6-2 LINEAR POLARIZED ABSORPTION SPECTROSCOPY

When a parallel beam of plane-polarized light is passed perpendicularly through a crystal plane, we observe the transition moments as projected in the directions of polarization inside the crystal. These directions are given by the principal dielectric axes of the crystal plane. Provided there is a symmetry axis lying in the plane of the plate, the dielectric axes are the symmetry axis and an axis normal to it. However, if there is no symmetry axis the polarization directions may change with the frequency.

A polarized crystal spectrum is therefore most easy to interpret in cases where we have a uniaxial crystal. The crystals can be either trigonal, tetragonal, hexagonal, or, rather trivially, cubic. One may distinguish, in these cases, between intensity induction by either magnetic-dipole or electric-dipole mechanisms.

One measures the axial or α spectrum when the unpolarized light is propagated along the unique axis **c**; in this instance, of course, both the electric vector **E** and the magnetic vector **H** are perpendicular to the optic axis. One next performs a measurement with plane-polarized light which is propagating perpendicular to **c**; two spectra may be obtained: the σ spectrum, for which $\mathbf{E} \perp \mathbf{c}$ and $\mathbf{H} \parallel \mathbf{c}$, and the π spectrum, for which $\mathbf{E} \parallel \mathbf{c}$ and $\mathbf{H} \perp \mathbf{c}$. If the axial spectrum agrees with the σ spectrum, then we have an electric dipole–allowed transition. Correspondingly, if the axial and π spectrum coincide, we have a magnetic-dipole transition.

The use of linear dichroism, that is recording the difference between a π- and a σ-polarized spectrum, may yield a more differentiated experimental spectrum than the more conventional method. A nice example is found in the rich line spectrum associated with the transitions ${}^4A_2 \rightarrow {}^2E, {}^2T_1$ met with in the complex $2Cr(en)_3Cl_3 \cdot KCl \cdot 6H_2O$. The lines had been analyzed using conventional techniques[4] (see section 6-1) but the linear dichroism spectrum adds considerably in the identification.[27]

The crystals of tris-ethylenediamine nickel(II) nitrate, $Ni(en)_3(NO_3)_2$, are hexagonal, with a nickel-site symmetry of D_3. The threefold axis of the complex unit $Ni(en)_3^{2+}$ is parallel to the **c** axis of the crystal. The measured coincidence of the σ spectrum with the α spectrum shows that for all the crystal-field bands the intensity-giving mechanism is an electric dipole transition.[28]

In ruby (Cr^{3+} in Al_2O_3) the symmetry of the distorted octahedron of O^{2-} surrounding each Cr^{3+} is C_3. The comparison of the σ and α spectra reveals[29] that the so called R lines (${}^4A_2 \rightarrow {}^2E$) are predominately electric dipole in character. When Cr^{3+} is doped into a spinel $MgAl_2O_4$, the chromium ions have a sixfold distorted octahedral configuration belonging to the D_{3d} point group. The site has thus a center of inversion. The accumulated evidence for the observed R-line fluorescence points towards a magnetic dipole transition.[30]

It turns out that in cases where we have a center of symmetry in the chromophore nearly all the intensity is either vibronic in origin or, in rare cases, magnetic dipole in nature. When, on the other hand, the symmetry is D_3 or C_3 a great part of the observed band intensity turns out to be electric dipole. Qualitatively the selection rules can be seen from the transformation properties of the states. In Fig. 6-4 we give the selection rules for electric dipole transitions of some low-lying states in Cr^{3+} complexes. Quantitatively the intensity ratio for σ and π polarizations may be estimated using a perturbation approach.

The ligand-field states are all even functions with respect to the inversion operator $\hat{i}$. In order to get electric dipole intensity in the transitions we need therefore a low-symmetry component in the hamiltonian operator which can mix

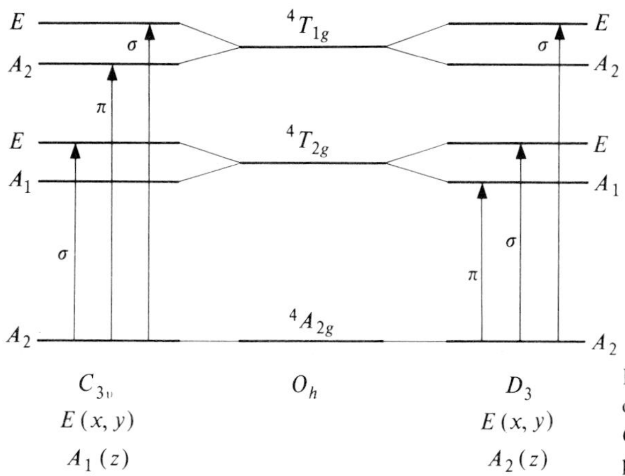

Figure 6-4 Low ligand field states of octahedrally coordinated Cr^{3+} in O_h and lower symmetries with polarization behavior indicated.

the even function with some odd functions. Calling the odd excited states $|t_u\rangle$ we have for the perturbed ligand field states $|o'\rangle$ and $|j'\rangle$

$$|o'\rangle = |o_g\rangle + \sum_t \frac{\langle t_u|\mathscr{H}^{(1)}|o_g\rangle}{W_0 - W_t} |t_u\rangle$$

$$|j'\rangle = |j_g\rangle + \sum_t \frac{\langle t_u|\mathscr{H}^{(1)}|j_g\rangle}{W_0 - W_t} |t_u\rangle$$

$\mathscr{H}^{(1)}$ must be invariant under the actual symmetry of the molecule and must change sign under inversion. If we expand $\mathscr{H}^{(1)}$ in powers of the electronic coordinates as expressed in the nonrotated coordinated system of Fig. 1-1, the first such term in C_{3v} symmetry is $\sum_n 1/\sqrt{3}(x_n + y_n + z_n)$. Such a function belongs to a t_{1u} manifold in O_h. As a tensor operator it is classified in C_{3v} as $\hat{V}(T_1 a_1)$. In D_3 symmetry the first term meeting our specification for $\mathscr{H}^{(1)}$ is $\sum_n 1/\sqrt{3}[x_n(y_n^2 - z_n^2) + y_n(z_n^2 - x_n^2) + z_n(x_n^2 - y_n^2)]$. In octahedral symmetry this function belongs to t_{2u}. The tensor operator is $\hat{V}(T_2 a_1)$.

The electric dipole transition moment $\mathbf{D}_{oj}$ is therefore

$$\mathbf{D}_{0j} = \sum_t \frac{\langle j_g|\sum_n \mathbf{r}_n|t_u\rangle\langle t_u|\mathscr{H}^{(1)}|o_g\rangle}{W_0 - W_t} + \sum_t \frac{\langle o_g|\sum_n \mathbf{r}_n|t_u\rangle\langle t_u|\mathscr{H}^{(1)}|j_g\rangle}{W_j - W_t} \tag{6-6}$$

Equation (6-6) can be simplified by using the closure approximation. To this approximation we may write dropping the g subscript

$$\mathbf{D}_{0j} \approx \frac{2}{\Delta W}\langle o|\mathscr{H}^{(1)}\sum_n \mathbf{r}_n|j\rangle \tag{6-7}$$

where the denominators in Eq. (6-6) have been replaced by an average ΔW.

The electric dipole operators can be represented by irreducible tensor operators of the type $\hat{V}(T_1x)$, $\hat{V}(T_1y)$, and $\hat{V}(T_1a_1)$ respectively. The products of the two tensor operators in Eq. (6-7) are reduced to linear combinations of tensor operators by using the coupling coefficients for octahedral symmetry on a trigonal basis as given in Table 1-3. We get in C_{3v} using Table 1-2

$$\hat{V}(T_1a_1)\hat{V}(T_1a_1) = \sqrt{\tfrac{1}{3}}\,\hat{V}(A_1a_1) - \sqrt{\tfrac{2}{3}}\,\hat{V}(T_2a_1) \tag{6-8}$$

$$\hat{V}(T_1a_1)\hat{V}(T_1y) = \sqrt{\tfrac{1}{3}}\,\hat{V}(Ex) + \sqrt{\tfrac{1}{2}}\,\hat{V}(T_1x) - \sqrt{\tfrac{1}{6}}\,\hat{V}(T_2x) \tag{6-9}$$

$$\hat{V}(T_1a_1)\hat{V}(T_1x) = -\sqrt{\tfrac{1}{3}}\,\hat{V}(Ey) - \sqrt{\tfrac{1}{2}}\,\hat{V}(T_1y) + \sqrt{\tfrac{1}{6}}\,\hat{V}(T_2y) \tag{6-10}$$

With $|o\rangle = {}^4A_{2g}$ and $|j\rangle = {}^4T_{2g}$ we have $A_{2g} \times T_{2g} = T_{1g}$. Hence in C_{3v} we get for the components of the transition moment $D_\parallel$ and $D_\perp$ as referred to the threefold axis of the complex, $D_\parallel = 0$ and $D_\perp = \pm\sqrt{\tfrac{1}{2}}\langle {}^4A_{2g} \| \sum_n \mathbf{r}_n \| {}^4T_{2g}\rangle$. The intensity ratio is therefore

$$\frac{|D^\pi_{A_2} \to T_2|^2}{|D^\sigma_{A_2} \to T_2|^2} = \frac{0}{1/2} \tag{6-11}$$

With $|o\rangle = {}^4A_{2g}$ and $|j\rangle = {}^4T_{1g}$ we get $D_\parallel = -\sqrt{\tfrac{2}{3}}\langle {}^4A_{2g} \| \sum_n \mathbf{r}_n \| {}^4T_{1g}\rangle$ and $D_\perp = \pm\sqrt{\tfrac{1}{6}}\langle {}^4A_{2g} \| \sum_n \mathbf{r}_n \| {}^4T_{1g}\rangle$. The intensity ratio is

$$\frac{|D^\pi_{A_2} \to T_1|^2}{|D^\sigma_{A_2} \to T_1|^2} = \frac{4}{1} \tag{6-12}$$

In D_3 symmetry the numerical results are the same, only the levels ${}^4T_2$ and ${}^4T_1$ are exchanged:

$$\frac{|D^\pi_{A_2} \to T_2|^2}{|D^\sigma_{A_2} \to T_2|^2} = \frac{4}{1} \tag{6-13}$$

$$\frac{|D^\pi_{A_2} \to T_1|^2}{|D^\sigma_{A_2} \to T_1|^2} = \frac{0}{1/2} \tag{6-14}$$

This simple perturbation treatment explains the qualitative features of the observed optical anisotropies. The C_{3v} symmetry is realized in Cr^{3+} dissolved in Al_2O_3, and the polarized spectrum as given by McClure[3] is in fair agreement with the above deductions. The polarized spectrum of $Cr(en)_3^{3+}$ published by McCarthy and Vala[4] shows on the other hand the expected D_3 behavior. Departures in the intensity ratios from the predicted ones may be due to the closure approximation, to vibronic-induced intensity, and to the use of a truncated expression for $\mathscr{H}^{(1)}$.

Without use of perturbation theory the intensity ratios can be evaluated by using the orbital transformation properties specific to the considered system. Characterizing the e orbitals in C_{3v} by a + or − according to their behavior when reflected in the $\tilde{y}\tilde{z}$ plane we have (consult Table 1-2 and Fig. 1-1)

$$t^{+} = \frac{-1}{\sqrt{6}}(yz + xz - 2xy) \qquad e^{+} = (3z^{2} - r^{2})$$

$$t^{-} = \frac{1}{\sqrt{2}}(yz - xz) \qquad e^{-} = (x^{2} - y^{2})$$

$$t^{0} = \frac{1}{\sqrt{3}}(xy + yz + xz)$$

For the one-electron transitions between the above orbitals we find using the symmetry elements relevant to C_{3v}:

$$\begin{aligned}
\langle t^{+}|\mathbf{r}|e^{+}\rangle &= 0\mathbf{i} + A\mathbf{j} + C\mathbf{k} \\
\langle t^{+}|\mathbf{r}|e^{-}\rangle &= A\mathbf{i} + 0\mathbf{j} + 0\mathbf{k} \\
\langle t^{-}|\mathbf{r}|e^{+}\rangle &= A\mathbf{i} + 0\mathbf{j} + 0\mathbf{k} \\
\langle t^{-}|\mathbf{r}|e^{-}\rangle &= 0\mathbf{i} - A\mathbf{j} + C\mathbf{k} \\
\langle t^{0}|\mathbf{r}|e^{+}\rangle &= 0\mathbf{i} - B\mathbf{j} + 0\mathbf{k} \\
\langle t^{0}|\mathbf{r}|e^{-}\rangle &= B\mathbf{i} + 0\mathbf{j} + 0\mathbf{k}
\end{aligned} \tag{6-15}$$

The unit vectors **i** and **j** are perpendicular to the C_3 axis while **k** is contained in it. The constants A, B, and C are given by

$$A = \langle t^{+}|\frac{1}{\sqrt{2}}(x - y)|e^{-}\rangle \tag{6-16}$$

$$B = \langle t^{0}|\frac{1}{\sqrt{2}}(x - y)|e^{-}\rangle \tag{6-17}$$

$$C = \langle t^{+}|\frac{1}{\sqrt{3}}(x + y + z)|e^{+}\rangle \tag{6-18}$$

and represent the magnitude of the three matrix elements of the transition moments which are different from zero by symmetry.

Writing down the wave functions for the low excited states of Cr^{3+} in O_h symmetry but quantized along the threefold axis we get the general result for Cr^{3+} in C_{3v} symmetry[3]

$$\frac{|D^{\pi}_{A_2} \rightarrow T_2|^2}{|D^{\sigma}_{A_2} \rightarrow T_2|^2} = \frac{0}{(A + B/\sqrt{2})^2} \tag{6-19}$$

$$\frac{|D^{\pi}_{A_2} \rightarrow T_1|^2}{|D^{\sigma}_{A_2} \rightarrow T_1|^2} = \frac{2C^2}{(A - B/\sqrt{2})^2} \tag{6-20}$$

Notice that only by assuming $\sqrt{2}A - B = C$ do we get the previous result, Eq. (6-12), back.

For Ti^{3+} ($3d^1$) doped into Al_2O_3 two far-infrared electronic absorption bands have been observed[31] at 37.8 cm^{-1} and 107.5 cm^{-1}. Both lines are σ polarized and the σ spectrum is identical with the axial (α) spectrum. The observed lines are therefore electric dipole transitions. The symmetry of the distorted octahedron of six oxygen around each Ti^{3+} ion is C_3.

The energy-level diagram is pictured in Fig. 5-1. The ground state can be either $2\bar{B}_{3/2}(^2E)$ or $\bar{E}_{1/2}(^2A)$. In C_3, z transforms like A and x, y as E. In the double group $\bar{E}_{1/2} \times \bar{B}_{3/2} = E$ and $\bar{E}_{1/2} \times \bar{E}_{1/2} = \{A\} + A + E$, where $\{A\}$ is an antisymmetric product. The absence of a π-polarized transition tells us that the ground state is $2\bar{B}_{3/2}$.

The fitting of the two infrared lines to the formulae of Eq. (5-34) yields $\Delta = -10\ cm^{-1}$ and $\zeta_M = 90\ cm^{-1}$. The extreme low value of $\zeta_M(Ti^{3+}$, $\zeta_{3d} = 175\ cm^{-1})$ shows that the diagram is only qualitatively correct. Indeed, the experimental results indicate the presence of a Jahn–Teller coupling.[32] We shall return to this system in the next section.

A good example of a vibrational electronic–induced dichroism is found in the absorption spectrum of *trans*-$[Co(en)_2Cl_2^+]$. We have seen, Eq. (4-140), that the intensity of a vibronically induced electronic transition should decrease with decreasing temperature. The temperature dependence of the spectrum of the complex[33] shows clearly that a considerable portion of the absorption band intensities is vibronic in nature. The possibility of a static intensity giving perturbation in the hamiltonian may thus be eliminated. The level scheme of the complex, as classified in D_{4h} symmetry, is given in Fig. 6-5. Note, however, that the actual molecular symmetry of the complex is C_{2h}, and the cobalt-ion site group symmetry in the crystal of *trans*-$[Co(en)_2Cl_2]Cl \cdot HCl \cdot 2H_2O$ is only C_i.

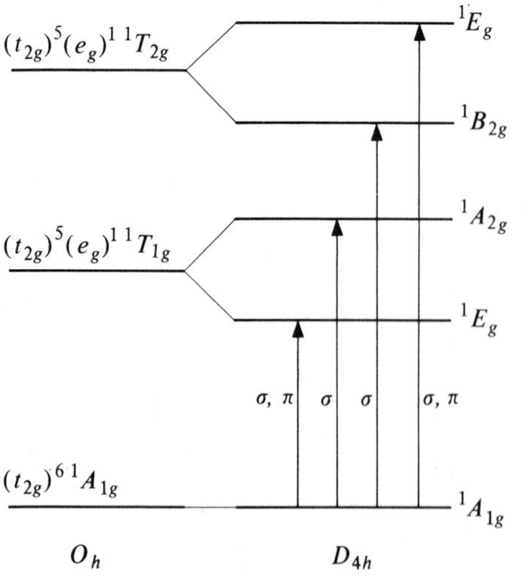

Figure 6-5 Vibronic selection rules for the low excited states of *trans*-$[Co(en)_2Cl_2^+]$ as classified in the point group D_{4h}.

In D_{4h} the electric dipole vector transforms as $A_{2u}(\pi)$ and $E_u(\sigma)$ as referred to the fourfold axis. For the transition ${}^1A_{1g} \rightarrow {}^1A_{2g}$ to be allowed in π polarization we must have (see Eq. (4-137))

$$(A_{1g}) \times (A_{2u}) \times \Gamma(\xi_i) \times (A_{2g}) = A_{1g}$$

which leads to $\Gamma(\xi_i) = \alpha_{1u}$. A similar consideration gives for the transition ${}^1A_{1g} \rightarrow {}^1B_{2g}$ that $\Gamma(\xi_i) = \beta_{1u}$. Neither of these vibrations is found in the cluster *trans*-[X_2MY_4]. The two excitations ${}^1A_{1g} \rightarrow {}^1A_{2g}$ and ${}^1A_{1g} \rightarrow {}^1B_{2g}$ are therefore also forbidden as vibronic transitions in D_{4h} symmetry. On the other hand, the vibrations which are present in the complex give us the selection rules pictured in Fig. 6-5. The agreement between the predicted and observed selection rules therefore allows us to identify the excited states.[33]

A vibronic analysis of band intensities can in favorable cases throw light on the equilibrium conformation of the excited states. Consider for instance the square planar complex $Ni(CN)_4^{2-}$. The molecular symmetry is D_{4h}, and the ground state transforms in this symmetry like ${}^1A_{1g}$. The low excited ligand-field spin singlet states are (see Table 6-1) ${}^1B_{1g}$, ${}^1E_g$, and ${}^1A_{2g}$. The odd vibrations of the complex are of species α_{2u}, β_{2u}, and ε_u. The electric dipole vector transforms in D_{4h} like $A_{2u}(\pi)$ and $E_u(\sigma)$. It happens that the three low crystal-field transitions ${}^1A_{1g} \rightarrow {}^1B_{1g}$, ${}^1E_g$, ${}^1A_{2g}$ are vibronically allowed in σ polarization, and that ${}^1A_{1g} \rightarrow {}^1B_{1g}$, ${}^1E_g$ are allowed in π polarization.

Experiments now show[34] that the band observed at some 23,000 cm^{-1} is seen in π polarization but not in σ polarization. The absence of this band in σ polarization indicates therefore that the D_{4h} vibronic selection rules may not be appropriate. However, if the equilibrium configuration of the excited state is D_{2d}, the vertical transition from the ground state ${}^1A_{1g}(D_{4h})$ will terminate on an excited state of ${}^1B_2$, ${}^1E$, or ${}^1A_2$ symmetry as classified in D_{2d} symmetry. In D_{2d} the z coordinate transforms as B_2 and x, y as E. The transition ${}^1A_{1g} \rightarrow {}^1B_2$ should therefore be allowed as an electric dipole transition and be π polarized. The band at 23,000 cm^{-1} is accordingly identified with such an electronic transition.

A correlation diagram[34] of the states of $Ni(CN)_4^{2-}$ in D_{4h} and T_d symmetry indicates that the ${}^1E_g$ ligand-field state likewise should have a D_{2d} equilibrium geometry. Combining the results of refs. 21, 22, 34, and 35, the potential curves for $Ni(CN)_4^{2-}$ probably look as pictured in Fig. 6-6.

Polarized electronic Raman transitions may also be of help in the assignment of excited states. Consider again Ti^{3+} doped into Al_2O_3. The site symmetry of Ti^{3+} in the unit cell is C_3. The symmetry of the scattering tensor in C_3 (see section 4-14) is $A(\alpha_{xx} + \alpha_{yy}, \alpha_{zz})$ and $E[(\alpha_{xx} - \alpha_{yy}, \alpha_{xy}), (\alpha_{yz}, \alpha_{xz})]$. In C_3 we have $\bar{E}_{1/2} \times \bar{B}_{3/2} = E$ and $\bar{E}_{1/2} \times \bar{E}_{1/2} = \{A\} + A + E$.

The Raman spectra are now recorded.[36] Let light polarized in the X direction, for instance, enter along the Z direction of the molecule. Collect the scattered light along the molecular X axis as polarized in the Y direction. The resulting spectrum is designated $Z(xy)X$. In this polarization signals were observed at 38 cm^{-1} and 109 cm^{-1}. The absence of an $X(zz)Y$ spectrum shows the observed transitions to be $2\bar{B}_{3/2} \rightarrow \bar{E}_{1/2}$ and $2\bar{B}_{3/2} \rightarrow \bar{E}_{1/2}$ (compare Fig. 5-1).

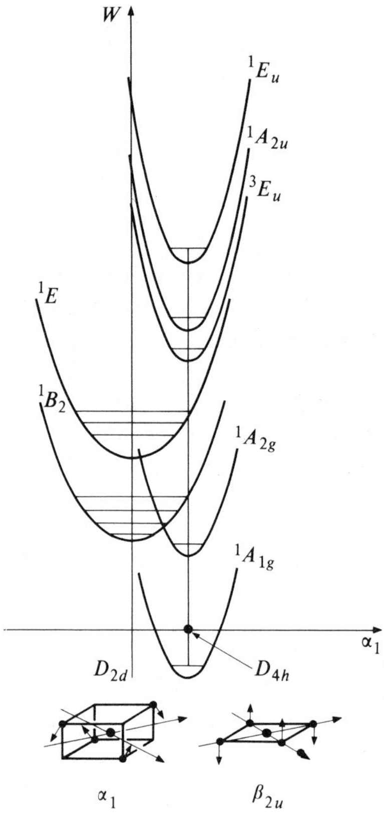

Figure 6-6 Some potential curves for $Ni(CN)_4^{2-}$. Note that the vibrational symmetry coordinate, which in D_{4h} transforms like β_{2u}, will be a totally symmetric vibration in D_{2d}. Reproduced with permission from *Semiempirical Methods of Electronic Structure Calculation*, Part B, Plenum Press, 1977.

Electric quadrupole transitions have been observed in single crystals of $Cs_2UO_2Cl_4$. Let $\boldsymbol{\kappa}$ be the propagation vector for the incident light and $\mathbf{u}$ the polarization vector. The electric quadrupole operator may then be expanded (compare Eq. (4-34))

$$\begin{aligned}(\boldsymbol{\kappa}\cdot\mathbf{r})(\mathbf{u}\cdot\mathbf{r}) &= \tfrac{1}{3}r^2(\boldsymbol{\kappa}\cdot\mathbf{u}) + \tfrac{1}{2}(x^2-y^2)(\kappa_x u_x - \kappa_y u_y)\\ &\quad + \tfrac{1}{3}(3z^2-r^2)(\kappa_z u_z - \tfrac{1}{2}\kappa_x u_x - \tfrac{1}{2}\kappa_y u_y)\\ &\quad + xy(\kappa_x u_y + \kappa_y u_x) + yz(\kappa_z u_y + \kappa_y u_z)\\ &\quad + xz(\kappa_x u_z + \kappa_z u_x) \qquad (6\text{-}21)\end{aligned}$$

The scalar product $\boldsymbol{\kappa}\cdot\mathbf{u}$ is zero. The remaining terms in Eq. (6-21) consist of the product of a component of the quadrupole tensor and a factor dependent upon the direction and polarization of the incident radiation. Using θ and ϕ as the polar and azimuthal angles referred to the molecular x axis we may take[37]

$$\begin{aligned}\boldsymbol{\kappa} &= (\sin\theta\cos\phi, \sin\theta\sin\phi, \cos\theta)\\ \mathbf{u}_1 &= (-\cos\theta\cos\phi, -\cos\theta\sin\phi, \sin\theta)\\ \mathbf{u}_2 &= (-\sin\phi, \cos\phi, 0)\end{aligned}$$

The angular dependence of, say, the xy quadrupole component for $\mathbf{u}_1$ polarization of the electric vector is $-\frac{1}{2}\sin 2\theta \cos 2\phi$; and for $\mathbf{u}_2$ polarization, $\sin\theta\cos 2\phi$. The intensity of the transition when $\theta = \pi/2$, that is with light propagation perpendicular to the molecular z axis, for $\mathbf{u}_1$ polarization is zero; and for $\mathbf{u}_2$ polarization, $\cos^2 2\phi$. For both electric dipole and magnetic dipole mechanisms the intensity is seen to be proportional to $\sin^2\phi$ or $\cos^2\phi$. The electric and magnetic dipole transitions vary therefore with a period of π compared with the period of $\pi/2$ for the electric quadrupole mechanism.

Using the notation $X(y)$ to indicate an experiment in which light propagates along the molecular X axis with polarization in the Y direction, various polarization experiments were performed on $Cs_2UO_2Cl_4$ crystals.[37] The molecular structure of the complex $UO_2Cl_4^{2-}$ is taken to be D_{4h}. The five components of the quadrupole tensor xy, (yz, xz), $(3z^2 - r^2)$, and $(x^2 - y^2)$ transform like b_{2g}, e_g, a_{1g}, and b_{1g}. The ground state of $UO_2Cl_4^{2-}$ is ${}^1A_{1g}$.

A line appears at 20,400 cm^{-1} with a vibration of 108 cm^{-1} superimposed on the electronic transition. From vibrational analysis the vibration associated with the "forbidden" transition must have $\beta_{1u}(D_{4h})$ symmetry. Since one quantum of β_{1u} appears in the $X(z)$ and $Y(z)$ spectra we have

$$A_{1g} \times A_{2u} \times \beta_{1u} \times \Gamma_j = A_{1g}$$

or $\Gamma_j = {}^1B_{2g}$. A magnetic dipole transition ${}^1A_{1g} \rightarrow {}^1B_{2g}$ is not allowed. The source of intensity for the pure electronic transition is therefore electric quadrupole, via the xy component of the tensor.

The electric quadrupole mechanism of the transition was then tested by studying its polarization properties. These measurements corroborated the fact that the line at 20,400 cm^{-1} was indeed electric quadrupole–allowed and that the terminating state was of B_{2g} symmetry.[37]

6-3 JAHN-TELLER COUPLINGS

Observations of weakly structured, broad absorption bands are not uncommon. If the assignment of the band involves an orbital-degenerate terminating state, the phenomenon is often interpreted as a manifestation of a Jahn–Teller coupling. A proof of the claimed effect is, however, not easily obtained. Fortunately, direct estimates of vibrational-electronic coupling strengths derived from careful analysis of detailed line absorption spectra are also available. We shall give here a few examples of both types of spectra. For a broad covering of the field we shall refer to the monograph by Englman.[38]

The complex $[Ti(urea)_6]I_3$ exhibits a broad double-peaked absorption band[39] at some 18,000 cm^{-1}. The spectrum is polarized, and the band in π polarization is about twice as intense as the σ spectrum. In this complex the Ti^{3+} ion occupies a D_3 site. The parent octahedral ground state is ${}^2T_{2g}(t_2)^1$. In D_3 symmetry this is split into a ${}^2A_1$ and ${}^2E$ state. The ligand-field excited state is of ${}^2E(e)^1$

Table 6-2 Selection rules for electric dipole transition in the group D_3

D_3	A_1	A_2	E
A_1	0	π	σ
A_2	π	0	σ
E	σ	σ	σ, π

symmetry. The selection rules for electric dipole transitions are given in Table 6-2. From the polarization behavior of the band we may conclude that the 18,000 cm^{-1} transition can be assigned as $^2E \rightarrow {}^2E$.

A ^2E state will experience a splitting in D_3 symmetry due to the combined action of a Jahn–Teller effect and the spin-orbit coupling. In the Jahn–Teller effect the active vibration is of ε symmetry and we can take (compare section 1-6)

$$\mathscr{H}^{(1)} = \left(\frac{\partial V}{\partial \xi_{2b}}\right)_0 \xi_{2a} + \left(\frac{\partial V}{\partial \xi_{2b}}\right)_0 \xi_{2b} + \zeta_M \mathbf{l} \cdot \mathbf{s}.$$

This perturbation leads to a coupling between the double group states $\overline{E}_{1/2}$ and $\overline{E}_{3/2}$, the coupling operator being

$$\begin{pmatrix} -\frac{1}{2}\zeta_M & -c\rho_2 \exp(\mathrm{i}\phi) \\ -c\rho_2 \exp(-\mathrm{i}\phi) & \frac{1}{2}\zeta_M \end{pmatrix} \tag{6-22}$$

where $\xi_{2a} = \rho_2 \cos\phi$, $\xi_{2b} = \rho_2 \sin\phi$, and c is the linear Jahn–Teller coupling constant. The two potential surfaces are given by

$$V(\rho_2, \phi) = \tfrac{1}{2}k_2\rho_2^2 \pm \sqrt{c^2\rho_2^2 + \tfrac{1}{4}\zeta_M^2} \tag{6-23}$$

The excited ^2E state will show no first-order spin-orbit splitting since the two orbitals $(x^2 - y^2)$ and (z^2) carry no orbital momentum. The dominating effect is thus the Jahn–Teller coupling. The two potential surfaces are seen from Eq. (6-23) to be separated by $2c_{ex}\rho_2$. The minimum of the lowest potential surface of the ^2E ground state manifold is found at

$$\rho_2^0 = \sqrt{\frac{c_0^2}{k_2^2} - \frac{\zeta_M^2}{4c_0^2}} \tag{6-24}$$

From a semiclassical point of view the splitting of the absorption band will therefore be given by

$$\Delta W \approx 2c_{ex}\sqrt{\frac{c_0^2}{k_2^2} - \frac{\zeta_M^2}{4c_0^2}} = 4\sqrt{\Delta V_{J-T}^0 \Delta V_{J-T}^{ex} - \frac{\zeta_M^2 \Delta V_{J-T}^{ex}}{16\Delta V_{J-T}^0}} \tag{6-25}$$

where ΔV_{J-T}^0 and ΔV_{J-T}^{ex} are the so-called Jahn–Teller energies (Eq. (1-85)) of the ^2E ground and excited states, respectively. The observed band splitting is[39] some 2,000 cm^{-1}. With $\Delta V_{\mathrm{J-T}}^{\mathrm{ex}} \approx \Delta V_{\mathrm{J-T}}^{0}$ and $\zeta_M \approx 150\ \mathrm{cm}^{-1}$ we find $\Delta V_{\mathrm{J-T}} \simeq$ 500 cm^{-1}.

Next we shall once more consider the ion Ti^{3+} doped into Al_2O_3. In O_h symmetry the ground state of the complex would be $^2T_{2g}$. Suppose now that this electronic state interacts strongly with a vibrational mode of ε symmetry. The investigation of the static Jahn–Teller effect gives us the three potential surfaces V_1, V_2, and V_3 of Eqs. (4-177), (4-178), and (4-179) pictured in Fig. 4-9. The equations which govern the dynamic Jahn–Teller problem are, using Eq. (4-176), with T_N representing the kinetic energy of the nuclei,

$$\left[T_N + \frac{1}{2}k_\varepsilon(\xi_{2a}^2 + \xi_{2b}^2) - \frac{c_\varepsilon}{2}(\xi_{2a} + \sqrt{3}\xi_{2b})\right]\chi_1 = W\chi_1 \tag{6-26a}$$

$$\left[T_N + \frac{1}{2}k_\varepsilon(\xi_{2a}^2 + \xi_{2b}^2) - \frac{c_\varepsilon}{2}(\xi_{2a} - \sqrt{3}\xi_{2b})\right]\chi_2 = W\chi_2 \tag{6-26b}$$

$$\left[T_N + \frac{1}{2}k_\varepsilon(\xi_{2a}^2 + \xi_{2b}^2) + c_\varepsilon\xi_{2a}\right]\chi_3 = W\chi_3 \tag{6-26c}$$

The solutions to the three differential equations (6-26a), (6-26b), (6-26c) are easily found to be two-dimensional harmonic oscillators, centered in the ξ_{2a}, ξ_{2b} plane at

$$\left(\frac{c_\varepsilon}{2k_\varepsilon}, \frac{\sqrt{3}c_\varepsilon}{2k_\varepsilon}\right), \left(\frac{c_\varepsilon}{2k_\varepsilon}, -\frac{\sqrt{3}c_\varepsilon}{2k_\varepsilon}\right), \left(-\frac{c_\varepsilon}{k_\varepsilon}, 0\right)$$

and with eigenvalues

$$W_{i,n} = -\frac{c_\varepsilon^2}{2k_\varepsilon} + (n_i + 1)\hbar\omega \tag{6-27}$$

$$i = 1, 2, 3 \quad \text{and} \quad n_i = 0, 1, 2, 3, \ldots$$

where ω is the angular frequency of the vibration.

Including the spin functions, the crude adiabatic ground state with $n_i = 0$ is sixfold degenerate. It can be represented by the functions

$$\psi_{xz}^{\alpha}\chi_{v''=0}^{xz}, \psi_{xz}^{\beta}\chi_{v''=0}^{xz}, \psi_{yz}^{\alpha}\chi_{v''=0}^{yz}, \psi_{yz}^{\beta}\chi_{v''=0}^{yz}, \psi_{xy}^{\alpha}\chi_{v''=0}^{xy}, \psi_{xy}^{\beta}\chi_{v''=0}^{xy}$$

Now let a combined spin-orbit coupling and trigonal distortion act as a perturbation. Such a perturbation is entirely off-diagonal. The energies of the perturbed system are given in Eq. (5-34), only the parameters ζ_M and Δ will have to be multiplied by the vibrational overlap $S_{00} = \langle\chi_{v''=0}^{xz} \mid \chi_{v''=0}^{yz}\rangle = \langle\chi_{v''=0}^{xz} \mid \chi_{v''=0}^{xy}\rangle = \langle\chi_{v''=0}^{yz} \mid \chi_{v''=0}^{xy}\rangle$. The vibrational overlap S_{00} can easily be evaluated. From Eq. (4-184) we have $S_{00} = \exp(-\frac{3}{2}\Delta V_{\mathrm{J-T}}/\hbar\omega_\varepsilon)$.

In the previous section we found that apparent values of $\zeta_M = 90\ \mathrm{cm}^{-1}$ and $\Delta = -10\ \mathrm{cm}^{-1}$ will fit the low-lying absorption bands of the Ti^{3+} system. Assuming an "unquenched" value of $\zeta_M = 175\ \mathrm{cm}^{-1}$, S_{00} is found to be 0.514. Therefore $\Delta \approx -20\ \mathrm{cm}^{-1}$ and taking $\hbar\omega \approx 200\ \mathrm{cm}^{-1}$ we get $\Delta V_{\mathrm{J-T}} \approx 90\ \mathrm{cm}^{-1}$. This example shows clearly how the Ham effect operates. However, our point of departure, that the $^2T_{2g}$ state is strongly coupled to an ε vibration and experiences a weak trigonal field in addition to the spin-orbit coupling, is not a realistic one. Second-order vibronic effects and more refined perturbation techniques will have to be introduced in order to give a meaningful description.[32,40]

A direct proof that the above simple treatment breaks down has been given by McClure.[3] The actual value of Δ can be found by measuring the absorption spectrum at temperatures high enough to populate the $^2A_1$ component of $^2T_{2g}$. The polarization ratio would change as the absorption from $^2A_1$ became significant, but no such effect was observed up to 800°K. The value of $|\Delta|$ must therefore be larger than 500 cm^{-1}. The ground state of Ti^{3+} in Al_2O_3 is therefore ^2E separated from $^2A_1$ by at least 500 cm^{-1}.

Concentrating on the ^2E ground state, a simple description assumes this state to experience a Jahn–Teller effect in an ε vibration and a spin-orbit coupling. Application of the coupling operator of Eq. (6-22) leads to a set of equations for the vibronic states

$$\begin{pmatrix} T_N + \frac{1}{2}k\rho^2 \mp \frac{1}{2}\zeta_M - W & -c\rho\exp(\mathrm{i}\phi) \\ -c\rho\exp(-\mathrm{i}\phi) & T_N + \frac{1}{2}k\rho^2 \pm \frac{1}{2}\zeta_M - W \end{pmatrix} \begin{pmatrix} \chi^+(\rho,\phi) \\ \chi^-(\rho,\phi) \end{pmatrix} = 0 \qquad (6\text{-}28)$$

In order to solve these[41,42] χ^+ and χ^- are expanded in terms of the two-dimensional harmonic oscillator functions $\chi_{n,m}$

$$\chi^+ = \sum a_{n,m}\chi_{n,m} \qquad (6\text{-}29)$$

$$\chi^- = \sum b_{n,m}\chi_{n,m} \qquad (6\text{-}30)$$

Introducing $A = \Delta V_{\mathrm{J-T}} \times \hbar\omega$ and using Eqs. (1-99) to (1-102), the eigenvalue problem then takes the secular form

$$[(n+1)\hbar\omega - \tfrac{1}{2}\zeta_M - W]\,a_{n,m} - \sqrt{A(n+m+2)}\,b_{n+1,m+1} + \sqrt{A(n-m)}\,b_{n-1,m+1} = 0 \qquad (6\text{-}31)$$

$$-\sqrt{A(n+m+2)}\,a_{n,m} + \sqrt{A(n-m+2)}\,a_{n+2,m} + [(n+2)\hbar\omega + \tfrac{1}{2}\zeta_M - W]\,b_{n+1,m+1} = 0 \qquad (6\text{-}32)$$

and the corresponding set of coupled equations with the sign of ζ_M changed. Because of the double degeneracy we can look at the solutions for which m is non-negative. W will then be given as the solutions to the infinite determinantal equations

$$\begin{vmatrix} (m+1)\hbar\omega \mp \frac{1}{2}\zeta_M - W & -\sqrt{A(2m+2)} & 0 & \cdot \\ -\sqrt{A(2m+2)} & (m+2)\hbar\omega \pm \frac{1}{2}\zeta_M - W & \sqrt{A\times 2} & \cdot \\ 0 & \sqrt{A\times 2} & (m+2)\hbar\omega \pm \frac{1}{2}\zeta_M - W & \cdot \\ 0 & 0 & -\sqrt{A(2m+4)} & \cdot \\ \cdot & \cdot & \cdot & \cdot \end{vmatrix} = 0 \qquad (6\text{-}33)$$

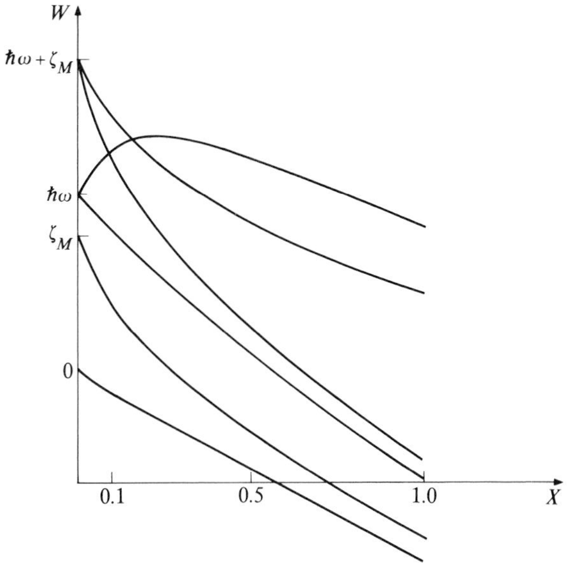

Figure 6-7 The variations in energies of the six lowest vibronic levels for a doubly degenerate vibration of a ^2E electronic state as a function of the Jahn–Teller coupling parameter

$$x = \frac{\Delta V_{\text{J-T}}}{\hbar\omega} \qquad \zeta_M = \tfrac{3}{4}\hbar\omega.$$

Defining $x = \Delta V_{\text{J-T}}/\hbar\omega$ and taking $\zeta_M = \frac{3}{4}\hbar\omega$, a machine solution to the six lowest vibronic states has been pictured in Fig. 6-7. For $x = 0$ the vibronic levels are given as two sets of $2(n + 1)$-degenerate groups of levels. The separation is ζ_M. As x increases, the degeneracy of each level is reduced to 2 and the apparent spin-orbit splitting decreases. This reduction is a manifestation of the Ham effect. The observed separation[30] of 37.8 cm^{-1} for the two lowest levels of Ti^{3+} in Al_2O_3 corresponds to $x = 0.6$ for $\hbar\omega = 200\ \text{cm}^{-1}$ on Fig. 6-7. Hence in this interpretation $\Delta V_{\text{J-T}} \approx 120\ \text{cm}^{-1}$.

As a final example of how a Jahn–Teller coupling can quench an expected spin-orbit splitting we return to the charge-transfer state $^2T_{2u}(t_{2u})^5(t_{2g})^6$ found in the octahedral complex $IrCl_6^{2-}$. A simple calculation (Eq. (6-4)) leads us to expect a spin-orbit splitting $W(G) - W(E_{5/2})$ of 440 cm^{-1}. However, the no-phonon lines of the two spin-orbit states are experimentally observed[23] to be separated by 5 cm^{-1} only. In the $^2T_{2u}$ state the Jahn–Teller and spin-orbit coupling effects are comparable. A perturbation approach considering only one of the couplings is therefore not feasible; they must be treated simultaneously. Both the τ_{2g} and ε_g vibrations are Jahn–Teller active in a $^2T_{2u}$ state and will be able to quench the spin-orbit coupling. A careful analysis of the observed line spectrum shows[23] that the experimental results can be accounted for using only a τ_{2g} Jahn–Teller active mode, and not on the basis of a coupling to the ε_g mode.

6-4 THE ZEEMAN EFFECT

Provided the lines in an absorption spectrum are sufficiently sharp, measurements of the optical Zeeman effect can be very useful in identifying the excited states. As a first example on its use we shall look at the complex $CoCl_4^{2-}$.

The electronic spectrum of the tetrahedral chromophore $CoCl_4^{2-}$ has been very carefully studied by Ferguson.[43] Some very narrow line absorptions were observed in a region near 17,400 cm^{-1}. From the correlation diagram he concluded that the lines must be due to spin-forbidden transitions $^4A_2 \rightarrow {}^2T_1, {}^2T_2$, but a closer identification was not possible.

The $CoCl_4^{2-}$ unit is also present in the Cs_3CoCl_5 crystal. The actual symmetry of the chromophore is D_{2d}. Spin-orbit coupling and the low symmetry make all states split into Kramers doublets of either $E_{1/2}$ or $E_{3/2}$ symmetry. Measuring at 1.6°K a very sharp line is now observed[44] at 17,308 cm^{-1}.

The $^4A_2$ ground state of tetrahedral $CoCl_4^{2-}$ is split into $E_{1/2}$ and $E_{3/2}$. The $E_{3/2}$ state is the absolute ground state, being separated from $E_{1/2}$ by[45] 8.6 cm^{-1}. The question is then whether the 17,308-cm^{-1} line corresponds to an $E_{3/2} \rightarrow E_{3/2}$ transition or to $E_{3/2} \rightarrow E_{1/2}$.

First of all we notice that formally the transition is spin-forbidden. However, the electric dipole–allowed transition $^4A_2 \rightarrow {}^4T_1$ is located[43] at some 13,000 cm^{-1}, and we take it that the spin-orbit coupling has scrambled the spin states, making the line allowed as an electric-dipole transition. Applying a magnetic field parallel to the axis of quantization, each Kramers doublet will split into two levels. The selection rules for transitions between the Zeeman levels of the ground and excited states can be found by using the coupling coefficients for D_{2d} symmetry (Table 6-3).

In D_{2d} symmetry z transforms like B_2 and (x, y) like E. Reference to Table 6-3 therefore gives us the polarizations as pictured in Fig. 6-8. Evidently the 17,308 cm^{-1} is split into four components if the excited state is $|E_{1/2}\rangle$ but only into two if it is a $|E_{3/2}\rangle$ state. A comparison of the experimental results with the theoretical predictions[46] identifies the excited state as $|E_{1/2}\rangle$.

A second example: the absorption spectrum of V^{3+} in Al_2O_3 was investigated by McClure.[3] An octahedral crystal field with a sizable trigonal field component was found to give a good overall description of the excited states with the exception of the sharp line structure seen at 15,876–15,890 cm^{-1}. The octahedral ground state $^3T_{1g}$ is strongly split in the actual C_3 trigonal field leaving a $^3A_2$ state

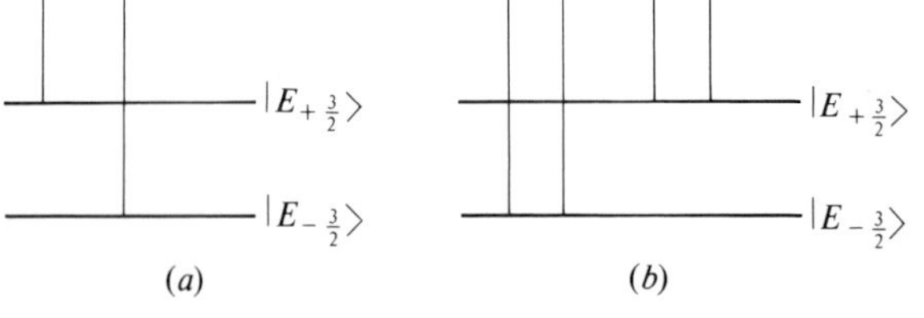

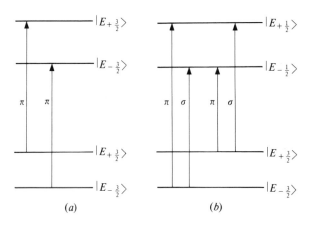

Figure 6-8 Zeeman splittings and transition patterns for (a) $|E_{3/2}\rangle \rightarrow |E_{3/2}\rangle$, and (b) $|E_{3/2}\rangle \rightarrow |E_{1/2}\rangle$. The magnetic field is applied along the axis of quantization.

Table 6-3 Some coupling coefficients for the double D_{2d}

(The phases are those of Koster, Dimmock, Wheeler, and Statz.[85])

$E_{1/2} \times E_{3/2}$		B_1	B_2	E	
		b_1	b_2	x	y
$\frac{1}{2}$	$\frac{3}{2}$	·	·	$\frac{1}{\sqrt{2}}$	$\frac{i}{\sqrt{2}}$
$\frac{1}{2}$	$-\frac{3}{2}$	$\frac{1}{\sqrt{2}}$	$-\frac{i}{\sqrt{2}}$	·	·
$-\frac{1}{2}$	$\frac{3}{2}$	$-\frac{1}{\sqrt{2}}$	$-\frac{i}{\sqrt{2}}$	·	·
$-\frac{1}{2}$	$-\frac{3}{2}$	·	·	$\frac{1}{\sqrt{2}}$	$-\frac{i}{\sqrt{2}}$

$E_{3/2} \times E_{3/2}$		A_1	A_2	E	
		a_1	a_2	x	y
$\frac{3}{2}$	$\frac{3}{2}$	·	·	$\frac{i}{\sqrt{2}}$	$\frac{1}{\sqrt{2}}$
$\frac{3}{2}$	$-\frac{3}{2}$	$\frac{1}{\sqrt{2}}$	$-\frac{i}{\sqrt{2}}$	·	·
$-\frac{3}{2}$	$\frac{3}{2}$	$-\frac{1}{\sqrt{2}}$	$-\frac{i}{\sqrt{2}}$	·	·
$-\frac{3}{2}$	$-\frac{3}{2}$	·	·	$-\frac{i}{\sqrt{2}}$	$\frac{1}{\sqrt{2}}$

lowest (Fig. 5-2). The sharp line structure was assigned by Scott and Sturge[47] to the zero-phonon transition $^3A_2 \rightarrow {}^3T_2$; the expected trigonal splitting and the spin-orbit coupling splitting of the octahedral $^3T_{2g}$ state was assumed to be nearly completely quenched by the Ham effect.

That this identification is correct was proved by Stephens and Lowe-Pariseau[48] by performing a Zeeman experiment on a very diluted crystal. Their measurements unambiguously identified eight excited states. The assumption of C_3 selection rules extended this number to nine, thereby accounting for all the components of the $^3T_{2g}$ state. The calculated Zeeman pattern was in good agreement with the experimental one, assuming the $\beta\mathbf{L}\cdot\mathbf{H}$ term in the Zeeman hamiltonian to be quenched.

Next, we shall again consider the ligand-to-metal charge-transfer transitions $E_{5/2}(^2T_{2g}) \rightarrow (E_{5/2} + G)(^2T_{2u})$ met with in $IrCl_6^-$. From hot-band experiments the line at 22,935.6 cm^{-1} can be shown to be associated with the no-phonon line $E_{5/2} \rightarrow E_{5/2}$. Assuming complete quenching of the orbital angular momentum, the Zeeman splittings of $E_{5/2}$ and G were then calculated.[49] Using a ground-state value of $g = 1.755$ the predicted splittings and polarizations of the expected six absorption lines were found to behave like a fully allowed quenched pair of no-phonon lines arising from the charge-transfer transition $E_{5/2}(^2T_{2g}) \rightarrow (E_{5/2} + G)$ $(^2T_{2u})$.

A final example on the use of the Zeeman effect: in the optical absorption spectrum of Cr_2O_3, five sharp lines were observed[50] at 13,743 cm^{-1} to 13,970 cm^{-1}. By analogy with the spectrum of Cr^{3+} doped into Al_2O_3 these lines are assigned to the $^4A_2 \rightarrow {^2E}$ transition (see Fig. 6-1). On the basis of their polarization behavior, four of the lines could be distinguished from line 5. Lines 1, 2, 3, and 4 are electric-dipole in nature, lines 1 and 4 being σ and 2 and 3 being π polarized. Only the σ lines are observed in the axial spectrum. They are separated by 183 cm^{-1}. Line 5, on the other hand, was observed in σ but not in axial polarization. This indicates a mixed electric and magnetic dipole character.

Cr_2O_3 is an antiferromagnetic crystal, containing four chromium ions per trigonal unit cell. The spins of the four chromophores lie along the crystal **c** axis and alternate up and down. The site symmetry of each chromium ion is C_3. This low symmetry in connection with the spin-orbit coupling and the local magnetic field will remove all electronic degeneracy. The parent cubic ^2E and $^4A_2$ states will therefore each split up into four levels. A coupled chromophore model will therefore give us sixteen excited states. Writing the wave function for a nondegenerate excited state Ψ^*, where the excitation runs over the four components, the coupled state functions will therefore be

$$|A\rangle = \tfrac{1}{2}(\Psi_1 + \Psi_2 + \Psi_3 + \Psi_4)$$

$$|B\rangle = \tfrac{1}{2}(\Psi_1 + \Psi_2 - \Psi_3 - \Psi_4)$$

$$|C\rangle = \tfrac{1}{2}(\Psi_1 - \Psi_2 - \Psi_3 + \Psi_4)$$

$$|D\rangle = \tfrac{1}{2}(\Psi_1 - \Psi_2 + \Psi_3 - \Psi_4)$$

with $\Psi_1 = |\Psi_{\rm I}^*\Psi_{\rm II}^0\Psi_{\rm III}^0\Psi_{\rm IV}^0|$, $\Psi_2 = |\Psi_{\rm I}^0\Psi_{\rm II}^*\psi_{\rm III}^0\Psi_{\rm IV}^0|$, $\Psi_3 = |\Psi_{\rm I}^0\Psi_{\rm II}^0\Psi_{\rm III}^*\Psi_{\rm IV}^0|$ and $\Psi_4 = |\Psi_{\rm I}^0\Psi_{\rm II}^0\Psi_{\rm III}^0\Psi_{\rm IV}^*|$. The energy separations between the coupled states are referred to as Davydov splittings.[51]

The lines 1 to 4 are now identified as allowed electric-dipole transitions to the Davydov-split levels arising from the lower two single-chromophore excited states. These eight states span representations in the factor group[51] D_3 of the crystal of (E, E) and $(2A_1, 2A_2)$. The absolute ground state transforms like A_1. Theory therefore predicts two σ-polarized and two π-polarized transitions. The separation of the two σ-polarized transitions of 183 cm^{-1} is seen to be a direct measure of the Davydov splitting.

The effect of a magnetic field applied parallel to the **c** axis is that each E level split with the same splitting factor. The calculated g factor is in nice accord with the experimental result, leaving no doubt as to the correctness of the assignments.[50]

6-5 MAGNETIC CIRCULAR DICHROISM

Most often the direct observation of a Zeeman effect in an excited state is impossible, owing to the width of the absorption band. However, an optical transition to the split electronic state will have a different absorption coefficient, according to whether one uses left- or right-circularly polarized light (section 4-6). Measurements of the magnetic circular dichroism, abbreviated MCD, can therefore be used to elucidate the g values of the excited state, thereby aiding in the assignment of the symmetry and electronic composition of the excited state.

As a first example of the utilization of MCD measurements for identification of excited states, we shall look at MnO_4^-. The absorption band found at 18,000–22,000 cm^{-1} was assigned as an allowed $^1A_1 \rightarrow {}^1T_2$ transition by Wolfsberg and Helmholz.[52] The electronic excitation was taken to be $t_1(\pi) \rightarrow t_2^*(d, \sigma, \pi)$. Later Ballhausen and Liehr[53] proposed that the excitation was mostly $t_1(\pi) \rightarrow e^*(d, \pi)$. The t_1 orbital in question is a nonbonding π-type orbital located entirely on the oxygens. The t_2^* orbital is antibonding with admixtures of both metal and π and σ ligand orbitals. The e^* orbital, on the other hand, is antibonding with admixtures of metal and π ligand orbitals.

In T_d symmetry an e orbital carries no angular momentum. It was recognized by Schatz et al.[54] that provided the transition in question was primarily $t_1 \rightarrow e^*$, the $\mathfrak{A}$ term, Eq. (4-97), in the expression for the MCD would depend only on the one-electron matrix elements of $\hat{l}_z$ inside the t_1 manifold of orbitals. The t_1 orbitals are pure, symmetry-determined linear combinations of oxygen π orbitals. The expression for the $\mathfrak{A}$ term would therefore contain no adjustable parameters. If, on the other hand, the transition corresponded mostly to $t_1 \rightarrow t_2^*$ the $\mathfrak{A}$ term would be dependent upon the composition of the molecular t_2^* orbital.

The electronic ground state of MnO_4^- is given by

$$|{}^1A_1\rangle = |\overset{+}{t}_{1x}\overset{-}{t}_{1x}\overset{+}{t}_{1y}\overset{-}{t}_{1y}\overset{+}{t}_{1z}\overset{-}{t}_{1z}| \tag{6-34}$$

where (see Fig. 2-2 and Table 2-2)

$$\begin{aligned} t_{1x} &= \frac{\sqrt{3}}{2} \times \frac{1}{2}(p_{x_1} + p_{x_2} - p_{x_3} - p_{x_4}) + \frac{1}{2} \times \frac{1}{2}(p_{y_1} + p_{y_2} - p_{y_3} - p_{y_4}) \\ t_{1y} &= -\frac{\sqrt{3}}{2} \times \frac{1}{2}(p_{x_1} - p_{x_2} + p_{x_3} - p_{x_4}) + \frac{1}{2} \times \frac{1}{2}(p_{y_1} - p_{y_2} + p_{y_3} - p_{y_4}) \\ t_{1z} &= -\frac{1}{2}(p_{y_1} + p_{y_2} + p_{y_3} + p_{y_4}) \end{aligned} \tag{6-35}$$

The excited ${}^1T_2$ components corresponding to the $(t_1)^5(e)^1$ configuration are

$$
\begin{aligned}
|T_{2x}\rangle &= \frac{1}{\sqrt{2}}\left[\left|\overset{+}{t}_{1y}\overset{-}{t}_{1y}\overset{+}{t}_{1z}\overset{-}{t}_{1z}\overset{+}{t}_{1x}\left(-\frac{1}{2}(x^2 \overset{-}{-} y^2) - \frac{\sqrt{3}}{2}(\overset{-}{z^2})\right)\right|\right. \\
&\quad \left. - \left|\overset{+}{t}_{1y}\overset{-}{t}_{1y}\overset{+}{t}_{1z}\overset{-}{t}_{1z}\overset{-}{t}_{1x}\left(-\frac{1}{2}(x^2 \overset{+}{-} y^2) - \frac{\sqrt{3}}{2}(\overset{+}{z^2})\right)\right|\right] \\
|T_{2y}\rangle &= \left[\left|\overset{+}{t}_{1z}\overset{-}{t}_{1z}\overset{+}{t}_{1x}\overset{-}{t}_{1x}\overset{+}{t}_{1y}\left(-\frac{1}{2}(x^2 \overset{-}{-} y^2) + \frac{\sqrt{3}}{2}(\overset{-}{z^2})\right)\right|\right. \\
&\quad \left. - \left|\overset{+}{t}_{1z}\overset{-}{t}_{1z}\overset{+}{t}_{1x}\overset{-}{t}_{1x}\overset{-}{t}_{1y}\left(-\frac{1}{2}(x^2 \overset{+}{-} y^2) + \frac{\sqrt{3}}{2}(\overset{+}{z^2})\right)\right|\right] \qquad (6\text{-}36) \\
|T_{2z}\rangle &= \frac{1}{\sqrt{2}}\left[\left|\overset{+}{t}_{1x}\overset{-}{t}_{1x}\overset{+}{t}_{1y}\overset{-}{t}_{1y}\overset{+}{t}_{1z}(x^2 \overset{-}{-} y^2)\right|\right. \\
&\quad \left. - \left|\overset{+}{t}_{1x}\overset{-}{t}_{1x}\overset{+}{t}_{1y}\overset{-}{t}_{1y}\overset{-}{t}_{1z}(x^2 \overset{+}{-} y^2)\right|\right]
\end{aligned}
$$

These components are easily found by a descent in symmetry to D_{2d} whereby the T_{2z} component is picked out, followed by a generation of T_{2x} and T_{2y} by the $\hat{C}_3$ symmetry operator.

The linear combinations

$$
\begin{aligned}
-\frac{1}{\sqrt{2}}[T_{2x} + \mathrm{i}T_{2y}] &\qquad \langle\hat{\mu}_z\rangle = \frac{\beta}{4} \\
T_{2z} &\qquad \langle\hat{\mu}_z\rangle = 0 \\
\frac{1}{\sqrt{2}}[T_{2x} - \mathrm{i}T_{2y}] &\qquad \langle\hat{\mu}_z\rangle = -\frac{\beta}{4}
\end{aligned}
$$

are seen to be diagonal in $\hat{\mu}_z = -|\beta|\hat{L}_z$ with the indicated expectation values. The calculation of the $\hat{L}_z$ values requires the evaluation of $\langle t_{1y}|\hat{l}_z|t_{1x}\rangle$. With $\hat{l}_z$ being referred to the metal coordinate system this is most easily performed directing $\hat{l}_z$ towards the ligands:

$$\hat{l}_z = \alpha_j\hat{l}'_z + \beta_j\hat{l}'_x + \gamma_j\hat{l}'_y \qquad (6\text{-}37)$$

where $(\alpha_j, \beta_j, \gamma_j)$ are the directional cosines to ligand j. For ligands 1 and 4, $\alpha = \sqrt{\frac{1}{3}}$; for ligand 2 and 3, $\alpha = -\sqrt{\frac{1}{3}}$. One gets $\langle t_{1y}|\hat{l}_z|t_{1x}\rangle = \frac{1}{2}\mathrm{i}$.

Evaluation of the $\mathfrak{A}$ term then gives using Eq. (4-97)

$$
\begin{aligned}
\mathfrak{A} = 3\cdot\frac{\beta}{4}\Big[&-\langle T_{2x}|\sum_n x_n|A_1\rangle\langle T_{2y}|\sum_n y_n|A_1\rangle + \\
&+ \langle T_{2y}|\sum_n x_n|A_1\rangle\langle T_{2x}|\sum_n y_n|A_1\rangle\Big] \qquad (6\text{-}38)
\end{aligned}
$$

Making use of the $\hat{S}_4$ symmetry operator found in T_d symmetry reduces this expression to

$$\mathfrak{A} = -\left(\frac{\beta}{4}\right)3\left[\left|\langle T_{2x}|\sum_n x_n|A_1\rangle\right|^2 + \left|\langle T_{2y}|\sum_n x_n|A_1\rangle\right|^2\right] \tag{6-39}$$

A calculation of the $\mathfrak{D}$ term, Eq. (4-100), gives

$$\mathfrak{D} = 3\left[|\langle T_{2x}|\sum_n x_n|A_1\rangle|^2 + |\langle T_{2y}|\sum_n x_n|A_1\rangle|^2\right] \tag{6-40}$$

Hence provided the transition ${}^1A_1 \rightarrow {}^1T_2$ is made up by a $t_1 \rightarrow e^*$ excitation we get

$$\mathfrak{A}/\mathfrak{D} = -0.25\beta \tag{6-41}$$

The value of $\mathfrak{A}/\mathfrak{D}$, obtained by moment analysis over the entire absorption band, was found[55] to be -0.22β, in good agreement with the calculated $t_1 \rightarrow e^*$ value. The identical calculation assuming a $t_1 \rightarrow t_2^*$ excitation leads to a much smaller value for $\mathfrak{A}/\mathfrak{D}$ which also has the wrong sign.[54]

In the diluted system MnO_4^- dissolved in KIO_4, integration over the band system led to a value[56] of $\mathfrak{A}/\mathfrak{D} = -0.259\beta$. Yet each member of the progression in the totally symmetric breathing frequency shows individual $\mathfrak{A}/\mathfrak{D}$ values very much smaller than -0.25β. Furthermore, a $\mathfrak{B}$ term appears, and the sign of $\mathfrak{B}/\mathfrak{D}$ changes along the progression.

The terminating ${}^1T_2$ state is degenerate in T_d and should experience a Jahn–Teller effect. The presence of the $\mathfrak{B}$ term is therefore due to the mixing under the Jahn–Teller coupling of the vibronic states. The apparent quenching of the angular momentum in each of the excited vibronic components was explained as a manifestation of the Ham effect. The magnitude of $\langle \hat{L}_z \rangle$ for each line is reduced due to the appearance of the vibrational overlap. However, the total $\mathfrak{A}/\mathfrak{D}$ value summed over the whole band system is preserved.

In $TiCl_4$ the first absorption band is in vapor phase found[57] to have a maximum at 35,600 cm^{-1}. It has a molar extinction coefficient of $\varepsilon_{max} = 6{,}900$. On the basis of an MCD experiment the transition is in analogy with the MnO_4^- band assigned[58] as ${}^1A_1 \rightarrow {}^1T_2(t_1 \rightarrow e^*)$. True as this assignment undoubtedly is, it shows the unreliability of using a parameterless CNDO-MO calculation[59] for absorption band identifications.

The square-planar $Ni(CN)_4^{2-}$ complex shows strong absorption bands in the near-ultraviolet region in addition to the ligand-field bands (Table 6-1). The charge-transfer band found at 36,700 cm^{-1} was assigned[60] as ${}^1A_{1g} \rightarrow {}^1E_u[e_g(d) \rightarrow a_{2u}(\pi^*)]$. The clear $\mathfrak{A}$ term seen in the MCD measurements definitely shows the terminating state to be degenerate.[61] The sign and magnitude of the $\mathfrak{A}$ term depend on the magnetic moment of the ${}^1E$ state. For an ${}^1A_{1g}(e_g^4) \rightarrow {}^1E_u(e_g^3 a_{2u})$ transition the sole contributor is the e_g level, the a_{2u} orbital carrying no orbital momentum. We now estimate the $\mathfrak{A}/\mathfrak{D}$ ratio.

The wave functions for $Ni(CN)_4^{2-}$ are for the ground and excited states of interest

$$|^1A_{1g}\rangle = |(\overset{+}{xz})(\overset{-}{xz})(\overset{+}{yz})(\overset{-}{yz})|$$

$$|^1E_u^x\rangle = \frac{1}{\sqrt{2}}\{|(\overset{+}{a_{2u}})(\overset{-}{xz})(\overset{+}{yz})(\overset{-}{yz})| - |(\overset{-}{a_{2u}})(\overset{+}{xz})(\overset{+}{yz})(\overset{-}{yz})|\}$$

$$|^1E_u^y\rangle = \frac{1}{\sqrt{2}}\{|(\overset{+}{xz})(\overset{-}{xz})(\overset{+}{a_{2u}})(\overset{-}{yz})| - |(\overset{+}{xz})(\overset{-}{xz})(\overset{-}{a_{2u}})(\overset{+}{yz})|\}$$

Diagonalizing ^1E under $\hat{\mu}_z = -|\beta|\sum_n \hat{l}_{zn}$ we find

$$|^1E_u^1\rangle = \frac{-1}{\sqrt{2}}(|^1E_u^x\rangle + i|^1E_u^y\rangle) \qquad \langle\mu_z\rangle = -\beta k_{\pi,\pi}$$

$$|^1E_u^{-1}\rangle = \frac{1}{\sqrt{2}}(|^1E_u^x\rangle - i|^1E_u^y\rangle) \qquad \langle\mu_z\rangle = \beta k_{\pi,\pi}$$

where $k_{\pi,\pi}$ is an orbital reduction factor.

For a $Ni(CN)_4^{2-}$ ion with the fourfold (Z) axis parallel to the propagation vector of the light Eq. (4-97) gives us

$$\mathfrak{A} = \sum_j \text{Im}\langle j|\sum_n x_n|^1A_{1g}\rangle^*\langle j|\sum_n y_n|^1A_{1g}\rangle\langle j|\hat{\mu}_z|j\rangle$$

We find easily, since $\langle{}^1E_u^x|\sum_n y_n|^1A_{1g}\rangle = \langle{}^1E_u^y|\sum_n x_n|^1A_{1g}\rangle = 0$, that

$$\mathfrak{A} = k_{\pi,\pi}\beta\langle{}^1E_u^x|\sum_n x_n|^1A_{1g}\rangle\langle{}^1E_u^y|\sum_n y_n|^1A_{1g}\rangle \tag{6-42}$$

Using the $\hat{C}_4$ symmetry operator

$$\mathfrak{A} = k_{\pi,\pi}\beta|\langle{}^1E_u^x|\sum_x x_n|^1A_{1g}\rangle|^2 \tag{6-43}$$

Relaxing the condition that the $Ni(CN)_4^{2-}$ ion be space-fixed, and performing an average over all orientations leads to

$$\mathfrak{A} = \frac{1}{3}k_{\pi,\pi}\beta\left|\langle{}^1E_u^x|\sum_n \mathbf{r}_n|^1A_{1g}\rangle\right|^2 \tag{6-44}$$

The averaged $\mathfrak{D}$ term is

$$\mathfrak{D} = \frac{2}{3}\left|\langle{}^1E_u^x|\sum_n \mathbf{r}_n|^1A_{1g}\rangle\right|^2 \tag{6-45}$$

yielding for the $^1A_{1g} \rightarrow {}^1E_u$ transition $\mathfrak{A}/\mathfrak{D} = \beta/2\,k_{\pi,\pi}$. With $k_{\pi,\pi} \approx 1$ the ratio is 0.5β. From gaussian band analysis the experimental ratio[61] is 0.8β in reasonable agreement with theory.

The above simple calculation neglects the scrambling of the ^1E_u and ^3E_u states due to the spin-orbit coupling.[62] However, the 36,700 cm^{-1} state seems to be predominantly ${}^1E_u$. Spin-orbit coupling appears to be more important in mixing $({}^1A_{2u}, {}^3E_u)$ as evidenced from the polarized absorption spectra.[22] The heavier the metal ion, the more important it is to incorporate the mixings due to the spin-orbit coupling if a proper assignment is to be reached. The MCD spectra of, for example, $ReCl_6^{2-}$ and $ReBr_6^{2-}$ bear witness to this.[63]

As a last example, consider the charge-transfer bands found in $Fe(CN)_6^{3-}$. The electronic ground state is ${}^2T_{2g}(t_{1u})^6(t_{2u})^6(t_{2g})^5$. Excited states are ${}^2T_{1u}(t_{1u} \rightarrow t_{2g})$ and ${}^2T_{2u}(t_{2u} \rightarrow t_{2g})$. The ground state is degenerate, and $\mathfrak{C}$ terms will appear in the MCD expression. We have for the metal t_{2g} orbitals

$$t_{2g}^1 = -\frac{1}{\sqrt{2}}(yx + ixz) \qquad \langle \hat{\mu}_z \rangle = \beta k_{\pi,\pi}$$

$$t_{2g}^0 = xy \qquad \langle \hat{\mu}_z \rangle = 0$$

$$t_{2g}^{-1} = \frac{1}{\sqrt{2}}(yz - ixz) \qquad \langle \hat{\mu}_z \rangle = -\beta k_{\pi,\pi}$$

For the t_{1u} and the t_{2u} ligand orbitals we use the linear combinations of Table 2-1. We have

$$t_{1u}^1 = -\frac{1}{\sqrt{2}}(\pi_x + i\pi_y)$$

$$t_{1u}^0 = \pi_z$$

$$t_{1u}^{-1} = \frac{1}{\sqrt{2}}(\pi_x - i\pi_y)$$

and

$$t_{2u}^1 = -\frac{1}{\sqrt{2}}(\pi_\xi + i\pi_\eta)$$

$$t_{2u}^0 = \pi_\zeta$$

$$t_{2u}^{-1} = \frac{1}{\sqrt{2}}(\pi_\xi - i\pi_\eta)$$

Writing down the determinantal wave functions and insertion in the formula for the $\mathfrak{C}$ term, Eq. (4-98), easily gives us, $j = 1$ or 2,

$$\mathfrak{C} = -\beta k_{\pi,\pi} \langle xz | x | t_{ju}^0 \rangle \langle yz | y | t_{ju}^0 \rangle \tag{6-46}$$

The orbitals t_{1u} and t_{2u} differ under the application of the symmetry operator $\hat{C}_4$. With $\hat{C}_4 t_{1u}^0 = t_{1u}^0$ and $\hat{C}_4 t_{2u}^0 = -t_{2u}^0$ we get easily

$$\mathfrak{C} = \mp \beta k_{\pi,\pi} |\langle xz | x | t_{ju}^0 \rangle|^2 \tag{6-47}$$

where the minus sign is to be used when $j = 1$, the plus sign for $j = 2$.

The calculation of $\mathfrak{D}$ encounters no difficulties. Using Eq. (4-100) we find

$$\mathfrak{D} = 2|\langle xz | x | t_{ju}^0 \rangle|^2 \tag{6-48}$$

and therefore $\mathfrak{C}/\mathfrak{D} = \mp\frac{1}{2}\beta k_{\pi,\pi} \approx \mp\frac{1}{2}$ in units of β, where the minus sign applies when the terminating state is of T_{1u} symmetry, the plus sign for a terminating T_{2u} state.[54]

In $Fe(CN)_6^{3-}$ the band with a maximum at 23,900 cm^{-1} has a $\mathfrak{C}/\mathfrak{D}$ ratio of -0.61. It is therefore assigned to the $^2T_{2g} \rightarrow {}^2T_{1u}$ transition. The charge-transfer band found at 33,400 cm^{-1}, on the other hand, has a $\mathfrak{C}/\mathfrak{D}$ ratio of 0.34. It is consequently assigned to the $^2T_{2g} \rightarrow {}^2T_{2u}$ transition.

6-6 NATURAL OPTICAL ACTIVITY

A molecule or ion which is not superimposable on its mirror image is called a *chiral system.* It is well known that such entities will show natural optical activity. For complexes we may distinguish between three types of optically active transitions: the ligand-field bands, the charge-transfer bands, and the transitions which are located entirely on the ligands.

As a measure of the optical activity we use the "rotatory strength", Eq. (4-85),

$$R_{ij} = \sum_{i,j} \frac{e^2}{2mc} \operatorname{Im} \langle j | \sum_n \mathbf{r}_n | i \rangle \cdot \langle j | \sum_n \mathbf{l}_n | i \rangle \tag{6-49}$$

for a transition $i \rightarrow j$.

Consider now a tri-bidentate complex, $M(\text{en})_3^{3+}$ (where en = ethylenediamine), having approximately D_3 symmetry.[64] Although such a complex has the molecular symmetry of D_3, the effective symmetry is really much higher and approximates the full regular octahedral symmetry O_h. The hamiltonian can therefore be written as consisting of two terms $\mathscr{H} = \mathscr{H}_0 + V_D$, where $\mathscr{H}_0$ spans the identical representation of O_h and V_D is invariant only under these symmetry operations which survive on going to D_3. We will consider V_D as a small perturbation superposed on the octahedral problem

$$\mathscr{H}_0 \Psi_k^0 = W_k^0 \Psi_k^0$$

The unperturbed functions are therefore given by Ψ_k^0, and the first-order corrections arising from the odd parts of V_D enable us to write the perturbed ligand-field states as

$$\Psi_k = \Psi_{kg}^0 + \lambda \Psi_{ku}^0$$

The low-lying ligand-field transitions are all due to transitions between "even" zero-order states. The electric dipole operator transforms under t_{1u} in O_h; the matrix element $\langle i | \sum_n \mathbf{r}_n | j \rangle$ is therefore proportional to λ. The magnetic-dipole operator further transforms like t_{1g}. It follows that if a given ligand-field transition is magnetically allowed in O_h, its rotational strength R_{ij} appears in first order in λ, whereas if it is forbidden as a magnetic-dipole transition in O_h, R_{ij} will be of order of magnitude λ^2. We may therefore construct a set of approximate selection rules that tell us the orders of magnitudes to be expected for the rotational strengths. As an example, the slightly perturbed band in $Cr(\text{en})_3^{3+}$ derived from the $^4A_{2g} \rightarrow {}^4T_{2g}$ transition found in octahedral Cr(III) complexes shows much stronger optical activity than, for instance, the band derived from the ${}^4A_{2g} \rightarrow {}^4T_{1g}$ transition. Measurements of the optical activity of chiral chromium(III) complexes can therefore be used for band identifications.

Quantitative estimations of the rotatory strength for ligand-field transitions are, however, still a moot point. Moffitt[64] tried to use a crystal-field model, in which the d → d transitions became allowed by admixture of 4p character under the odd trigonal field. However, Sugano[65] subsequently showed that Moffitt's mechanism must be incorrect.

Let us consider Moffitt's problem a little closer. In order to look at the one-electron rotatory strengths we consider the augmented wave functions

$$|t_{2g}\rangle = |t_{2g}^0\rangle - \sum_{\gamma_u} |\gamma_u^0\rangle \frac{\langle \gamma_u^0 | V_D | t_{2g}^0 \rangle}{\Delta W_{t\gamma}} \tag{6-50}$$

$$|e_g\rangle = |e_g^0\rangle - \sum_{\gamma_u} |\gamma_u^0\rangle \frac{\langle \gamma_u^0 | V_D | e_g^0 \rangle}{\Delta W_{e\gamma}} \tag{6-51}$$

We are interested in the transformation properties of V_D. Inserting Eqs. (6-50) and (6-51) in Eq. (6-49) and performing the closure over $|\gamma_u^0\rangle$ lead to

$$R_{t_{2g}, e_g} \propto \sum_{e_g, t_{2g}} \langle e_g^0 | V_D \mathbf{r} | t_{2g}^0 \rangle \cdot \langle e_g^0 | \mathbf{l} | t_{2g}^0 \rangle \tag{6-52}$$

Now $e_g \times t_{2g} = t_{1g} + t_{2g}$. The actual decomposition is given by the coupling coefficients for the octahedral group. The orthogonality properties of the coupling coefficients then require the two operators $V_D\mathbf{r}$ and $\mathbf{l}$ to transform the same way if $\sum \langle e_g^0 | V_D\mathbf{r} | t_{2g}^0 \rangle \cdot \langle e_g^0 | \mathbf{l} | t_{2g}^0 \rangle$ is to be different from zero. With $\mathbf{l}$ transforming like t_{1g} and $\mathbf{r}$ as t_{1u}, V_D is seen to transform like a_{1u}. Sugano[65] pointed out that for the group O_h the a_{1u} term does not show up before the ninth spherical harmonic. Assuming $|t_{2g}\rangle$ and $|e_g\rangle$ to be pure d orbitals, such a potential cannot mix d and p orbitals. Moffitt's erroneous conclusions were found to rest on an error in sign in the evaluation of $\langle t_{2g}^0 | \mathbf{l} | e_g^0 \rangle$.

If we now consider the $|t_{2g}\rangle$ and $|e_g\rangle$ functions to be molecular orbitals, we can see from Eqs. (6-50) and (6-51) that $V_D(a_{1u})$ may couple $|t_{2g}\rangle$ to $|t_{2u}\rangle$ and $|e_g\rangle$ to $|e_u\rangle$. No e_u orbitals are readily available in O_h symmetry, but filled πt_{2u} ligand orbitals are at hand.[66]

The molecular t_{2g} orbital can be written

$$|t_{2g}\rangle = c_1 |dt_{2g}\rangle + c_2 |\pi t_{2g}\rangle + c_3 |\pi t_{2u}\rangle$$

It was argued by Hamer[67] that the coupling of $|\pi t_{2u}\rangle$ to $|dt_{2g}\rangle$ under the V_D perturbation term must be extremely small if not completely absent. With $\hat{F}$ being the Hartree–Fock operator for the perturbed system, the secular equation is therefore

$$\begin{vmatrix} F_{dd} - W & F_{d\pi} - G_\pi W & 0 \\ F_{d\pi} - G_\pi W & F_{\pi\pi} - W & \beta \\ 0 & \beta & F_{\pi\pi} - W \end{vmatrix} = 0$$

where $\beta = \langle \pi t_{2g} | \hat{F} | \pi t_{2u} \rangle$. With $\Delta = F_{dd} - F_{\pi\pi}$ being sufficiently large and positive we get to good approximation,[68]

$$|t_{2g}\rangle = |dt_{2g}\rangle - G_\pi |\pi t_{2g}\rangle - \frac{\beta G_\pi}{\Delta} |\pi t_{2u}\rangle$$

Using the real, trigonal orbitals of Table 1-3 we find for the rotational strengths r_v of the one-electron transitions:

r_v	e_x	e_y
t_{2x}	$\frac{1}{2}A$	$\frac{3}{2}A$
t_{2y}	$\frac{3}{2}A$	$\frac{1}{2}A$
t_{2a}	A	A

with

$$A = -\frac{e^2\hbar}{2mc} k_{\pi,\sigma} \frac{\beta G_\pi}{\Delta} \frac{1}{\sqrt{3}} \langle t_2 \| T_1 \| e \rangle$$

$\langle t_2 \| T_1 \| e \rangle$ is a reduced matrix element and $k_{\pi,\sigma}$ the orbital reduction factor.

By the invariance of the trace[64] the rotational strength of the $(t_{2g})^3\,{}^4A_{2g} \rightarrow (t_{2g})^2\,(e_g)^1\,{}^4T_{2g}$ of dihedral Cr^{3+} complexes is therefore $\sum r_v = 6A$. Expansion of the reduced matrix element shows A to be proportional to $G_\sigma G_\pi$. Hence A is not dependent upon a linear term in the overlap integrals. In this it follows the expression for the orbital energy, Eq. (2-68).

Provided we do not ignore the splittings of the octahedral levels under the trigonal field, Moffitt's original model may also give a net rotation. That was the solution to the problem proposed by Hamer[67] and Piper and Karipides.[69] A splitting at the vertical transition of ~ 500 cm^{-1} for the excited $^1T_1$ state in $Co(en)_3^{3+}$ was shown[67] to account for the observed rotatory dispersion. On the other hand Liehr[70] introduced an angle of cant α between the ligand orbitals and the metal orbitals, to account for a metal-ligand orbital mismatch. He found that a zero-order molecular orbital theory could explain the observed d → d rotatory strengths of the trigonal dihedral complexes, and that R was proportional to sin α. The number of parameters used by Liehr unfortunately makes it rather hard to draw more than qualitative conclusions. In spite of the many ingenious models found in the literature we may say that a conclusive calculation of R for d–d transitions has not been performed. This is, however, not surprising. As we have seen, estimations of R demand sophisticated perturbation theory, and it is probably impossible to pinpoint one dominating mechanism.

Turning our attention toward the electron-charge transfers, the situation is somewhat better. Using a coupled chromophore model, the rotatory strengths were calculated for some optically active metal-to-ligand π^* orbital transitions in the tris-*o*-phenanthroline iron(II) ion.[71] For the electron-transfer transitions found in $Fe(phen)_3^{2+}$ the 21,400 cm^{-1} band was assigned to $^1A_1 \rightarrow {}^1A_2$ and the 19,600 cm^{-1} band as due to $^1A_1 \rightarrow {}^1E$. These identifications follow the proposals by

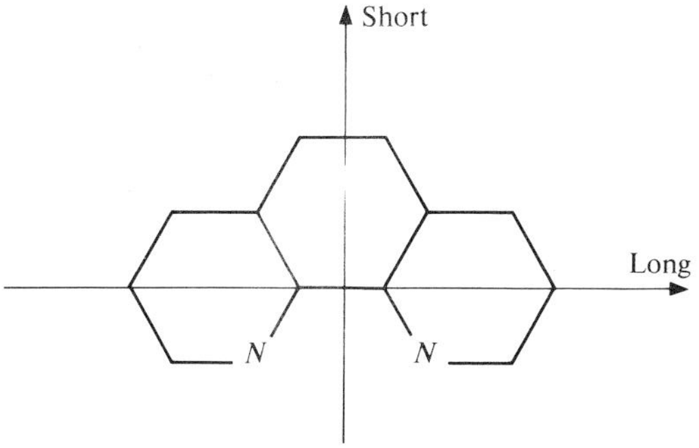

Figure 6-9 Reference axes for *o*-phenanthroline.

Orgel.[72] The absolute configuration of $Fe(phen)_3^{2+}$ can be obtained from the sign of R as associated with the electronic transitions. The calculated value of R is, however, greater than the experimental one by a factor of four.

For molecules or ions possessing an extended π system of orbitals the low-lying $\pi \rightarrow \pi^*$ electronic transitions found in the uncomplexed ligand can often also be identified in the complexed ligand. If that is the case the sign of R for these bands can be used to establish the absolute configuration of the complex. As an example we shall look at an $M(phen)_3^{n+}$ complex.[73,74]

In a planar, conjugated system all electric dipole $\pi \rightarrow \pi^*$ transitions are polarized in the plane of the molecule. This follows because a reflection in the plane of the molecule reverses the signs of all the π orbitals. Thus a $\pi \rightarrow \pi^*$ transition in (phen) will be either long- or short-axis polarized (Fig. 6-9). On the other hand the magnetic dipole transitions $\pi \rightarrow \pi^*$ will be polarized perpendicular to the molecular plane.

Each of the three ligands in $M(phen)_3^{n+}$ can be excited, and we use excitation theory to construct suitable wave functions. Denoting the three state wave functions for the ground state of the (phen) ligand Ψ_1^0, Ψ_2^0, and Ψ_3^0, the totally symmetric ground state will be $|A_1\rangle = \Psi_1^0\Psi_2^0\Psi_3^0$. The wave function for an excited state of one of the (phen) ligands will be denoted by Ψ_j^*. Choosing the orientations of the long-axis transition dipoles as shown in Fig. 6-10, we have the three zero-order excited state functions

$$\Psi_{\text{I}} = \Psi_1^*\Psi_2^0\Psi_3^0 \qquad \Psi_{\text{II}} = \Psi_1^0\Psi_2^*\Psi_3^0 \qquad \Psi_{\text{III}} = \Psi_1^0\Psi_2^0\Psi_3^*$$

Taking proper linear combinations of Ψ_{I}, Ψ_{II}, and Ψ_{III}, we construct three linear combinations which serve as a basis set for the irreducible representations A_2 and E in a D_3 point-group symmetry

$$|A_2\rangle = \frac{1}{\sqrt{3}}(\Psi_{\text{I}} + \Psi_{\text{II}} + \Psi_{\text{III}}) \tag{6-53}$$

$$\begin{aligned} |E^a\rangle &= \frac{1}{\sqrt{2}}(\Psi_{\text{II}} - \Psi_{\text{III}}) \\ |E^b\rangle &= \frac{1}{\sqrt{6}}(2\Psi_{\text{I}} - \Psi_{\text{II}} - \Psi_{\text{III}}) \end{aligned} \tag{6-54}$$

We now treat each ligand as if it were an electrically neutral system. The interaction between the coupled chromophores is then given by Eq. (4-220). With $\mathbf{D}_i = \langle \Psi_i^* | \sum_n \mathbf{r}_n | \Psi_i^0 \rangle$, $i = 1, 2, 3$, being the transition moment on ligand i and Δ the distance between the centers of the ligands, the interaction energy between two ligands is given by

$$V_{1,2} = \frac{\mathbf{D}_1 \cdot \mathbf{D}_2 - 3(\mathbf{m} \cdot \mathbf{D}_1)(\mathbf{m} \cdot \mathbf{D}_2)}{\Delta^3} \tag{6-55}$$

with

$$\mathbf{m} = \frac{\mathbf{\Delta}}{|\Delta|} \tag{6-56}$$

Using the wave functions of Eqs. (6-53) and (6-54) the perturbed energies are easily calculated to be

$$W^{(1)}(A_2) = \tfrac{1}{2}|D|^2/\Delta^3 \tag{6-57}$$

$$W^{(1)}(E) = -\tfrac{1}{4}|D|^2/\Delta^3 \tag{6-58}$$

The coupled chromophore model therefore predicts that the unperturbed $\Psi^0 \rightarrow \Psi^*$ transition will be split, and that $W(A_2) > W(E)$. The $A_1 \rightarrow A_2$ transition will be polarized along the C_3 axis of the complex and the $A_1 \rightarrow E$ transition perpendicular to the C_3 axis. We can calculate the intensity ratio by projecting the transition moments upon the C_3 axis and get $I_{\parallel} : I_{\perp} = 2|D|^2 : \frac{1}{2}|D|^2 = 4$.

For the calculation of the rotational strengths of the transitions we make use of Eq. (4-85), which is written in the form

$$R_{0j} = -\frac{\hbar e^2}{2mc} \langle j | \sum_n \mathbf{r}_n | 0 \rangle \cdot \langle j | \sum_n \mathbf{r}_n \wedge \nabla_n | 0 \rangle \tag{6-59}$$

For the transition $A_1 \rightarrow A_2$ we get therefore

$$R_{A_1 A_2} = -\frac{\hbar e^2}{6mc}(\mathbf{D}_1 + \mathbf{D}_2 + \mathbf{D}_3) \cdot$$

$$\left(\langle \Psi_1^* | \sum_j \mathbf{r}_j \wedge \nabla_j | \Psi_1^0 \rangle + \langle \Psi_2^* | \sum_j \mathbf{r}_j \wedge \nabla_j | \Psi_2^0 \rangle + \langle \Psi_3^* | \sum_j \mathbf{r}_j \wedge \nabla_j | \Psi_3^0 \rangle \right) \tag{6-60}$$

Moving the origin of the operator $\mathbf{r} \wedge \nabla$ from the metal ion to the center of ligand i (see Fig. 6-10) we have

$$\langle \Psi_i^* | \sum_j \mathbf{r}_j \wedge \nabla_j | \Psi_i \rangle = \langle \Psi_i^* | \sum_j (\boldsymbol{\rho}_{ij} + \mathbf{r}_j) \wedge \nabla_j | \Psi_i \rangle = \sum_j \boldsymbol{\rho}_{ij} \wedge \langle \Psi_i^* | \nabla_j | \Psi_i \rangle \tag{6-61}$$

where in the case $\langle \Psi_i^* \sum_j \mathbf{r}_j \wedge \nabla_j | \Psi_i \rangle \neq 0$ we have incorporated this magnetic dipole transition into an "effective" value of $\boldsymbol{\rho}_i$. The vector $\boldsymbol{\rho}_i$ therefore has a given (unknown) length ρ, going from the center of the coordinate system perpendicular through $\mathbf{D}_i$. Using Eq. (4-29), Eq. (6-60) is transformed into

$$R_{A_1A_2} = \frac{e^2\omega_{j0}}{2c}\cdot\frac{1}{3}(\mathbf{D}_1 + \mathbf{D}_2 + \mathbf{D}_3)\cdot(\rho_1 \wedge \mathbf{D}_1 + \rho_2 \wedge \mathbf{D}_2 + \rho_3 \wedge \mathbf{D}_3) \qquad (6\text{-}62)$$

This expression shows us that only transitions polarized along the long axis of (phen) can be optically active since for short-axis polarizations $\rho_i \wedge \mathbf{D}_i = 0$ (see Fig. 6-10). Looking at the directional cosines we see by expanding Eq. (6-62) that

$$R_{A_1A_2} = -\frac{e^2\omega_{j0}}{2c}\rho|\mathbf{D}|^2\sqrt{2} \qquad (6\text{-}63)$$

Similarly we get

$$R_{A_1E^a} = R_{A_1E^b} = \frac{e^2\omega_{j0}}{2c}\rho|\mathbf{D}|^2\frac{1}{\sqrt{2}} \qquad (6\text{-}64)$$

Had we taken the mirror image of the complex, the A_2 component would of course still have had higher energy than the E component but the rotational strength of the A_2 state would now have been positive and the E state negative.

In the uncomplexed ligand *o*-phenanthroline the $\pi \rightarrow \pi^*$ band at about 38,000 cm^{-1} is long-axis polarized. This transition remains largely unchanged in energy during complex formation. With the absolute configuration pictured in Fig. 6-10, the measured circular dichroism around 38,000 cm^{-1} should therefore first show a positive and then a negative CD curve for increasing energies. The reverse is of course true for the mirror conformation of the complex. We conclude that measurement of the CD curve can determine the absolute configurations of chiral systems.

Complications such as overlapping bands may be met in actual situations, but in principle the method is sound. The results depend only on the symmetry of the molecule and the level order of A_2 and E. In order to calculate the sequence of the excited states, we assumed that the dipole–dipole interaction between the

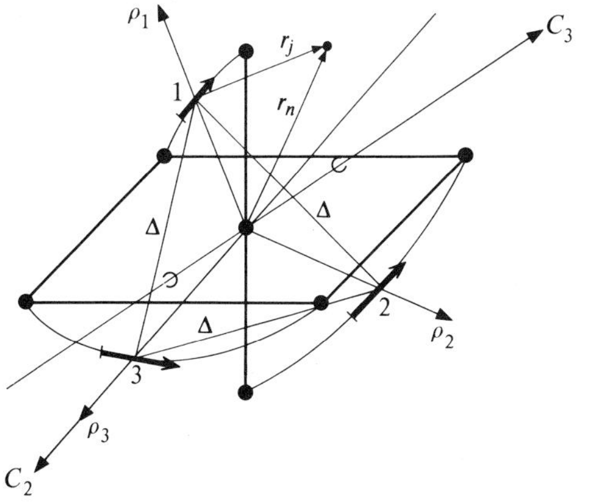

Figure 6-10 Numbering and orientation of the three long-axis transition dipole vectors in a M(phen)$_3^{n+}$ complex. The coordinate system is the same as used in Fig. 1-1.

ligands was the dominant factor. However, the most convincing determination of absolute configuration using CD methods should also include an independent, experimental determination of the positions of the A_2 and E levels. Polarized absorption-band measurements are of course the natural choice here.

6-7 THE STARK EFFECT

Applying an electric field to a single molecule may remove some symmetry elements. This in turn can lead to a splitting of symmetry-determined degeneracies. Such a phenomenon is called a *Stark effect*. In a molecular crystal we can furthermore have orientation degeneracy in case more than one chromophore is found in the unit cell. The electric field may again destroy two or more symmetry elements in the unit cell, resulting in a lifting of the orientational degeneracy. This is called a *pseudo* Stark effect.

No case of the first effect has been recorded for inorganic complexes. The pseudo Stark effect has, for instance, been observed in the emission lines of ruby[75] and in the charge-transfer transitions in CrO_3Cl^-. In the latter case[76] four chromophores are found in the unit cell. These are mutually related through the symmetry operations of the crystal space group C_{2h}^5 and are thus all equivalent. Each chromophore has C_{3v} symmetry, and the lowest observed electronic transition is the allowed $^1A_1 \rightarrow {}^1E$ correlating to the $^1A_1 \rightarrow {}^1T_1$ transition in the parent tetrahedral ion, CrO_4^-. The site symmetry of Cr is only C_1, and the two lines found at 5,476 Å and 5,379 Å represent 0–0 transitions to the split $^1E(C_{3v})$ state.

The C_3 axis of each CrO_3Cl^- ion is lying in the **ab** crystal plane. The permanent electric dipole $\boldsymbol{\mu}$ of each chromophore in the ground state is aligned along the C_3 axis. When the electric field is applied parallel to the crystal axis **a**, the electronic transition energy for chromophores 1 and 2 will experience a Stark shift $\Delta W = \mathbf{F}_{\text{local}} \cdot \Delta\boldsymbol{\mu}$ and the chromophores 3 and 4 experience a shift of equal magnitude but opposite direction. A splitting of the original orientational fourfold degenerate level into two levels will take place. This is indeed observed in the absorption spectrum.

The unit cell of $KCrO_3Cl$ possesses a center of symmetry. Therefore, if the unit cell is considered as the molecular entity no Stark splitting should occur. The observation of a strong linear Stark effect makes it possible to conclude that no resonance interaction takes place between the four chromophores. From the directions of $\Delta\boldsymbol{\mu}$ it was further inferred that for the two split excited states $\boldsymbol{\mu}^*$ is not parallel to the ground state $\boldsymbol{\mu}$. The equilibrium conformation of these excited states is therefore not C_{3v}.

When a static electric field F is applied, the medium no longer has isotropic refractive properties. By virtue of the Kramers–Kronig relations a birefringence n_a–n_b is associated with a difference in the absorption coefficients $\varepsilon_a - \varepsilon_b$. An electro-optical Kerr effect is observed, which has much in common with magnetic rotation spectroscopy. So far no applications to inorganic complexes seem, however, to have been performed.[77]

6-8 FORBIDDEN ELECTRONIC TRANSITIONS

Studies of optical absorption spectra are the most powerful technique we possess in elucidating the electronically excited states of molecules. Nevertheless, a common drawback is that the method fails to detect some of the excited states due to some selection rule. A number of techniques exist, however, which enable us to observe optical transitions to ordinarily inaccessible excited states.

The performance of pressure experiments is one such method. As an example consider the $^1A_1 \rightarrow {}^1T_1$ transition found in MnO_4^-. Assuming T_d symmetry, this transition is orbitally forbidden. In the mixed crystal $LiClO_4 \cdot 3H_2O/MnO_4^-$ the site group of MnO_4^- is C_{3v}. Polarization studies[78,79] revealed that the 0–0 line of the Teltow system at 13,772 cm^{-1} is polarized perpendicular to the crystallographic **c** axis (**E** $\perp$ **c**). An external pressure applied along the (110) direction lowers the site group symmetry of the MnO_4^- ion to C_s. This in turn causes the perpendicularly polarized line to split into two lines[80] which are polarized with the electric vector parallel and perpendicular, respectively, to the applied pressure **P**. At the same time the application of pressure brings forward a line located approximately 10 cm^{-1} below the 0–0 line. This new line, which has zero intensity in zero pressure, we take to be the third component of the $^1T_1$ state in the parent tetrahedral point group.

Provided a molecule has an excited state of different spin-multiplicity from that of the ground state, located not too high in energy, the lifetime of such a state may be sufficiently long to perform an excited-state absorption measurement. A fraction of molecules is optically "pumped" by means of a flash or a pulsed laser into the excited state, and the absorption spectrum is measured using the pumped state as the ground state. In this way we do away with the spin-selection rule, which normally prevents us from localizing the higher-lying states having a different spin-multiplicity from that of the ground state of the molecule.

As an example, we can take Cr(III) complexes.[81,82] It is possible here to pump the complex from the $^4A_2$ ground state up into the excited ^2E state, and to locate by means of absorption techniques the higher-lying spin-doublet states. These doublets will show up with the band intensities to be expected of spin-allowed ligand-field transitions.[83]

Finally we shall mention two-photon absorption experiments, which have become feasible due to the dye laser (see section 4-13). This technique has so far only been used on organic molecules,[84] but for transition-metal complexes it should be of special interest in locating the 0–0 transitions in the parity-forbidden ligand-field bands.

REFERENCES

1. C. J. Ballhausen, *J. Pure Appl. Chem.*, **44**: 13, 1975.
2. A. D. Liehr, *J. Phys. Chem.*, **67**: 1314, 1963.
3. D. S. McClure, *J. Chem. Phys.*, **36**: 2757, 1962. (See also for more details *Solid State Physics*, **9**: 488, 1959.

4. P. J. McCarthy and M. T. Vala, *Mol. Phys.*, **25**: 17, 1973.
5. M. D. Sturge, H. J. Guggenheim, and M. H. L. Pryce, *Phys. Rev.*, **B2**: 2459, 1970.
6. A. D. Liehr and C. J. Ballhausen, *Ann. Phys. N.Y.*, **6**: 134, 1959; and *Mol. Phys.*, **2**: 123, 1959.
7. E. I. Solomon and C. J. Ballhausen, *Mol. Phys.*, **29**: 279, 1975.
8. M. H. L. Pryce, G. Agnetta, T. Garofano, M. P. Palma-Vittorelli, and M. U. Palma, *Phil. Mag.*, **10**: 477, 1964.
9. J. C. Eisenstein, *J. Chem. Phys.*, **34**: 1628, 1961.
10. P. B. Dorain and R. G. Wheeler, *J. Chem. Phys.*, **45**: 1172, 1966.
11. L. A. Woodward and M. J. Ware, *Spectrochim. Acta*, **20**: 711, 1964.
12. W. Moffitt, G. L. Goodman, M. Fred, and B. Weinstock, *Mol. Phys.*, **2**: 109, 1959.
13. B. Weinstock and G. L. Goodman, *Adv. Chem. Phys.*, **9**: 169, 1965.
14. J. C. Eisenstein and M. H. L. Pryce, *Proc. Roy. Soc.*, **A255**: 181, 1960.
15. C. J. Ballhausen, *Theoret. Chim. Acta*, **24**: 234, 1972.
16. H. Johansen, *Chem. Phys. Lett.*, **17**: 569, 1972.
17. I. H. Hillier and V. R. Saunders, *Chem. Phys. Lett.*, **9**: 219, 1971.
18. M. H. Wood, *Theoret. Chim. Acta*, **36**: 309, 1975.
19. H. Hsu, C. Peterson, and R. M. Pitzer, *J. Chem. Phys.*, **64**: 791, 1976.
20. K. H. Johnson, *Adv. Quant. Chem.*, **7**: 143, 1973.
21. J. Demuynck and A. Veillard, *Theor. Chim. Acta*, **28**: 241, 1973.
22. C. D. Cowman, C. J. Ballhausen, and H. B. Gray, *J. Am. Chem. Soc.*, **95**: 7873, 1973.
23. W. C. Yeakel and P. N. Schatz, *J. Chem. Phys.*, **61**: 441, 1974.
24. J. C. Green, M. L. H. Green, P. J. Joachim, A. F. Orchard, and D. W. Turner, *Phil. Trans. Roy. Soc. Lond.*, **A268**: 111, 1970.
25. B. D. Bird and P. Day, *J. Chem. Phys.*, **49**: 392, 1968.
26. J. C. Rivoal and B. Briat, *Mol. Phys.*, **27**: 1081, 1974.
27. R. Gale, A. J. McCaffery, and R. Shatwell, *Chem. Phys. Lett.*, **17**: 416, 1972.
28. R. Dingle and R. A. Palmer, *Theoret. Chim. Acta*, **6**: 249, 1966.
29. D. F. Nelson and M. D. Sturge, *Phys. Rev.*, **137**: A1117, 1965.
30. D. L. Wood, G. F. Imbusch, R. M. Macfarlane, P. Kisliuk, and D. M. Larkin, *J. Chem. Phys.*, **48**: 5255, 1968.
31. E. D. Nelson, J. Y. Wong, and A. L. Schawlow, *Phys. Rev.*, **156**: 298, 1967.
32. R. M. Macfarlane, J. Y. Wong, and M. D. Sturge, *Phys. Rev.*, **166**: 250, 1968.
33. R. Dingle, *J. Chem. Phys.*, **46**: 1, 1967.
34. C. J. Ballhausen, N. Bjerrum, R. Dingle, K. Eriks, and C. R. Hare, *Inorg. Chem.*, **4**: 514, 1965.
35. J. P. Dahl, R. Dingle, and M. T. Vala, *Acta Chem. Scand.*, **23**: 47, 1969.
36. B. F. Gächter and J. A. Koningstein, *Solid State Comm.*, **14**: 361, 1974.
37. R. G. Denning, T. R. Snellgrove, and D. R. Woodwark, *Mol. Phys.*, **30**: 1819, 1975.
38. R. Englman, *The Jahn–Teller Effect in Molecules and Crystals*, Wiley-Interscience, 1972.
39. R. Dingle, *J. Chem. Phys.*, **50**: 545, 1969.
40. P. J. Stephens, *J. Chem. Phys.*, **51**: 1995, 1969.
41. W. Moffitt and W. Thorson, *Calcul des Fontions D'onde Moléculaire*, Edition du Centre National de la Recherche Scientifique, Paris, 1958, p. 141.
42. H. C. Longuet-Higgins, U. Öpik, M. H. L. Pryce, and R. A. Sack, *Proc. Roy. Soc.*, **244A**: 1, 1958.
43. J. Ferguson, *J. Chem. Phys.*, **39**: 116, 1963.
44. N. Pelletier-Allard, *Compt. Rend. Acad. Sci. Paris*, **256**: 115, 1963.
45. R. P. van Stapele, H. G. Beljers, P. F. Bongers, and H. Zijlstra, *J. Chem. Phys.*, **44**: 3719, 1966.
46. B. Judd, *Proc. Phys. Soc.*, **84**: 1036, 1964.
47. W. C. Scott and M. D. Sturge, *Phys. Rev.*, **146**: 262, 1966.
48. P. J. Stephens and M. Lowe-Pariseau, *Phys. Rev.*, **171**: 322, 1968.
49. W. C. Yeakel, J. L. Slater, and P. N. Schatz, *J. Chem. Phys.*, **61**: 4868, 1974.
50. J. W. Allen, R. M. Macfarlane, and R. L. White, *Phys. Rev.*, **179**: 523, 1969.
51. A. S. Davydov, *Theory of Molecular Excitations*, McGraw-Hill, 1962.
52. M. Wolfsberg and L. Helmholz, *J. Chem. Phys.*, **20**: 837, 1952.
53. C. J. Ballhausen and A. D. Liehr, *J. Mol. Spectroscopy*, **2**: 342, 1958; and **4**: 190, 1960.

54. P. N. Schatz, A. J. McCaffery, W. Suëtaka, G. N. Henning, A. B. Ritchie, and P. J. Stephens, *J. Chem. Phys.*, **45**: 722, 1966.
55. P. J. Stephens, *Chem. Phys. Lett.*, **2**: 241, 1968.
56. P. A. Cox, D. J. Robbins, and P. Day, *Mol. Phys.*, **30**: 405, 1975.
57. D. S. Alderdice, *J. Mol. Spect.*, **15**: 509, 1965.
58. B. Briat, J. C. Rivoal, and R. H. Petit, *J. Chim. Phys.*, **67**: 463, 1970.
59. C. A. L. Becker, C. J. Ballhausen, and I. Trabjerg, *Theoret. Chim. Acta*, **13**: 355, 1969.
60. H. B. Gray and C. J. Ballhausen, *J. Am. Chem. Soc.*, **85**: 260, 1963.
61. P. J. Stephens, A. J. McCaffery, and P. N. Schatz, *Inorg. Chem.*, **7**: 1923, 1968.
62. S. B. Piepho, P. N. Schatz, and A. H. McCaffery, *J. Am. Chem. Soc.*, **91**: 5994, 1969.
63. J. C. Collingwood, S. B. Piepho, R. W. Schwartz, P. A. Dobosh, J. R. Dickinson, and P. N. Schatz, *Mol. Phys.*, **29**: 793, 1975.
64. W. Moffitt, *J. Chem. Phys.*, **25**: 1189, 1956.
65. S. Sugano, *J. Chem. Phys.*, **33**: 1883, 1960.
66. C. J. Ballhausen and H. B. Gray, *Molecular Orbital Theory*, W. A. Benjamin, Inc., 1964.
67. N. K. Hamer, *Mol. Phys.*, **5**: 339, 1962.
68. C. J. Ballhausen, *Chem. Phys. Lett.*, **49**: 405, 1977.
69. T. S. Piper and A. Karipides, *Mol. Phys.*, **5**: 475, 1962; and *J. Chem. Phys.*, **40**: 674, 1964.
70. A. D. Liehr, *J. Phys. Chem.*, **68**: 665, 1964.
71. M. Král, A. Moscowitz, and C. J. Ballhausen, *Theoret. Chim. Acta*, **30**: 339, 1973.
72. L. E. Orgel, *J. Chem. Soc.*, 3683, **1961**.
73. A. J. McCaffery and S. F. Mason, *Proc. Chem. Soc.*, 211, **1963**.
74. A. J. McCaffery, S. F. Mason, and B. J. Norman, *Proc. Chem. Soc.*, 259, **1964**.
75. W. Kaiser, S. Sugano, and D. L. Wood, *Phys. Rev. Lett.*, **6**: 605, 1961.
76. J. H. Høg, C. J. Ballhausen, and E. I. Solomon, *Mol. Phys.*, **32**: 807, 1976.
77. A. D. Buckingham, C. Graham, and R. E. Raab, *Chem. Phys. Lett.*, **8**: 622, 1971.
78. J. Teltow, *Z. phys. Chem.*, **B40**: 397, 1938; and **B43**: 198, 1939.
79. L. W. Johnson, E. Hughes, and S. P. McGlynn, *J. Chem. Phys.*, **55**: 4476, 1971.
80. C. J. Ballhausen and I. Trabjerg, *Mol. Phys.*, **24**: 689, 1972.
81. T. Kushida, *J. Phys. Soc. Japan*, **21**: 1331, 1966.
82. R. A. Krause, I. Trabjerg, and C. J. Ballhausen, *Chem. Phys. Lett.*, **3**: 297, 1969.
83. M. Shinada, S. Sugano, and T. Kushida, *J. Phys. Soc. Japan*, **21**: 1342, 1966.
84. A. Bergman and J. Jortner, *Chem. Phys. Lett.*, **15**: 309, 1972.
85. G. F. Koster, J. O. Dimmock, R. G. Wheeler, and H. Statz, *Properties of the Thirty-two Point Groups*, MIT Press, Cambridge, Mass., 1963.

INDEX